Twentieth-Century New England Land Conservation

Twentieth-Century New England Land Conservation

A Heritage of Civic Engagement

Edited by Charles H.W. Foster

Harvard Forest • Petersham, Massachusettts
Harvard University
2009
Distributed by Harvard University Press
Cambridge, Massachusetts, and London England

ISBN

Please address inquiries to quote to: Harvard Forest
324 North Main Street
Petersham, MA 01366

FIRST EDITION

ISBN 978-0-674-03289-7

Published in the United States of America by Harvard Forest, Petersham, Massachusetts.

Distributed by Harvard University Press

PRINTED IN THE UNITED STATES OF AMERICA

CONTENTS

FOREWORD

This journey back through the twentieth century began in earnest on October 14, 2005. At the invitation of the Massachusetts Historical Society, five New Englanders met in Boston to consider compiling the first-ever, contemporary, conservation history of the New England region. The original cooperators included former Maine state planning director and former gubernatorial candidate Richard E. Barringer, an economist; long-time president/forester Paul O. Bofinger of the Society for the Protection of New Hampshire Forests; Vermont historic preservation specialist, planner, and historian Robert L. McCullough; former Massachusetts commissioner of natural resources and environmental secretary Charles H.W. Foster; and Brown University anthropologist Ninian Stein. Much-respected Connecticut corporate lawyer and citizen activist Russell L. Brenneman could not make the first meeting, but pledged his personal support for and subsequent involvement in the enterprise. Like all such ventures, there was already a history behind the history.

In 1999, at the suggestion of Foster and former Trustees of Reservations Director Gordon Abbott, Jr., and with the enthusiastic support of then-director William M. Fowler, Jr., the Massachusetts Historical Society (MHS) set about to explore the potential for a New England environmental archive that would house important nongovernmental records and personal papers. By the summer of 2002, the MHS was sufficiently convinced of the merits to launch such an initiative. Its first acquisitions were the historic records of the Massachusetts Forest & Park Association, later followed by those of the Massachusetts Audubon Society and its director, Allen H. Morgan, thereby significantly augmenting the personal papers already in its possession such as those of Theodore Lyman III, an early New England fisheries conservationist, and Charles Francis Adams, a driving force behind the creation of Boston's Metropolitan Park Commission. The MHS also sought a commission from the Massachusetts Environmental Trust to take oral histories from the first eight of the Massachusetts secretaries of environmental affairs.

In 2004, the MHS received a grant from the National Historical Publications and Records Commission, an office of the National Archives, to engage a project archivist. Cheryl Beredo's initial effort was to develop an implementation plan for a New England environmental archive. In the course of her work, it soon became evident that a history of the region would be needed to identify particular priorities for records collection and preservation. An important part of any such effort would be a network of cooperating institutions, interconnected electronically, such that papers could be accessed readily regardless of where they might be physically archived. During a series of inquiries and regional meetings, Beredo was assured of interest and support from existing repositories.

Meeting in Boston in October 2005, the first cooperators were faced with the daunting prospect of not only the geographic sweep of the region, but the breadth of what might be included in conservation. After much discussion, it was decided that the history would focus primarily on the twentieth century, and that the subject matter would be limited, where possible, to land conservation. This would provide a manageable organizational scope for the project. By land conservation were meant actions affecting the ownership or use of land in order to safeguard one or more of the public values arising therefrom.

What is at stake here is that in New England, where land is 75 percent in private ownership, these are the properties that actually produce the bulk of the public benefits (e.g., drinking water, air quality, open space, recreation habitat for fish and wildlife, climate amelioration, greenhouse gas reduction). We have come to expect to receive the benefits at no cost to ourselves. In the future, in some instances, outright acquisition will be the answer, but because of the increased price of land today as contrasted with the early part of the twentieth century, most of the actions taken will have to be more inventive (e.g., purchase of certain rights in the land rather than the land itself). Regrettably, our space limitations only allow us to allude to these problems.

It was further agreed that the account would be predicated upon individual land-conservation overviews for each state, prepared as the authors and their advisors so elected, and fitted to particular audiences in those states. And since there was no apparent source of financial support, the regional history would have to be a largely volunteer effort.

The first cooperators also tried their hands at identifying a set of characteristics of New England conservation to be tested. These went beyond the conventional caricature of the native New Englander as unsophisticated, stubborn, thrifty, laconic, rural, and cantankerous. The six themes settled on are described in the introduction and revisited in the concluding chapter. By and large, there does, indeed, seem to be a special quality to New England conservation.

In the course of preparing the state overviews, the cooperators encountered another common experience — how New England conservationists dealt with representations from the federal government over the years. This has led to an additional chapter written by the National Park Service documenting how one federal agency has operated in New England, and its perceptions of what it takes to accomplish successful land conservation within our region.

What has emerged is a set of stories written by people who were there for an appreciable part of the century under examination. As the interest grew and a set of supplemental advisors became involved, the project was able to engage their collective three hundred years of personal involvement and experience. That in itself is an extraordinary accomplishment.

In addition to the three original cooperator/authors — Robert L. McCullough (Vermont), Charles H.W. Foster (Massachusetts), and Russell Brenneman (Connecticut) — we extend our thanks to the three additional authors who arrived later — former Maine Audubon Society Director Thomas A. Urquhart (Maine), *Yankee* magazine writer Jim Collins (New Hampshire), and *The Providence Journal* environmental reporter Peter B. Lord (Rhode Island) — all accomplished writers long before our project.

The six state overview chapters to follow remain faithful to the original concept. They are purposely individualistic. Like New England conservation itself, they follow their own instincts, are presented in their own styles, and come to their own conclusions. The final chapter attempts to draw together a consensus on what conservation has accomplished over the years within the region as a whole, and how history might be used to inform the activities of the next generation of leaders.

Since that first meeting in Boston, the New England conservation history project has been able to attract modest support. The New England Natural Resources Center, a bridge-building institution founded during the early, turbulent, environmental 1970s — now simply a facilitating nonprofit for unusual but worthwhile regional endeavors — agreed to serve as the institutional sponsor. Four of its trustees are actively engaged in the project. The Betterment Fund and Quimby Family Foundation have supplied funds for the work in Maine; the New Hampshire Charitable Foundation for activities in New Hampshire and other parts of the region.

The National Park Service has supplied a matching Civic Engagement grant to help pay for operating expenses and the convening of a review conference at the New England Center in Durham, New Hampshire. For these tangible forms of support, we are most grateful. Despite these welcome fund sources, the project has remained faithful to its early concept as a largely volunteer endeavor.

In addition to the network of individual cooperators, institutional cooperators have been recruited in several of the states: the Muskie School of Public Service at the University of Southern Maine, the Rubenstein School of Environment and Natural Resources at the University of Vermont, the New Hampshire Historical Society, the Massachusetts Historical Society, the Center for Environmental Studies at Brown University, and the Connecticut Forest & Park Association, Their involvement presages the likelihood of a continuing use of the material once the New England environmental history project itself has been completed.

One last contribution is especially noteworthy. As we began wondering how this extraordinary story of people and the land over so many years could be shared with others, the Lincoln Institute of Land Policy stepped forward with an offer to help publish the account. This was meaningful for several reasons, none the least the institute's historic role in conceiving of the idea of a national land-trust movement and then supporting its early, New England-based efforts. As former Land Trust Alliance director and Lincoln trustee Jean Hocker has documented ("The Land Trust Alliance and the Modern American Land Trust Movement," Forest History Today, Spring/Fall 2007), the seminal moment was a National Consultation on Local Land Conservation held at Lincoln House in Cambridge on October 15-16, 1981, and attended by forty conservation leaders from local and regional land trusts across the country. Lincoln followed up this event by convening a series of Land Conservation in New England study groups charged with designing the needed information clearinghouse, service bureau, and training programs. The results speak for themselves—nearly 1,700 land trusts now operating nationwide with an estimated 37 million acres already protected and an average of 1.2 million acres being added annually. This remarkable record of civic engagement, inspired first by New Englanders and replicated later throughout America, confirms our extraordinary heritage of self-reliance, local leadership, place-based innovation, and civic action.

As the director of the project since it inception, I can only conclude with a heartfelt expression of appreciation to my colleagues for making a dream come true. Theirs has been a significant contribution not just to history, but to the fundamental importance of the human connection with the environment and the ability people have to make a significant difference when so involved. Let this account similarly inspire those who may follow in our footsteps.

Charles H.W. Foster
Needham, Massachusetts

PREFACE

For many years, people have been lamenting the so-called homogenizing of America. The distinctive qualities of the American character, the landscape and, yes, even our language have been slowly suffocating, they say, in an ever-advancing sea of shopping malls, housing developments, fast-food restaurants, highways and, of course, all the varied ramifications of the centerpiece of the high-tech revolution, television. Lately, the movement toward globalization, exciting as that is in many ways, has contributed as well. You might say it's like a stew, originally consisting of meat, potatoes, carrots, onions, celery, and gravy, being ground into a squishy, brown mush.

There exists today, however, a permanent — and I emphasize that word — obstacle in the path of that seemingly relentless trend. I'm referring to a beloved, admired, and rather small six-state region occupying the northeast corner of the United States known as New England. New England is not only a sort of homeplace for things we hold to be uniquely American, but it is also where our American history is more readily available — and visible — that anywhere else. Sure, it has not been spared the invasion of commercial growth but, the mysterious part of it is that at the same time it has not only maintained its regional identity but even strengthened it through persistent efforts in both historical preservation and land conservation.

Dismissed as irrelevant by some uninformed occupants of other less definable regions (*of course*, I'm prejudiced!), New England over the years has also created institutions and innovations truly profound in their influence on American society. In fields as diverse as law, religion, medicine, education, literature, industry, agriculture and, for sure, protecting our natural environment, New England's contributions have played—and continue to play—pivotal roles, roles so often on the cutting edge of vital, national concerns. When, for instance, during the 1960s Rachel Carson's *Silent Spring* was echoing throughout laboratories of chemical industries and board rooms of industrial giants on both sides of the Atlantic, with everyone trying to figure out how to kill bugs and weeds without damaging humans and wildlife, New England was already showing the way. It would, for example, require the cooperation between government, business, and private individuals so dramatically demonstrated by the preservation of 45,000 acres north of New Hampshire's White Mountain National Forest, under the auspices of the Trust for New Hampshire Lands.

In New England, this and so many other programs to do with conservation danced around the theme that the land as well as our historic-built environment is for the greater good rather than for the greater profit. The irony here is what New England commenced to demonstrate early on: namely that the greater profit almost always springs from the greater good. In other words, preserving our natural and built environment is also—surprise!—good business.

"Thank you, New England," people across the country might say today, "Guess you were right all along."

Of course, in a genealogical sense, New England stretches across the country from coast to coast. The Mathers, for instance, founded Cleveland, Ohio; William Greenleaf Eliot (grandfather of T. S. Eliot) established the public school system of St. Louis; the Adamses helped develop Kansas City, Kansas, as well as Denver, San Antonio and Houston, Texas; Sherburne County, Minnesota is named for Moses Sherburne of Mt. Vernon, Maine, who helped write the Minnesota state constitution; Wisconsin's first, third, seventh, eleventh, twelfth, and eighteenth governors were New Englanders . . . well, the list could go on and on.

Still, many would agree with me that the area specifically occupied by Connecticut, Rhode Island, Massachusetts, Vermont, New Hampshire, and Maine is distinct: different from the rest of the country. People arriving from elsewhere often say they can "feel" the difference but there are also plenty of studies, using statistics, to support the notion. One, for instance, was recently conducted by the University of Iowa that revealed the rankings of all fifty states in what's referred to as "civic culture." That is the ability to work together for the public good on projects such as, of course, land conservation as well as a myriad of others. Criteria used included participation in public affairs, a strong belief in equality, a willingness to trust one another, a high percent of the population joining community groups, and so forth.

Well, it so happens Vermont scored first in the nation when it comes to "civic culture." Massachusetts was second, Maine fourth, New Hampshire sixth, and Connecticut ninth. So New England placed five of its six states in the top ten. Rhode Island didn't do too badly either — coming in sixteenth out of fifty.

Can natives of those six states take all the credit? Hardly. Pearl Buck once wrote that to survive a region must "treasure the image of itself." I think those who treasure New England's image most are often those who have moved to New England from somewhere else, the image firmly in their minds. Then, by joining planning boards, historical societies, land-trust boards, and so on, they proceed to work toward making the image real. As a result, New England is, in my opinion, more New England today than it was years ago. "Assume the virtue if you have it not," Shakespeare's Hamlet told his mother, "for use can change the stamp of nature." So it is that outsiders have assumed New England's virtues, as they once visualized them from afar.

The results of all this over the past fifty years is quite dramatic. Environmentalists, although far from enjoying total victory, are no longer looked upon as the village eccentrics. When I was a young man, it took a brave citizen to stand up to a town's major employer for, say, dumping chemicals into the river. Today, it takes a brave (and foolish) factory manager to keep polluting in the face of community opinion. The idea of the earth as a spaceship with a limited supply of food, fuel, pure air, and water — an idea that was dimly occurring to a few thoughtful people fifty years ago — is familiar to every third grader today. Now, *that's* progress.

So I don't believe it to be overreaching to say that *all* Americans can take pride in and feel a sense of ownership in New England, the birthplace of so many things, including the environmental movement about which you'll read in this book. Yes, to be sure, the winters are harsh, the spring is muddy, and the mosquitoes and blackflies can be ferocious but, you know, all of these things inexplicably increase our devotion. Maybe our legendary contrariness has something to do with it — the same contrariness that enabled New Englanders to buck uncontrolled land development long before anyone else. As former American poet laureate Donald Hall once wrote, "There's no reason to live here [in New England] except for love."

Well . . . reason enough.

Judson D. Hale, Sr.
Dublin, New Hampshire

Twentieth-Century New England Land Conservation

LAND CONSERVATION IN NEW ENGLAND

AN INTRODUCTORY OVERVIEW

CHARLES H.W. FOSTER

This is the story of land conservation in New England. It has been written by and about New Englanders. By New England, we mean the northeastern sector of the United States traversed in 1614 by soldier, explorer, and author Captain John Smith—the area he came to call "new" England. Since then, the region has been considered to be the six states of Maine, New Hampshire, Vermont, Massachusetts, Connecticut, and Rhode Island. By conservation, we mean the wise use and management of natural resources, a concept embracing not only the concern for a sustainable, quality, natural evironment but also the direct social, cultural, and even economic values that derive from its air, water, and land resources. As we will see later on, New Englanders' commitment to the environment, although deeply rooted, has always recognized the pragmatic contributions made by its natural resources.

As a topographic entity, New England is relatively distinct. It covers approximately 62,000 square miles of land area and is home to some 14 million residents. To this should be added another 5,000 square miles of inland and coastal water subject to state control. The region is drained by 28 major river and coastal basins of which the four-state Connecticut River, Massachusetts-New Hampshire's Merrimack River, and Maine's Kennebec River are the largest. The entire region was shaped by the great Pleistocene glacier 10,000 years ago, which left behind smooth-sided mountains, U-shaped notches and valleys, and associated plains. Except for a portion of the Connecticut valley, the grain of New England runs roughly north and south. In most places, with bedrock as little as twenty feet below the surface, the soils are shallow and thin.

Viewed from the air, the region is seemingly one dominated by woods and water. Two-thirds of the land is in forest, most of it owned privately. The forest occurs in six distinct forest zones ranging from spruce-fir to the north to pitch-pine-oak to the south. The generally shallow groundwater basins, augmented by a plentiful precipitation averaging 44 inches annually, evenly distributed throughout most years, support an abundance of lakes, ponds and, wetlands — what Thoreau once described, standing at the top of Maine's Katahdin, as "a mirror broken into a thousand fragments."

Perhaps the earliest natural history of New England, "a true, lively, and experimental description of that part of America called New England," was written by William Wood in 1634 to whet the appetites of potential investors and settlers. The account sold briskly in England. Wood described an environment where the trees, with the exception of the riverine white pines, "were not very thick" but grew "straight and tall," a landscape that was surprisingly open as the result of Indian-set fires. Extensive lists of birds, mammals, and fish were provided, most of them still with us today. The three "annoyances" — wolves, rattlesnakes, and mosquitoes — were more than offset by a climate, Wood said, that left the complexions of men and women "fresh and ruddy." As for New England's changeable weather, his characterization was *nullum violentum est perpetuum* (no extremes last long).

Despite the well-meaning effort over the years to keep things the way they were, New England's past history has been one of almost constant change. Environmentally, it has gone through periods where as much as half of its forests were cleared for settlement and agriculture, only to have the land abandoned and revert to forest when the original settlers moved west. The forest was stripped several times for lumber, paper, fuel, charcoal, and railroad ties, massively clearcut and burned, only to rebound and now record its highest standing biomass since the time of settlement.

In the early centuries, fish and wildlife were pursued relentlessly for their hides, pelts, and flesh to the point where populations of deer, beaver, turkey, and salmon were virtually extinguished. In recent years, most have returned. Despite an accepted common-law principle that natural bodies of water (the so-called Great Ponds) and the ocean "foreshore" should remain forever open to the public for fishing and fowling, compliant legislatures saw to it that the region's water resources were increasingly dammed, developed, and debilitated in the interest of the early industrial development for which New England became renowned.

Today's conservation efforts, although still manifest in specific programmatic areas, are now concentrated on safeguarding basic ecological processes so that the region's resources can maintain their resilience and continue to deliver their important social, economic, and quality-of-life values and services. At a time when environmental impacts have been found to be as much global as local, this has become a daunting proposition.

The role of New England as a distinct social, cultural, political, and economic region, the cornerstone of our inquiry, is a matter of some debate given the influences of New York to the south and west, Boston

Wildlife was a direct beneficiary of the abandonment of New England farms in the late nineteenth and early twentieth century as shown by artist and sportsman A. Lassell Ripley's scene of ruffed grouse enjoying wild apples at the site of an overgrown farmstead cellar hole.
(The Guild of Boston Artists, Boston, 1972)

and Canada to the north and east, and a landscape that reflects a mix of rural, urban, inland, and coastal land uses. The matter of cultural homogeneity is particularly debatable in the face of the recent in-migration from many sources. For example, early writers extolled New England as "the seedbed of the great American Republic" (Harriet Beecher Stowe, 1869), "the first American civilization to be finished, to achieve stability in the conditions of its life" (Bernard DeVoto, 1932), and "America's palladium of truth, justice, freedom, and learning (Van Wyck Brooks, 1940). Contemporary observers have been more skeptical, calling the region "an imagined world of pastoral beauty, rural independence, virtuous simplicity, and religious and ethnic homogeneity" (Dona Brown, 1995). In numbers, the traditional Yankee is now clearly an endangered species. For a larger treatment of these varying impressions, we commend to the reader the recent compendium, *The Encyclopedia of New England* (Yale University Press, 2005).

NEW ENGLAND AS A REGION

As a natural resources concept, the term "region" encompasses many different forms. Most regions display distinct geographic features — in the words of the 1935 National Resources Committee report "homogeneity in one or more of its aspects." These can be topographic, hydrologic, or biotic. Regions are also defined by the economic services they provide, such as reconciling competing proposals for jobs and development, stimulating growth, and distributing resources equitably among component jurisdictions. Still a third characterization of a region is its social context — what the early southern regionalist, Howard Odum, described as "the sense of the people and their culture being close to the soil and their resources" (Jensen, 1951). A final category is what might be called administrative regionalism accomplished through mechanisms where, in our case, natural resources activities and services have been decentralized within circumscribed areas. New England is blessed by examples of all four regional forms. As we have seen, it is geographically distinct and, at times in the past (and even today), in a number of discrete areas of concern, it has attempted to operate cohesively as a single economic, social, economic, and/or political region. Take for example the case of its administrative forms.

Administrative regionalism dates back to the concept of *imperium in imperio* (an empire within an empire) used by the Romans, and later the British, to control their far-flung holdings. Its first manifestation in New England was James II's appointment of Sir Edmund Andros in 1686 as colonial governor of the Consolidated Dominion of New England, the one and only time New England has been governed as a single political unit. New England's unrest over Andros's heavy-handed limitation of town meetings and restriction of their right to impose local taxation brought about an early end to the practice of all-powerful colonial governors and an enduring cultural aversion by New Englanders to governance from afar.

Modern regionalism has been the subject of many inquiries. The first notable event was the December 1935 report of the National Resources Committee, *Regional Factors in National Planning and Development*, a product of the turbulent economic times of the New Deal and the first and only time America would have (briefly) a centralized national planning agency. In the interest of investigating the potential use of regions for social and economic development, twelve prominent regionalists were asked for their opinions as to what they conceived a region to be, how regions should be sized and bounded, especially with respect to state lines, and what might be the best type of region for use in regional planning and development. Although opinions were divided, natural resources were cited frequently as candidates for viable regions (Meyer and Foster, 2000).

In April of 1972, the Congressionally established Advisory Commission on Intergovernmental Relations issued a special report on multistate regionalism, examining the experience with specific regional instruments for economic development, water resource management, interstate compacts, and federal-multistate regionalism. The bottom line of its analysis was the sense conveyed by its summary chapter title: Let the Experiment Continue.

On January 23, 1996, under the joint auspices of the New England Natural Resources and Environmental Policy Centers, more than fifty leaders were convened at Harvard's John F. Kennedy School of Government for a day-long symposium revisiting the past and exploring the prospect for regionalism in New England (Reidel, 2001), The New Deal's basic criteria for a region — space, territory, and activity — were first confirmed. New England regionalists were then urged to select scales that were appropriate to the "many New Englands" (e.g., the

North Woods, rural uplands, coastline, offshore waters and islands, river valleys, and urban centers) and to take advantage of areas where coalitions of interests already existed or could be formed readily. Regionalists were warned not to assume the presence of a governmental entity as evidence of a viable region.

In 1999, stimulated by the earlier Natural Resources Committee and Intergovernmental Relations reports, researchers at Harvard's John F. Kennedy School of Government undertook a further examination of forms of natural resources regionalism, seeking advice from forty-eight North American theorists and practitioners on what the best form of an environmental region should be (Foster and Meyer, 2000). Most respondents urged flexibility and variety in regional arrangements, counseling against seeking a single, empirically determined, environmental region. The pragmatists among the respondents favored regions specific to one or more problems, thereby echoing the earlier assertion that "regions do not have truth, only utility" (Raitz and Ulack, 1984).

In the meantime, quite opportunistically, New England had been accumulating substantial experience with the transboundary potential of interstate compacts relating to natural resources.

The interstate compact, a negotiated agreement that enjoys the legal status of both a contract among the participating states and the benefits of statutory law, has been hailed by American Bar Association experts as the only mechanism in the U.S. Constitution permitting states to alter their relationships without running afoul of the federal government (Brown, Caroline, et al., 2006). As jurists Felix Frankfurter and James Landis observed in 1925, the compact clause is "the equivalent of the age-old treaty-making power of the independent sovereign nations." Because measures were needed promptly to adjudicate boundaries between the colonies and later the states, the compact device, they observed, has "roots deep in Colonial history."

During the period 1940–1975, some 100 interstate compacts came into being — not only an increase in sheer numbers, but also in types of usage. In 1983, the Council of State Governments recognized 25 distinct compact categories. By then, more than 200 formal interstate agreements were in place, 123 involving one or more of the New England states. Most of the compacts were bilateral, but 48 were multistate/regional in nature (13 of these in New England). But further experimentation was yet to come, and New England was to be in the vanguard.

In 1961, after a bruising series of political confrontations reportedly settled personally by New England-born President John F. Kennedy, the Delaware River Basin multipurpose regional water resources compact was approved by Congress. A so-called federal-interstate compact, what made this different was the inclusion of the federal government as an actual member of the commission and thus potentially bound by a majority of the state members (Delaware, New Jersey, New York, and Pennsylvania). Seven years later, a similar Susquehanna River Basin Compact came into being with Maryland, New York, and Pennsylvania as its state members.

As will be seen later in this conservation history, New England's role in the federal-interstate compact experimentation turned out to be bittersweet, for its own Northeastern Water and Related Land Resources Compact, framed in 1959 with the federal government to be equally bound, was never approved by Congress. Despite this rejection, New England's water resources leaders refused to abandon their insistence on a co-equal role for the states with the federal government in water resources planning and related land-resources management. Not only did they lend their support to the Delaware and Susquehanna compacts, but they saw to it that the Water Resources Planning Act approved by Congress in 1965, and the actions of all such river basin commissions authorized under Title II of the Act, would contain an arrangement whereby only two votes would be cast: one by the federally appointed chairman on behalf of the federal agencies, and one by the elected vice chairman on behalf of the states. In practice, all commission decisions would have to be by consensus.

New England also helped pave the way for other imaginative and creative forms of interstate arrangements. For example, its three flood control compacts for the Connecticut (1953), Merrimack (1957), and Thames (1958) river basins, forged during the controversy over the Corps of Engineers' role in flood control, contained a unique provision. Not only were the downstream states to reimburse the upstream states for the tax losses of providing sites for dams and reservoirs, but the compacts included compensation for "economic damages."

One flood control commissioner still recalls the experience of listening to local people estimate their economic losses when a mill was acquired to make way for a flood control reservoir. It was both an object lesson in interconnected community economics, and a tribute to the decency and fairness of New Englanders when given a chance to document their own needs. So rarely were the estimates inflated that settlement could usually proceed without delay.

The Northeastern Forest Fire Protection Compact, approved by Congress in 1949, after the great Bar Harbor, Maine forest fire of 1947, was pathbreaking for its inclusion of the eastern Canadian provinces. Not only did these provisions across the border echo the early constitutional discussion about inviting Canada to join the Union, but it would anticipate a move by the New England governors in 1973 to form the nonprofit New England Governors/Eastern Canadian Premiers Conference, a policy body now involving six states and five provinces (Rausch, 1997). In addition to their compact involvement, the role of the governors in regional affairs is itself worth reviewing.

THE GOVERNORS

As delegates gathered in Carpenters Hall, Philadelphia, in 1776 to debate the constitutional and other provisions of the new Union, a curious question came up — what to call the head of the new United States. President was selected, not because it was the most prestigious but because it signified less power than governor. Even the states recognized that distinction. New Hampshire, for example, briefly designated its chief executive as president before adopting the conventional term governor.

As distrust of outside authority grew, the New England governors became increasingly instruments of their state legislatures, given short and fixed terms of office and often constrained in their actions by independent executive councils (Sabato, 1983). Vermont and New Hampshire still have two-year terms for their governors, and Massachusetts and New Hampshire still retain separately elected executive councils. By the time of the early nineteenth century and de Tocqueville's *Democracy in America*, the governor was derided as "counting for absolutely nothing."

As of 1994 (Congressional Quarterly, Inc.), the New England states could list 441 individuals who had served since 1789 as governors. Before the Revolution, governors were appointed by the Crown — seemingly all-powerful, but in actuality of limited influence due to their lack of enforcement powers. In Colonial and immediate post-Colonial

Convened by President Theodore Roosevelt on May 13, 1908, the national conference of governorsdid much to inaugurate the national conservation movement. New Englanders attending included Governor Rollin S. Woodruff of Connecticut, Governor Charles M. Floyd of New Hampshire, Governor Fletcher Proctor of Vermont, Governor James H. Higgins of Rhode Island, and former Governor John F. Hill of Maine. Although not attending the White House Conference, Massachusetts Governor Curtis Guild, Jr. later took steps to convene in Boston on November 23, 1908, the first conservation conference ever called by the governors of the New England states and the occasion for which this book is the centennial. (Forest History Society)

times, the governors were at least prominent individuals, meriting memorial chairs in the Beacon Hill reading room of the New England Historic Genealogical Society. By the 1920s many governors had become, in Larry Sabato's colorful terminology, little more than "Good Time Charlies." By mid-century, it was clearly time for further reforms at the state level to counter James Reston's view of the state capitals as "over their heads in problems and up to their knees in midgets" (*New York Times*, October 5, 1962). At that time, the governors were persuaded to use more extensively the one broad power allowed them by their legislatures — the discretion to enter into agreements with other jurisdictions (Zimmerman, 2002).

With regard to collective actions taken by the governors for the region as a whole using their executive powers, one of the earliest natural resources manifestations was what Massachusetts governor Curtis Guild, Jr. advertised as the First New England Conference Called by the Governors of the New England States. Held in Boston on November 23–24, 1908, at the suggestion of Herbert Myrick, editor of the influential journal, *New England Homestead*, it included every governor and governor-elect as well as prominent citizens chosen by the governors in numbers of two per representative and two per senator in Congress. Agriculture, forestry, and fisheries dominated the official agenda topics. Gifford Pinchot, chief forester of the United States, was among the keynote speakers.

Pinchot's appearance was no accident because it came on the heels of the landmark May 13–15, 1908, White House Conference of the Governors of the United States called by President Theodore Roosevelt, himself a former governor. Natural resources issues, especially the nationwide concern over forests and their importance to river headwater regions, dominated the discussions there and, later, in Boston. They were given added credence by Pinchot's new assignment as president of the National Conservation Commission created by Roosevelt.

New England's fledgling state forest agencies, and their nonprofit forest conservation organization supporters, were measurably assisted by these developments, for waiting in the wings was Massachusetts Congressman John Weeks with legislation that Congress would enact in 1911 authorizing a new system of eastern national forests to protect such headwater areas.

When the Conference concluded, Governor Guild assured his audience that he had no intention of creating a permanent regional organization that would displace the statutory responsibilities of Congress or the state legislatures, but he did remind the delegates that the states have constitutional rights of their own, and urged the assembled officials to develop and promote uniform laws and actions on the subjects discussed. As a reminder of the importance of conservation, each delegate's badge contained a representation of New England's historic white pine superimposed upon a white silk ribbon.

Mindful of Guild's advice, in November 1925 the governors exercised their executive powers to convene a conference of business and industry leaders. Faced with the imminent departure of the textile industry to the South, the New England governors urged the business community to unite. At the governors' suggestion, there was subsequently formed the New England Council for Economic Development (now simply the New England Council), self-described as the "town meeting of business that never adjourns." The council was comprised of eighteen-member councils in each of the states and, in 1961, opened a lobbying office in Washington headed by Charles W. Colson, later of Watergate prominence. The council maintained an especially active committee and staff devoted to vacation travel and natural resources.

In the summer of 1933, the New Deal National Planning Board approached the New England governors with an offer of assistance to help assess the region's natural resources and plan for their wisest and best use. At that time, New England had no state planning boards. Only 172 city and town planning agencies were in existence. New Hampshire became the first state in the nation to set up a state planning board. The remaining New England states followed suit a year later. In March 1934, the governors consented to the establishment of a New

Delegates attending the New England conservation conference included two nominees proposed by each New England member of Congress. Such was the concern over the state of the region's forests that the official badge for attendees contained the image of a historic New England white pine placed on the background of a white silk ribbon.
(Conference Proceedings, 1908, Wright & Potter Printing Co., Massachusetts State Printers, Boston)

England Regional Planning Commission, the first such commission in the country. The three state planning boards then in existence (New Hampshire, Maine, and Connecticut) were officially represented; the governors of Vermont, Massachusett, and Rhode Island appointed special representatives.

For the next ten years, the commission would do remarkable work using a small staff and consultants operating out of Boston. It would claim some 1,500 New Englanders involved in its programs. Not only would the first system-wide information on water resources be assembled, but there would be active committees formed and studies undertaken for other natural resources, including recreation and forestry. The commission's New England-wide regional planning conference held in Boston in May 1940 included business leaders from the New England Council and citizen attendees from the Federation of Womens Clubs, an indication of the breadth of interest and support at that time.

We now know that the choice of New England as a pioneer in regional planning was not by accident, for the executive director of the National Planning Board was none other than New Englander Charles W. Eliot, II, grandson of the great Harvard president by that name and the nephew of Charles Eliot, the founder of the Massachusetts Trustees of (Public) Reservations, the region's and the nation's first land trust. A member-at-large of the commission for much of its existence was a redoubtable lady, Mrs. Charles Sumner Bird of Walpole, Massachusetts, wife of the politically connected businessman and philanthropist. She would come to play a leading role in the founding and advancement of the New England town forest movement.

With much happening within the region, the New England Council in 1937 returned the favor by encouraging the governors to create a formal regional association of their own, the New England Governors Conference (NEGC), which would become the first and oldest such organization in the nation. For almost three decades thereafter, the council would provide the actual secretariat for the governors, materially assisting them in formulating and advancing their agendas. Suitably organized and staffed, the New England governors were now in a position to utilize their executive powers and influence in a number of fields including natural resources.

For example, in the late 1940s, caught up in the fervor for multipurpose river valley authorities in Congress, New England was confronted with proposals for natural resources-based federal authorities in the Connecticut and Merrimack River valleys. The measures called for unified water control, resource development and, especially, low-cost publicly generated, hydroelectric power — all hot-button items in New England. In addition, many resource study proposals were pending in Congress.

As has been documented (Foster, 1984), New England's governors, congratulating themselves on having maneuvered various legislative proposals into oblivion, awakened in 1950 to discover that by presidential directive a New England–New York Interagency Committee (NENYIAC) had been created to carry out a comprehensive study of the region's water and related land resources. What made the action especially galling was the fact that the NENYIAC investigation would be chaired by the Corps of Engineers, carried out exclusively by federal agencies, and only advised by governors' representatives.

In action reminiscent of the American Revolution, the governors' appointees rose up in 1951 to demand an actual voice in all decisions. A responsive federal engineer, Colonel Benjamin B. Talley, sympathetic to the states' position, created an executive council in which the six federal agency representatives and the six state advisors would each have one vote.

Four years later, forty-six volumes of inventory information had been assembled, bound within gold covers, and subjected to public hearings in each state — the first comprehensive account of water and related land resources in New England's history. The consensus recommendation was that this document would serve only as a guideline for future program activity. What to do next would occupy the governors' representatives and their now-cordial federal colleagues for a decade and a half.

The preferred action by the states was an interstate compact that would solidify the principle of state-federal co-equality pioneered by NENYIAC, an effort that ultimately failed, despite the best efforts of the governors and a supportive New England Council. The proposed Northeastern Water and Related Land Resources Compact, though endorsed by a majority of the New England state legislatures, never received congressional consent due to opposition from Washington-level federal bureaucrats and the unrelenting hostility of Vermont Senator George D. Aiken, a man who never forgot his earlier battles with the

Corps of Engineers as governor over flood control projects in his state.

But, in 1967, New England's day would finally arrive. The governors' request for the establishment of a New England River Basins Commission (NERBC) was approved by the federal Water Resources Council under provisions of Title II of the 1965 Water Resources Planning Act. Wary Washington deliberately selected a non-New Englander, Coloradoan R. Frank Gregg, to serve as federal chairman, a choice that ultimately proved to be a wise one, for under Gregg's leadership the NERBC went on to become the premier national river basin commission of them all.

Picking up where NENYIAC left off with the help of some fifty bright young professionals, the NERBC built and analyzed important resource data bases and prepared comprehensive plans for all of southeastern New England, Long Island Sound, and the Connecticut River valley. It also undertook special resource investigations for areas and issues identified by the states. But, by far, the greatest contribution made by the NERBC was its service as a forum for state, federal, and public participants.

As Foster later wrote, the NERBC was a place where there was the capacity to pull people together to discuss mutual needs and concerns. It was a place where an outsider could usually find out what was going on. It was also a place where insiders at high policy levels could share experiences and get to know each other better. When the commission was terminated in 1987, as one participant put it, the region lost "an important instrument for the cooperation of man."

As the 1970s neared, times would change again for the New England governors. By then an additional regional agency was in place — the New England Regional Commission (NERCOM) — a federal-state agency established under Title V of the Economic Development and Public Works act of 1965 with program responsibilities for economic development, labor, transportation, and energy. The governors themselves, not their designees, served as the commissioners because unlike the NERBC, which was supported primarily by special project funds, NERCOM received block grants of as much as $5 million annually for the governors/commissioners to allocate as they wished. NERCOM was headed by an appointed chairman who was recognized as the senior federal official in the region.

As times and priorities changed, so would the governors' involvement in natural resources affairs. The remarkable bi-partisanship on regional water resources matters during the 1950s and 1960s began to fray under the pressure of worsening economic conditions and polarizing issues such as the 1970s oil shocks and the issue of nuclear power plants. To assure their independence, the governors chose to cast off their historic dependency on the business community and create their own nonprofit organization, the New England Governors Conference, Inc., an arrangement that worked well as long as there were NERCOM funds to pay for staff and program services.

Despite efforts to encourage harmony between NERBC and NERCOM, each headed by experienced, aggressive, and ambitious officials, such as having the current chairman-governor's staff person serve as common liaison and co-officing the two agencies, relations remained cool. There were occasional projects where the region's growing environmental awareness and NERCOM's economic priorities were in consonance, such as the $300,000 grant made in 1970 to the New England Natural Resources Center (NENRC). With funds channeled through the states' leading nonprofit conservation organizations, the NENRC-led effort was designed to identify and offer transportation and business interests advance notice of natural areas to be avoided. It was the forerunner of today's state-based Natural Heritage and Endangered Species programs. But, for the most part, the two regional organizations went their separate ways. By the advent of the Reagan administration in 1981, and its decision to terminate both the Title V and the Title II commissions, it was time for New England to go back to the drawing board.

One trigger of renewed regional activity was the action taken by the governors and others in 1988 when European investor Sir James Goldsmith suddenly acquired Diamond International's near million-acre New England forestland holdings and put them on the market, thereby fulfilling the warning in 1987 by economist-forester Perry R. Hagenstein that it was only a matter of time before the pattern of New England's vast industrial holdings would change. Accustomed to ready access and use of industrial lands, a shocked environmental community turned first to Congress for a preliminary study of forest lands in northern New York and New England, conducted by the Forest Service, with the states in the usual position of advisors.

Much like the earlier NENYIAC period, the states then rose up in protest to insist on the formation of a subsequent Northern Forest Lands Council composed of governors' appointees with the federal presence in the minority. Echoing the experience of the earlier New England River Basins Commission, former members reconvening in 2004 for a retrospective look at their accomplishments gave the highest marks to the role the council played in serving as an open forum and sponsoring "listening sessions" throughout the region to enable individuals, organizations, and agencies to address contentious issues of forestland conservation and usage. Although the council declined to recommend a continuing organizational presence, it did stress the need for "new, imaginative thinking and doing" among individuals, communities, and agencies and set the stage for the creation in 1997 of the nonprofit, Concord, New Hampshire-based, Northern Forest Center. By then, the dominant issue had become not the future of the region's forest, but the future of its forest-dependent communities.

Another innovation was the governors' decision in 1973 to create a second nonprofit organization—the New England Governors/Eastern Canadian Premiers Conference (NEG/ECP)—which reflected the emerging relationship between northeastern Canada and the United States in the areas of agriculture, energy, trade, and natural resources. Quarterly meetings of both organizations now occur concurrently, guided by co-secretariats in Boston and Halifax and the governors/premiers' region-wide, cabinet level Committee on the Environment. Important natural resources initiatives have included those relating to air pollution, acid rain, mercury, climate change, and ocean resources. Canadian researcher Ulrike Rausch, the author of a candid assessment of the status of the Northeast's regional affairs, has applauded the uniqueness of this relationship, but has found the region increasingly "torn between the need to solve regional problems and the regime of the global market."

But as New England Futures experts Neal Peirce and Curtis Johnson warned recently (*Boston Globe* Op-Ed, April 20, 2006), this "proverbially smart" region of America had better not continue sitting on its hands and letting the quality of its environment decline. There is precedent in history, they wrote, for imaginative and creative governors setting to work to make their region both survive and excel. "Only the governors have the popular mandate, the cache, the recognition, and respect to get agendas moving quickly and clearly."

Thus, as the millenium turned, much like the situation a century ago, New England and its states would be in need of a new, strategic vision for land conservation accompanied by a critical mass of political will to carry it out. Only time would tell whether this objective would come to be realized.

OVERARCHING AND UNDERLYING THEMES

From our perspective, the rationale for land conservation in New England is less important than its reality for, over the years, New Englanders have compiled an enviable record of successful actions relating to the natural environment that merit commendation. While ours can only be an overview, it will illustrate several characteristics that seem to distinguish New England's approach from those of other regions. They are set forth below. In the states' overview portion of our account, we will demonstrate how these themes have become manifest over the years in particular programs. In our concluding section, we will offer thoughts on what these actions and events portend for the future, and how our experience can be instructive for other regions.

The first common theme has been the New Englanders' commitment to self-determination — that is, not letting others design and dictate their future. We see this first in the legacy of the Revolutionary War for Independence and, later, in the history of state-federal relations within the region. This characteristic helps explain New England's reputation for occasionally contentious political and programmatic independence.

A second common theme has been that of innovation. For years, ours has been a region of inventors and tinkerers. The list of New England firsts is simply staggering, ranging from manufacturing products and processes (e.g., clocks, guns, shoes, electronics, mills), financial services (e.g., banking, insurance), to the social inventions (e.g., public schools, hospitals, research, higher education, and nonprofits, to name just a few). As you will see, New Englanders have been equally creative in addressing natural resources and environmental issues and needs.

The third characteristic of New England conservation has been a reliance on individual leadership as a first resort, much of it nongovernmental in nature. This was particularly evident at the turn of the twentieth century before conventional conservation institutions were in place. But, even later, those in office have continued to affix an individual stamp on conservation affairs despite the formal constraints of their organizations and agencies.

A fourth theme over time has been the New Englanders' strong commitment to place — in the case of conservation, actions taken to secure specific locations with special meanings to people.

A fifth theme has been the long history of civic engagement. One example is the town meeting, still the primary local governance device for the region. Thomas Jefferson once applauded the New England town as "the wisest invention ever devised by the wit of man" and Alexis de Tocqueville called it "the best way to bring (liberty) within the people's realm." Despite the constraints of the town format, the principles of bringing people together to consider common problems, and requiring government to seek the consent of those governed, are as valid today as they were three centuries ago. Bottom-up solutions remain the ones most likely to succeed in our region.

The final theme has been the New Englanders' moral and ethical concerns for the environment, which include a willingness to have the region's natural resources contribute directly to their individual or collective economic benefit. Today's environmentalists are quick to criticize the pragmatic land-use and natural resources practices of the past. In so doing, they fail to use the long lens of history. In the beginning, for example, when sheer survival depended upon subduing the wilderness, and the legal imperatives of the new plantations lay in producing an economic return to their sponsors, natural resources were understandably to be developed and even exploited. But even in Colonial times, it was often deemed sensible, and morally desirable, to temper the extraction of materials and the use of resources in order to make provision for the future. New England has clearly been a national leader in devising ways to both conserve and utilize its natural resources.

When we are done, there will be a record of a conservation heritage in which all New Englanders can take pride. It will be an account not just of individuals, events, and accomplishments but also the fundamental relationship between humans and their environment over time — what writer Thomas Urquhart has properly termed "finding the proper balance between artifact and nature" (Urquhart, 2004). But before doing so, the six overarching and underlying themes expressed above need to be described in more detail.

SELF-DETERMINATION

Given its historic roots, it is no surprise that New Englanders have developed a strong antipathy to being told what to do by others. Much can be traced to the origins of the American Revolution when citizens rose up to oppose the dictates of the Crown, but the seeds of self-reliance and self-determination had actually been sown years before when colonists resisted similar domination by their religious leaders. The sense of community represented by the founding of villages and towns, and the development of the machinery of representative government, led to an early separation of church and state.

But more than anything, their town meeting heritage fortified the New Englanders' determination to be heard, a characteristic poet and novelist Archibald MacLeish described as the "sea-captain's habit of speaking out in the wind's teeth" or, in colloquial Vermont terms, "becoming as independent as a hog on ice" (*New England In a Nutshell*, 2002). The region's well-known penchant for individualism (Lauter and Zagarell, 2006), was part of the seventeenth century Protestant concept of personal salvation — the need for each individual to experience personally God's "pure grace."

In the New World, the ethos of individualism soon became manifest in more tangible ways, such as the decision by the Mayflower community, three years after its founding, to abandon the concept of communal ownership of land in favor of individually owned plots. The early settlers found it not inconsistent to exist within a community where all were equal, but allowance was made for the exercise of self-determination and individualism (Moakley, 2006). For conservation, this has led to the seeming Yankee contradiction of individuals acquiring great personal wealth, but also retaining a strong commitment to public service, good government, and the well-being of the commonwealth as a whole.

Whether cause or effect, the town meeting tradition has had much to do with the New Englander's insistence on remaining self-determinative. Settled by small, close-knit, compactly organized groups, not by adventurers, towns constituted a useful way to balance individualism and conformity (Bond, 2006). Yet, town meetings, both then and now, remain strongly human occasions — inquisitive, assertive, gossipy, and contentious. Towns still serve as the primary governance institutions for the 62,000 square mile, six-state region of the Northeast. Having existed for more than three hundred years, the basic concept of the citizen assembly continues to retain much of the vitality and popular support it enjoyed three centuries ago.

As we will see, New England conservationists have made effective use of this long tradition of self-reliance and grassroots governance during the twentieth century. They have employed town-based entities successfully as instruments for policy and program formulation, as forums to weigh the acceptability of regional and national proposals, and even ways to carry out actual projects. In fact, it is local entities that are now bearing most of the burden of land-conservation investments during a time of declining support for the environment at state and federal levels.

INNOVATION

As mentioned earlier, the New England qualities of ingenuity, pragmatism, and thrift have given the region a national reputation as a place where ideas and inventions readily arise and prosper. The list of important firsts is impressive — many but not all the result of the New England region being the first in the country to be extensively settled. As we will see, among these firsts have been significant conservation innovations. Before turning to what has made New England conservationists successful innovators, one needs to describe what innovation is, how it manifests itself, and the means by which innovative ideas and practices come about.

As Foster and Levitt advise, an innovation, by definition, is simply something new (*nova*). An environmental innovation is a new idea applied in an environmental setting. To be a true innovation, the outcome must represent a significant departure from the norm. The result will be a fundamental transformation — at least additive and often disruptive. Further, an innovation should be rare, new to the organizational scene, substantial in size or impact, and durable over time (Lynn, 1997). Innovation, in general, has a positive connotation but, in practice, it can lead to mixed results — hazardous as well as successful.

Conservation innovators are encountered in both the public and private sectors. Public innovators benefit from their generally broad license to meet statutory environmental objectives, but are constrained by the lack of flexibility of their organizations and the largely reactive nature of public service. In an environment where public servants tend to be judged more by what may be lost rather than what might be gained, there are few incentives for innovation. The culture of such agencies can make departure from the norm appear to be an assault on the very essence of their organizations and lead to internal resistance to change. There are also political time constraints working against public innovation. For example, the high-level administrator is usually there for four years at best — with luck, long enough to get an innovation authorized but too short to see it implemented in practice.

On the private side — especially in what is referred to as the nonprofit sector — a host of other challenges can militate against innovation. Although their relatively small size, constant needs, and high aspirations can create a favorable climate for innovation, nonprofits are usually better at conceiving new ideas and approaches than carrying them out. The obvious answer is collaboration with the public sector or other private organizations, but that risks losing the distinction of the innovation and diminishing the perceived ownership of the entrepreneurship that each organization seeks for its own prominence. These problems can be mitigated if the nonprofit is of a size to sustain its own activities, but then the equivalent baggage of a public agency may surface — conservatism, caution, and inertia.

In conservation, innovation is frequently the result of nonprofit organizations who "shadow" their governmental counterparts, often sponsoring initiatives that the latter are then asked to carry out. The reverse is true also — programs advanced by governmental agencies that the nonprofit community and the public are called upon to support. Another variant occurs when a legislative body enters into a working partnership with both public and private conservation entities, such as the famous "iron triangle" of legislative leaders, federal executives, and nonprofit leaders that has been responsible for so many pathbreaking, national environmental program authorizations. But what about the human side of innovation in practice? Here we are indebted to the observations of a number of outside specialists.

Although innovation may arise internally from within a program or organization, there is generally a trigger or what some call a "precipitating jolt" (Poole and Van der Ven, 2004) that disturbs existing practice and sets the stage for change. In governmental programs, this can be an unplanned event such as a well-publicized exposé of alleged inadequacy or malfeasance, or an orderly occurrence such as a change in individual leadership or elective office. On the non-governmental side, a similar precipitating event might be the unexpected termination of operating support by a major donor or philanthropy or a decline in membership. Among planned, change devices, a common practice is what the Ford Foundation's Innovation in American Government Program has called "scaffolding" (Gilmore and Krantz, 1997) — the creation of outside study and advisory structures to search out and recommend opportunities for change.

When it comes to implementation, some administrators will feel more comfortable simply adopting innovations that have proven successful elsewhere (Kanter, 1983), but most will want to apply their own stamp and modify them to meet their own needs. The small, relatively informal, and noncomplex organization will generally be the best one to innovate, but this can lead to what Altshuler and Behn (1997) call "quandaries" — the fact that dispersed power helps conceive innovations but concentrated power is required for their adoption. In most instances, incremental change will be preferable to wholesale change. In such cases, the best type of implementation will be an evolving, not an episodic occurrence, extending over a period of time.

But how can one recognize an innovation when it occurs? To sum up, we will use James Levitt's suggested criteria for a landmark conservation innovation (Levitt, 2002):

> *novelty*, to the extent that the innovation demonstrates a spark of creativity;
> *significance*, to the degree that it addresses an issue of public concern;
> *effectiveness*, to the degree that it delivers tangible, quantifiable results;
> *transferability*, to the degree that it can be replicated by other organizations;
> *ability to endure*, to the extent that the innovation has

> demonstrated, or shows strong promise of demonstrating, a lasting impact over a significant span of time.

Finally, just who are innovators? They are usually singular creatures with well-developed egos. While many will cast their expectations in terms of apparent economic and social returns, the truth is that most just like to tinker, invent, and enjoy the limelight. In their case, the mountain is to be climbed simply because it is there. Pure innovators tend to be a lonely lot requiring patience, understanding, encouragement, and support from others. They are often difficult to work with. To preserve creativity, they need a large measure of freedom from competing responsibilities. But at the critical juncture when the innovation is ready to be applied, those with requisite people skills must be ready to step forward so that the creations can be brought to fruition in a timely and meaningful manner.

LEADERSHIP

A third characteristic of New England conservation has been its long tradition of leadership by individuals.

In the early part of the twentieth century when institutions for conservation were in their infancy, there was simply no alternative to individual and largely private leadership. Since the pool of leaders was relatively small, one encountered individuals who performed that role in a variety of settings, thereby cross-pollinating a number of needed societal initiatives and collaborating at prominent and influential levels. One finds the same individuals who were founding museums and hospitals, building universities, and launching social programs involved in conservation as well.

Later on, despite the growth of agencies with statutory responsibilities for conservation and the environment, the tradition of leadership by individuals continued, often at the governmental level. At times, the distinction between public and private initiative has been blurred. Viewed historically, one could even assert that the current concepts of public-private collaboration and partnership, if not originating in New England, have at least risen to particular prominence in this region. Before offering illustrations of such individual conservation actions in each of the New England states, it would profit us to examine leadership in both principle and concept.

The word leader seems to derive from the Anglo-Saxon word *laedon*, which means to travel, to go, to move in some way, or to set a direction (Langton, 1984). In its simplest form, leadership is the capacity of an individual, in New England political scientist and environmentalist James MacGregor Burns's view (1978), to arouse, engage, and satisfy the needs of others. This definition raises the obvious need for a leader to have followers and to bring about end results that are perceived by others to be useful.

Leaders come in all shapes and sizes (Foster, 1993). They are both born and made. Some have characteristic styles, such as numinous (godlike) or inspirational qualities. Others operate largely through familial processes predicated upon bonds between equals or their perceived expertise. So-called charismatic leaders exert a pure form of leadership based upon shared values and the singular devotion of followers. The results of their leadership tend to be transformational in nature. In contrast, there are also transactional types of leaders who simply make things happen. In the increasingly complex world of conservation, such leaders play a crucial, if often, unheralded role.

At its best, leadership is morally purposeful. In practice, it is invariably dissensual, generating conflict through sharpened demands and challenges to the status quo. To constitute true leadership, however, the act must always be causative — that is, something must happen.

What does it take to be a leader? The individual, beyond being knowledgable, must be perceptive, able to grasp information readily, have good intuition and interpersonal skills, be empathetic, and enjoy a substantial measure of inner security and confidence. Moreover, the situation at hand must be one amenable to an application of these qualities. Equally important is what leadership is not. For example, leaders are not dependent upon status or office. They need not have or exercise authority. They may not even be popular.

Mumford, et al. (2006) offers three common pathways to outstanding leadership. *Charismatic* leaders rely upon follower-based appeals. Their objectives often include what is or appears to be, self-aggrandizement. *Ideological* leaders are particularly skilled at conceptual integration. In contrast to the charismatic leader, they are often content to allow their objectives to be advanced by trusted associates. *Pragmatic* leaders operate through expertise, logical appeals, and rational persuasion, relying on their ability to craft viable solutions to problems arising as crisis or change occur.

Finally, different organizational contexts and different times will require different kinds of leaders. As an example, a personalized style of leadership, undertaken at the wrong time, can become a significant disruption to an organization's overall performance, especially in a governmental setting. For that reason, the most successul conservation programs are often started by a charismatic or ideological leader and then brought to prominence through those with pragmatic knowledge and skills.

SENSE OF PLACE

As Springfield College's Paige W. Roberts has observed ("Images and Ideas," Encyclopedia of New England, 2005), the perceptions, values, and meanings New Englanders have attached to their region over time represent a distinct sense of place. Before seeing how these have played out in practice within the various states, a clearer understanding of place itself needs to be set forth. For this, we are indebted to two main sources: the early classic by Canadian geographer Edward Relph (*Place and Placelessness*, 1976), and the recent summary of key geographic concepts (*Place: A Short Introduction*, 2000) written by the University of Wales's Tim Cresswell and dedicated to the legendary humanist, Yi-Fu Tuan.

As we are advised, place differs from physical space by being invested with human meaning. It usually contains one or more of the following elements (Henri Lefebvre, 1991): an area that can be physically bounded and mapped; an area that is more subjective and imagined; and an area that is practiced and lived in. In New England can be found all three of these elements. What is called home is a special kind of place where people feel a particular sense of attachment and rootedness.

Place is far from a static concept. For a moment in time, it may have a distinct identity — even a uniqueness — but this can change as interactions take place in conjunction with outside processes and events. In all cases, however, the result is an accumulating storehouse of human memories and experiences. It is these that practicing land conservationists so frequently access and put to effective use.

The modern erosion of place has been much decried, but it is a fact that even in an increasingly global environment, humans still find ways to develop distinct place connections—often at multiple locations of work, play, and even home. As some have argued (Foster, 1995), the modern wonders of ready transportation and communications, far from threatening place, have actually expanded its boundaries. People can now attach themselves to many different places, becoming persons of several places in several dimensions of time. Humans can develop a strong connectedness to places even when living thousands of miles distant. Despite the absence of propinquity, these can become powerful and influential attachments.

Relph's concern for authenticity—an awareness and acceptance of responsibility for a given place—has been modified by the growing importance of what French anthropologist Marc Auge (1995) calls empirical "nonplaces"—spaces where people coexist and even cohabit without actually living together for any length of time. Examples are banks and post offices, supermarkets, malls, neighborhood eating places and gas stations, vacation and retirement sites and, of course, the Internet. These can all become potential meeting places to share and articulate ideas and form commitments.

The most common use of place connectedness in New England land conservation has been to protect a particular area from development or other encroachment. Indeed, the most successful examples of land conservation in practice have been those focused on a particular place. But another prominent manifestation occurs when place is largely socially constructed—situations where physical location is neither a necessary nor sufficient condition for action (Relph, 1976). This makes it possible for people to associate in a programmatic context—say, for parks, open space, or environment in general—and to focus their intentions within a larger dimension.

Working conceptually with place has spawned a variety of approaches (Foster, 1995) but, generally speaking, place reflects three prominent characteristics: a landscape setting, a group of associated activities, and a significance to people. Each of these must be assessed and measured carefully. Mapping can identify the topographic characteristics of a place, but personal interviews are also needed to reveal its human dimensions. Where the physical setting is plausible and a commonality of concern appears to exist, the stage is set for successful, place-based conservation.

Carrying forward place connectedness to a regional level is more complicated, especially in New England where the sense of local community is so strong. But as we saw earlier, the six most northeastern states, viewed collectively and broadly, have sufficient topographic and historic similarities to warrant an assertion that there really is a place worth saving called New England.

CIVIC ENGAGEMENT

Elements of leadership, innovation, and place connectedness often merge into acts of what is variously called public participation, civic engagement or, in our case, civic environmentalism. All reflect de Tocqueville's observation of the ready willingness of Americans to associate with one another to state a position or advance a cause. The manifestation is usually an organization or association centered around a specific issue, often institutionalized later on to provide an enduring presence. New England has an extraordinary array of such environmental citizenship institutions, some more than a century old.

As Bowdoin College environmental studies director DeWitt John has observed (John, 1994), civic environmentalism has the virtue of being able to engage local citizens directly in conservation matters. By doing so, authorities can access specialized information and social networks and identify a range of options that are likely to make it easier for citizens to accept responsibility and support remedial actions.

The principles of civic engagement are as old as New England itself. The famous Mayflower Compact, entered into by the Pilgrims in Provincetown Harbor on November 11, 1620, established as a basic principle of American government that law rests on the consent of the governed. The compact signers established a "civil body politic" to guide their actions and later replicated it at the local level in their town meeting and governance mechanisms. As has been true of so much of New England conservation, the origins of the idea were partly religious (faithful to the settlers' Congregationalist principles), partly cultural (familiar to those emigrating from England), but also eminently practical (the most effective way to cope with the unknown circumstances of the New World).

Emerging in the New Deal era of the 1930s, and strengthened in the 1960s, was the group-pluralist approach variously called "participatory democracy" or "creative federalism," all based on the same principle of governmental accountability to the public. No longer constrained by the limits of particular jurisdictions, the key operational questions were what citizens, acting in what ways, through what means, and fulfilling what roles in the public participation process (Priscoli, 1995).

Since then, civic engagement has become an accepted part of American conservation. As has been observed (Landy, 1999), it seeks to replace coercion from afar with the exercise of responsibility at home, thereby increasing peoples' capacities to govern themselves. The approach cuts across the political spectrum, appealing equally to conservatives who prefer nonfederal and market-based approaches and to liberals who favor creative, democratic action. The common thread is empowerment at the local level.

Civic engagement helps build what Putnam (2000) has termed "social capital"—the connections between people that provide the support, cooperation, and trust essential for institutional effectiveness. These connections are usually of two types: those that band individuals together in common cause, and those that bridge differences in order to gain additional assets and support. Examples of both in New England conservation will be offered later.

But the record of civic engagement applied to conservation in New England has been a mixed one. In earliest times, rugged individualism was the celebrated way to go—what de Tocqueville described as the "calm and considered feeling which disposes each citizen to isolate himself from the mass of his fellows . . . (leaving) the greater society to look after itself." As others have noted (e.g., Putnam, 2000), this myth still pervades the American psyche. Over time, as problems proliferated, necessity forced an uneasy accommodation between community and individual actions.

During the Progressive Era at the turn of the twentieth century, grassroots and national leadership joined forces to carry out a remarkable agenda of social invention and political reform that included the founding of the American conservation movement and its organizational manifestations. The leaders in this instance were an unlikely combination of naturalists, fashionable sportsmen, and a few men of power in government (Brooks, 1980). In the aftermath of World War II, there was another burst of environmental experimentation built upon the civic mindedness of the wartime period of national unity and patriotism.

By the late 1950s, however, as Putnam has documented, the commitment to community-based environmentalism had begun to wane, replaced by the artificial conservation organization membership increases resulting from skillful marketing and direct mail approaches. There were exceptions, of course, such as the birth of the locally based land-trust movement but, in general, support for the environment was found to be achieved more efficiently, and at less cost, by simply seeking individual contributions rather than personal participation. The most influential organizations were Washington-based, professionally rather than citizen-operated, and long removed from their original, place-based, participatory roots.

As environmental issues became more complex, it was clear that single contexts or levels of concern — national, state, regional, and local — would not be the answer. Multidimensional actions were required. It was recognized that competition between organizations and jurisdictions must give way to a new era of mutual concern and cooperation supported by a combination of classic civic engagement and more contemporary cybertechnology.

As we will see, New England has blazed an important trail through these shifting times, contributing meaningfully to the development of the new national environmental ethos. New Englanders have played numerous roles — variably staunch individualists, effective if somewhat reluctant cooperators, institutional innovators and operators, and iconic action and thought leaders. In these senses, they have broken new ground for, as well as mirrored, one of the most important social movements in the history of the United States.

THE ETHICAL IMPERATIVE

New England's historic place-related consciousness and activism profited much from the remarkable group of scientists, writers, and artists who emerged in the early 1800s to set the stage for the region's later conservation movement. Influenced by, but appreciably removed from, what was happening on the European continent, the American version of the Enlightenment saw a surge of inquiry accompanied by a growing popular interest in science. This was followed by the late nineteenth century rejection of materialism in favor of the moral, aesthetic, and ethical virtues of the environment as expressed by the successor Romanticists.

The scientific inquiries of the early period were appreciably different from those of today since they still retained a strong commitment to the supposed divinity of the organic world. However, the extensive field studies carried out by explorer-naturalists such as Alexander Wilson and John James Audubon (Judd, 2006) were beginning to reveal the diversity of the natural world and its fundamental interconnectedness, albeit still within the concept of the purposiveness of the Creator and the prevailing belief that nature needed to be tamed and altered in the interest of humankind. The later primacy of the ideal over material reality would later give rise to the proposal to transcend the material world of everyday experience through the intellectual effort and self-reliance advocated by the Concord transcendentalists Ralph Waldo Emerson, Henry David Thoreau, and Nathaniel Hawthorne.

At the turn of the twentieth century, the writings of Henry David Thoreau were finally published, giving rise to the reputation and the myth of what the world came to believe was Thoreau's unreserved advocacy of unspoiled wildness and nature. In fact, Thoreau found nature devoid of human influence often "repellent and fearsome" (Lowenthal, 2000), and Thoreau's remarkable natural history observations were often tinged by a genuine interest in the way natural resources were being utilized pragmatically by his neighboring Concord-area residents and farmers (e.g., Foster, 1999).

Close on the heels of the explorer-settlers and explorer-naturalists, and coincident with the rise of Romanticism, came the so-called Hudson River School of landscape painting led by Thomas Cole, Asher B. Durand, and Thomas Doughty—artists based first in New York's Catskills but later successful portrayers of New England's White Mountains and other regions.

As historian Christopher Johnson has written (Johnson, 2006), the growing nationalistic pride in the American wilderness encouraged these artists to employ three basic concepts: the *sublime* (powerful and threatening images of nature that inspired awe and terror); the *beauty* (features that were smooth and delicate and invoked a feeling of empathy); and the *picturesque* (using light, surface, texture, and other qualities to produce a set of images that would often be allegorical or even quasi-imaginary). So popular were these artworks that, in addition to the Bible, a New England household of that era would typically contain a landscape print hanging over the kitchen table.

The literary and artistic contributors to the Romantic Era enjoyed significant cross-connections with one another, with the explorer-settlers, and with the explorer-naturalists. They also came under the influence of the religious evangelists who emerged in the 1820s and 1830s. Thus, it was no accident that the works of scientists, writers, and artists often contained Christian metaphors, symbols of the presence of God in nature, and evidence of spiritual transcendence.

One further ethical characteristic of New England conservation emerged at the turn of the twentieth century—the advocacy of a balanced use of the environment that was both utilitarian and conservative. This stemmed in part from the need to arrive at an accommodation between the exploitative approach to natural resources, whose effects were captured so graphically by the pioneer photographers of the late nineteenth century, and the need to accept the growing scientific awareness of nature operating through long-term, interconnected systems. In all aspects of science, literature, and art, questions began to arise as to the wisdom of a policy of human dominance over nature, lacking a corresponding commitment to actions that would be thoughtful, knowledgeable, productive, and custodial for the environment as a whole. The professional, it was asserted, should be the one to chart the way.

By the mid-twentieth century, an important advocate of this policy was discovered (actually rediscovered), Vermonter George Perkins Marsh (Lowenthal, 2000). The 1955 Princeton symposium, *Man's Role in Changing the Face of the Earth*, brought Marsh's views into particular prominence, providing a national platform to underscore his fundamental views of the roles of human agency and enlightened self-interest. So irrevocably overlain was the natural world by human agency, Marsh wrote in *Man and Nature*, that prudent stewardship would be required in order to reverse the effects of environmental misuse. He was optimistic that such stewardship would provide the answers.

This utilitarian view of nature would dominate until the new wave of environmentalism appeared in the 1960s and 1970s. As ethicist Thomas Dunlap has observed (Dunlap, 2004), it then fell to trained professionals such as Aldo Leopold and Rachel Carson to serve among the first in New England and the nation to successfully preach the overarching duty of humankind to assure a secure place for all life on earth.

With this thematic structure in mind, we can now turn our attention to preparing an account of New England conservation over time. The most meaningful elements will be those we offer that have occurred on a state by state basis. However, case examples will also be offered of innovations that expanded beyond individual state boundaries to encompass most, if not all, of the region. We will also track elements that originated at regional or national levels. The twentieth century will be our primary time frame. In the interest of preparing a manageable account, land conservation will be our principal area of illustration. When we are done, there will be a record of a conservation heritage in which all New Englanders can take pride. It will be an account not just of individuals, events, and accomplishments but also the fundamental relationship that developed between humans and their environment in our region over time, and what it may portend for the future.

SELECTED REFERENCES

Introduction

Brooks, Van Wyck. 1950. *New England: Indian Summer*. Dutton. New York, NY.

Brown, Dona. 1995. *Inventing New England: regional tourism in the nineteenth century*. Smithsonian Institution Press. Washington, D.C.

DeVoto, Bernard. March 1932. "New England, there she stands," *Harper's*.

Feintuch, Burt and David H. Watters (eds.). 2006. *The Encyclopedia of New England*. Yale University Press. New Haven, CT.
"Captain John Smith" (Karen Ordahl)
"Governors" (Scott Harris)

Foster, Charles H.W. 2006. "Bridging borders: the prospect for environmental regionalism in New England and Atlantic Canada." In *Regionalism In a Global Society* (Stephen G. Tomblin and Charles S. Colgan, eds.). Broadview Press. Peterborough, ON.

Stowe, Harriet Beecher. 1869. *Oldtown Folks*. Houghton, Mifflin. Boston, MA.

Wood, William. 1634. *New England's Prospect* (1977 edition, Alden T. Vaughan, ed.). University of Massachusetts Press. Amherst, MA.

New England as a Region

Advisory Commission on Intergovernmental Relations. April 1972. "Multistate regionalism." U.S. Government Printing Office. Washington, DC.

Brown, Caroline N., et al. 2006. "The evolving use and the challenging role of interstate compacts: a practitioner's guide." American Bar Association. Chicago.

Conforti, Joseph A. 2001. *Imagining New England: Exploration of Regional Identity from the Pilgrims to the Mid-twentieth Century*. University Press of New England. Hanover, NH.

Feintuch, Burt and David H. Watters (eds.). 2006. *The Encyclopedia of New England*. Yale University Press. New Haven, CT.
"Introduction — images and ideas" (Paige W. Roberts)

Foster, Charles H.W. and William B. Meyer. 2000. The Harvard environmental regionalism project. Discussion Paper 2000-11. JFK School of Government, Harvard University. Cambridge.

Jensen, Merrill (ed.). 1951. *Regionalism in America.* University of Wisconsin Press. Madison.
Meyer, William B. and Charles H.W. Foster. 2000. New Deal regionalism: a critical review. Discussion Paper 2000-02. JFK School of Government, Harvard University. Cambridge.
National Resources Committee, December 1935. Regional factors in national planning and development. U.S. Government Printing Office. Washington, DC.
Raitz, Karl B. and Richard Ulack. 1984. *Appalachia, a Regional Geography.* Westview Press. Boulder, CO.
Reidel, Carl H. and Charles H.W. Foster. 2001. New England's prospect: environmental regionalism in New England. Discussion Paper M-96-01. JFK School of Government, Harvard University. Cambridge.
Zimmerman, Joseph F. 2002. *Interstate Cooperation: Compacts and Administrative Agreements.* Praeger. Westport, CT.

Interstate Compacts

Brown, Caroline N., et al. 2006. "The evolving use and the challenging role of interstate compacts: a practitioner's guide." American Bar Association. Chicago, IL.
Council of State Governments. 1983. *Interstate compacts* (a revised compilation). Lexington, KY.
Foster, Charles H.W. 1984. *Experiments in Bioregionalism: The New England River Basins Story.* University Press of New England. Hanover, NH.
Frankfurter, Felix and James N. Landis. 1925. "The compact clause of the Constitution: a study in interstate adjustments." *Yale Law School Journal*: 34(7): pp. 685–758.
Hardy, Paul T. 1982. *Interstate compacts: the ties that bind.* Institute of Government, University of Georgia. Athens.
Rausch, Ulrike. undated. *The potential of transboundary cooperation: still worth a try.* An assessment of the Conference of New England Governors and Eastern Canadian Premiers. Center for Foreign Policy Studies, Dalhousie University. Halifax, NS.

Governors

Blanchard, Newton C. (ed.). 1909. Proceedings of a conference of governors in the White House, May 13–15, 1908. U.S. Government Printing Office. Washington, DC.

Congressional Quarterly, Inc. 1994. "American leaders 1789–1994: a biographical summary" (governors). Washington, DC.

de Tocqueville, Alexis. 1831. *Democracy in America*. 1979 edition (J. P. Mayer, ed.). Anchor Books, Doubleday & Co. New York, NY.

Foster, Charles H.W. 1984. *Experiments in Bioregionalism: the New England River Basins Story*. University Press of New England. Hanover, NH.

Hagenstein, Perry R. 1987. "A challenge for New England: change in large forest ownerships." New England Natural Resources Center. Boston, MA.

New England Conference. November 23, 1908. Proceedings of the first New England conference called by the governors of the New England states. Wright and Potter. Boston.

New England Regional Planning Commission. June 1943. *A decade of regional planning in New England*. Publication No. 72, National Resources Planning Board. Boston.

Peirce, Neal and Curtis Johnson. April 20, 2006. "A call to action for New England's governors." Opinion A-13, *Boston Globe*.

Rausch, Ulrike. undated. *The potential of transboundary cooperation: still worth a try*. An assessment of the Conference of New England Governors and Eastern Canadian Premiers. Center for Foreign Policy Studies, Dalhousie University. Halifax, NS.

Rosenholm, Glenn. Autumn 2005. "Ten years later, common ground is easier to find." *Northern Woodlands*: Autumn 2005.

Sabato, Larry. 1983. *Goodbye to Good-Time Charlie: the American Governorship Transformed*. Congressional Quarterly Press. Washington, DC.

Zimmerman, Joseph F. 2002. *Interstate Cooperation: Compacts and Administrative Agreements*. Praeger. Westport, CT.

Self-determination

Bold, Christine. 2006. *Writers, Plumbers, and Anarchists: the WPA Writers' Project in Massachusetts*. University of Massachusetts Press. Amherst.

Commonwealth Editions (eds.). 2002. *New England In a Nutshell: Quotations about the People, Places & Particulars of Life in the Six New England States*. Commonwealth Editions. Beverly, MA.

de Tocqueville, Alexis. 1831. *Democracy in America*. 1979 edition (J. P. Mayer, ed.). Anchor Books, Doubleday & Co. New York, NY.

Feintuch, Burt and David H. Watters (eds.). 2005. *The Encyclopedia of New England*. Yale University Press. New Haven, CT.

"Introduction: politics" (Maureen F. Moakley)

"Introduction: literature" (Paul Lauter and Sandra A. Zagarell)

Foner, P.S. (ed.). 1949. *The Basic Writings of Thomas Jefferson*. Wiley Book Co. New York, NY.
Webster, Clarence M. 1945. *Town Meeting Country*. Duell, Sloan & Pearce. New York, NY.

Innovation

Altshuler, Alan A. and Robert D. Behn (eds.). 1997. *Innovation in American Government*. Brookings Institution. Washington, DC.
Berry, Joyce K. and John C. Gordon (eds.). 1993. *Environmental Leadership: Developing Effective Skills and Styles*. Island Press. Covelo, CA.
Foster, Charles H.W. and James N. Levitt. 2001. Reawakening the beginner's mind: innovation in environmental practice. Kennedy School Discussion Paper 2001-7, Harvard University. Cambridge.
Gilmore, Thomas N. and James Krantz. 1997. "Resolving the dilemmas of ad hoc processes." In *Innovation in American Government* (Altshuler and Behn, eds.). Brookings Institution. Washington, DC.
Kanter, Rosabeth Moss. 1983. *The Changemasters: Innovation and Entrepreneurship in the American Corporation*. Simon & Schuster. New York, NY.
Lynn, Lawrence E., Jr. 1997. "The dilemmas of innovation in American government." In *Innovation in American Government* (Altshuler and Behn, eds.). Brookings Institution. Washington DC.
Poole, Marshall Scott and Andrew Van de Ven (eds.). 2004. *Handbook of Organizational Change and Innovation*. Oxford University Press. New York, NY.

Leadership

Burns, James MacGregor. 1978. *Leadership*. Harper & Row. New York, NY.
Foster, Charles H.W. 1993. "What makes a leader?" In *Environmental Leadership* (Joyce K. Berry and John C. Gordon). Island Press, Covelo, CA.
Gardner, John W. 1990. *On Leadership*. Free Press. New York, NY.
Langton, Stuart. 1984. *Environmental Leadership*. Lexington Books. Lexington, MA.
Mumford, Michael D. 2006. *Pathways to Outstanding Leadership*. Lawrence Erlbaum Associates. Mahwah, NJ.
Zalesnik, Abraham. 1966. *Human Dimensions of Leadership*. Harper & Row, New York.

Sense of Place

Aberley, Doug. 1993. *Boundaries of Home: Mapping for Local Empowerment*. New Society Publishing. Gabriola Island, BC.

Auge, Marc. 1995. *Non-places: Introduction to an Anthropology of Supermodernity*. Verso. London, UK.

Cresswell, Tim. 2004. *Place: a Short Introduction*. Blackwell Publishing, Ltd. Oxford, UK.

Feintuch, Burt and David H. Watters (eds.). 2005. *The Encyclopedia of New England*. Yale University Press. New Haven, CT.
"Introduction — images and ideas" (Paige W. Roberts)

Foster, Charles H.W. 1995. The environmental sense of place: precepts for the environmental practitioner. Research Paper WP95CF1. Lincoln Institute of Land Policy. Cambridge.

Gallagher, Winifred. 1993. *The Power of Place: How our Surroundings Shape our Thoughts, Emotions, and Actions*. Poseidon Press. New York, NY.

Glassberg, David. June 1997. "Sense of place: a report to the Massachusetts Foundation for the Humanities." Northampton.

Hiss, Tony. 1990. *The Experience of Place*. Alfred A. Knopf. New York.

Kemmis, Daniel. 1990. *Community and the Politics of Place*. University of Oklahoma Press. Norman.

LeFebrve, Henri. 1991. *The Production of Space*. Blackwell Publishing, Ltd. Oxford, UK.

Relph, Edward. 1976. *Place and Placelessness*. Pion Ltd. London, UK.

Civic Engagement

Bold, Christine. 2006. *Writers, Plumbers, and Anarchists: The WPA Writers' Project in Massachusetts*. University of Massachusetts Press. Amherst.

Brooks, Paul. 1980. *Speaking for Nature: How Literary Naturalists from Henry Thoreau to Rachel Carson Have Shaped America*. Houghton Mifflin Co. Boston.

Commonwealth Editions (eds.). 2002. *New England in a Nutshell*. Commonwealth Editions. Beverly, MA.

de Tocqueville, Alexis. 1831. *Democracy in America*. J. P. Mayer (ed.), 12th Edition (1969). Anchor Books, Doubleday & Co., Inc. New York.

Foster, Charles H.W. 1968. "The New England town meeting." Unpublished working paper.

John, DeWitt. 1994. *Civic Environmentalism*. Congressional Quarterly Press. Washington, DC.

Landy, Marc C., et al. January 1999. *Civic Environmentalism in Action: a Field Guide to Regional and Local Initiatives*. Progressive Policy Institute. Washington, DC.

Priscoli, Jerome Delli. Fall 1995. "Twelve challenges for public participation practice." Interact 1(1): pp. 77–95. International Association of Public Participation Practitioners.

Putnam, Robert D. 2000. *Bowling Alone: The Collapse and Revival of American Community*. Simon & Shuster. New York.

The Ethical Imperative

Cronon, William. 1983. *Changes in the Land: Indians, Colonists, and the Ecology of New England*. Hill and Wang. New York.

Dunlap, Thomas R. 2004. *Faith in Nature: Environmentalism as a Religious Quest*. University of Washington Press. Seattle.

Foster, David R. 1999. *Thoreau's Country: Journey Through a Transformed Landscape*. Harvard University Press. Cambridge.

Johnson, Christopher. 2006. *This Grand and Magnificent Place: the Wilderness Heritage of the White Mountains*. University Press of New England. Hanover, NH.

Judd, Richard W. January 2006. "A wonderfull order and ballance: natural history and the beginnings of forest conservation in America, 1730–1830." *Environmental History* 11: pp. 8–36.

Levitt, James N. December 2002. Conservation innovation in America: past, present, and future. Institute for Government OPS-02-03, Kennedy School, Harvard University. Cambridge.

Lowenthal, David. 2000. *George Perkins Marsh: Prophet of Conservation*. University of Washington Press. Seattle.

Merchant, Carolyn. *Ecological Revolutions: Nature, Gender, and Science in New England*. University of North Carolina Press. Chapel Hill.

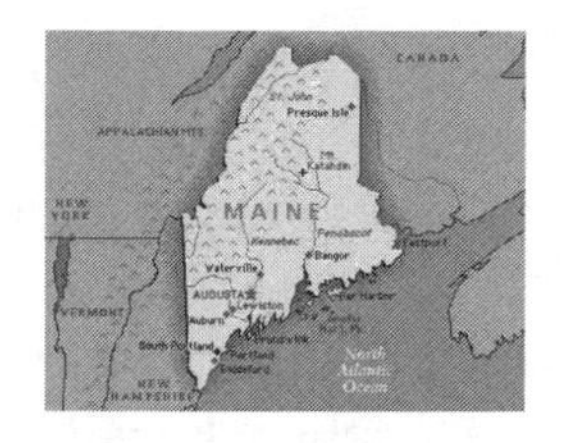

A CERTAIN PERSISTENCE OF CHARACTER

LAND CONSERVATION IN MAINE — 1900–2000

THOMAS A. URQUHART

The establishment of an approximately fixed ratio between the two most broadly characterized distinctions of rural surface — woodland and plough land — would involve a certain persistence of character in all the branches of industry, all the occupations and habits of life, which depend upon or are immediately connected with either, and would thus help us to become, more emphatically, a well-ordered and stable commonwealth, and, not less conspicuously, a people of progress.

— *George Perkins Marsh*[1] *1864*

WHOSE WOODS THESE ARE — UP TO 1900

Stopping by woods on a snowy evening, Hugh J. Chisholm was struck by an impressive set of waterfalls. He saw a source of power that "would be the largest east of Niagara. . . . From the winter of 1882, when I first saw the Androscoggin River tumbling over Rumford Falls, dropping 180 feet in all in the course of a half mile, ... it was eleven years before I had acquired the 1,100 acres of land which were considered necessary

1. *Man and Nature*, by George Perkins Marsh, p. 280.

to control the falls."[2] Up and running in 1893, the Rumford mill would become the lynchpin of the International Paper Company.

A Bangor businessman was doing the same thing on the West Branch of the Penobscot. By 1900, Charles Mullen's Great Northern Paper Company was producing paper under the direction of a former Chisholm employee, Garrett Schenck. Millinocket was transformed from a few farm buildings into "a duly organized and well-governed town of about 3,000 inhabitants . . . The history of the development of Millinocket, like that of Rumford Falls, reads more like romance than reality."[3]

The times favored these entrepreneurs. New processes to make paper from wood pulp were coming on line just as saw-log production was declining. The state had finished selling off "[t]he magnificent heritage of public land which Maine received on its separation from Massachusetts. . . . The value of permanent forest was not understood, while the extension of farming was the one thing which was looked to as meaning industrial progress. Further, the State was weak against individual interest and push."[4]

Chisholm and Mullen "understood;" they had "individual interest and push." Their genius was in establishing control over the entire process from raw material to finished product. Where a link might be missing, they created it: the "magic cities in the wilderness" for the workforce, the railroads to link mill to market. They bought up the forests for wood and managed the rivers for hydropower.

By 1900, some thirty pulp mills were spread throughout Maine. The towns that sprang up around them were company towns, but they offered one particular perk. Workers had free access to company land for hunting and camping. Thus evolved Maine's unique tradition of public access to private land.

The North Woods were soon discovered by "sports." The railroad, which now went as far as Maine's northern border, brought increasing numbers of visitors in and transported pulp and paper out. Neither over-

2. *Pulp and Paper Industry in Maine*, Hugh Chisholm, p. 163.

3. Maine Commission of Industrial and Labor Statistics, 1903, Quoted by Rolde, *The Interrupted Forest*, p. 280–1.

4. *Commissioner of Industrial and Labor Statistics*, p. 182; Maine Forests, Their Preservation, Taxation, and Value, Austin Cary.

Hugh J. Chisholm of the International Paper Company, a turn-of-the-century industry entrepreneur, was among the first with the genius to visualize Maine's need to establish control over its natural resources all the way from raw material to finished product. Inspired by a winter's sight of the Androscoggin River at Rumford Falls dropping 180 feet in the course of a half mile, Chisholm went on to create an integrated system of forests, mills, railroads, water power, and workforce centered around what were later described as "magic cities in the wilderness."
(MAINE PAPER & HERITAGE MUSEUM)

fishing nor overhunting stopped the "carnival barker"[5] mentality that promoted tourism as a numbers game.

Wealthy "rusticators" had discovered the coast even earlier. Like high-class advertisements, luminous landscapes painted by Thomas Cole and Frederic Church lured sophisticated New Yorkers to Mount Desert Island to escape the un-air-conditioned city for the summer. With the coming of rail service, tourists who could stay for only a week or two began to disrupt this civilized lifestyle. The rusticators bought estates of their own.

As the twentieth century began, therefore, a few companies controlled the roughly half of Maine that has never been organized into towns. Lacking local governance, ignored by the state, and at half a penny an acre virtually untaxed, the Unorganized Territory was a corporate fiefdom. The forests produced paper; rivers powered the mills and served as sewers. Entrenched in the State House, Maine's industrial forest economy ensured that Republican political power, unchallenged since the Civil War, would persist for another fifty years.

Along the coast, the threat was different: more and more vacationers brought in by ever-improving means of transportation. The oligarchs — rusticators "from away" — became the conservationists.

CONSERVATION BY PERSONAL APPOINTMENT ONLY — 1900–1950

In 1903, fire destroyed over 200,000 acres of Maine forest. In response, the Maine Forest Commission, established in 1891, added a warden service.[6] Fire was the primary conservation issue for the large industrial owners, and they bolstered their case for public assistance by pointing to their open-land policy for residents and visitors.

Otherwise, Hugh Chisholm "forged a close relationship with Yale University's forestry program, . . . paving the way for the industry standard of sustainable forest management principles."[7] That was as far as corporate conservation efforts went.

5. Rolde, op. cit., p. 296, referring to the Bangor and Aroostook Railroad's magazine *In the Maine Woods*.

6. Rolde, *The Interrupted Forest*.

7. From Paper Industry Hall of Fame www.paperhall.org.

Nonetheless, Forest Commissioner Forrest H. Colby, in his annual report of 1919, gave this urgent rationale for forest conservation: "that our forests serve as a never ending source of clean, potable water; that they are closely affiliated with stream flow, favorable climatic conditions and rainfall; and, of no little consequence, that they harbor and sustain a variety of wildlife which constitutes in itself a great natural resource."

Colby argued for "a broad and permanent policy of acquiring lands for the State," and considered, overoptimistically, that "[p]ublic ownership of timberlands should be a popular measure with the citizens of Maine." His call for protection, public ownership, and reforestation concluded with a nice touch of Maine pragmatism. "For a beginning I would rather see these three basic principles diligently applied, than to undertake a forestry program with more complicated, and, in a way, experimental details which might serve perhaps to distract from the main issue, namely: to stop wasting what we have and to add what we can to our forest resources." [8]

Protecting land in Maine, nevertheless, remained the province of a few private individuals.

One of Nature's Wonder Spots

That same year, Percival Baxter — not yet governor — called for a Mount Katahdin Centennial Park as a fitting celebration of Maine's hundredth anniversary as a state in 1920.[9] His bill was defeated, but the legislature approved a measure to allow the state to accept gifts of land. It also appropriated $5,000 — Colby had asked for $20,000 — for "General Forestry Purposes," although none of it was spent on land purchase.[10]

As governor two years later, Baxter continued to press for the park. But the Great Northern Paper Company owned the land, and it had no intention of selling. A speech Baxter made had not endeared him to its chief, Garrett Schenck. Forest experts had told him, Baxter said, "that Maine's supply would last but 20 or 30 years ... I realized that the days

8. *Forest Protection and Conservation in Maine*, Forrest H. Colby, Forest Commissioner, 1919.

9. Baxter's struggle to acquire and give to Maine the park that bears his name has been well described in many works, notably Neil Rolde's *The Baxters of Maine*.

10. Philip T. Coolidge, *History of the Maine Woods*, p. 653.

Shown with Maine's tallest mountain, Katahdin, as a backdrop, Percival Baxter devoted much of his life as legislator, governor, and philanthropist to acquiring and donating more than 200,000 acres to form the area now known as Baxter State Park. Arguably the state's pioneering conservationist, Baxter's vision is expressed best on the plaque placed on a boulder in Katahdin Stream: "Man is born to die. His works are short-lived. Buildings crumble, monuments decay, and wealth vanishes, but Katahdin in all its glory forever shall remain the mountain of the people of Maine." (COURTESY BAXTER STATE PARK)

of Maine's forests were numbered unless something was done to check the destruction and to replenish the supply ... Someday I want the State of Maine to own Mount Katahdin. It is one of nature's wonder spots."[11]

Only when Schenck died did Great Northern become more cooperative. In 1930, the company agreed to sell 5,760 acres, including

11. Rolde, *Baxters* op. cit., p. 219.

Mount Katahdin. There was one snag: Great Northern's ownership was held as a common undivided interest.[12] The other owner had fought the idea of a park from the beginning. A court had to divide the land before, in 1933, Baxter State Park would be officially proclaimed, to be "forever wild."

But Baxter's struggle was not over. Louis Brann, Maine's first Democratic governor in nearly twenty years, wanted a national park around Katahdin because of the federal funds that would come with it. Four years later, Ralph Owen Brewster and Myron Avery launched another effort to establish a national park. Brewster, Baxter's old political foe, was now in the U.S. Congress. Avery, had brought the Appalachian Trail to Maine. It was said, "Myron left two trails from Maine to Georgia. One was hurt feelings and bruised egos. The other was the A.T."[13] The former Governor fought them all, tooth and nail.

Throughout the1940s, Baxter added parcels of land almost annually. As well as a wilderness lover, he was an ardent animal rights advocate, and hunting was banned in the expanding state park. When 150,000 acres had been set aside, local sportsmen became concerned. In 1954, Baxter bowed to their pressure, and allowed a section of the park to be open to hunters.

By the time Percival Baxter died, "his" park had just over two hundred thousand acres. It was an entirely individual triumph. Not an acre had been purchased with public funds.

A Unique and Noble Tract of Land

The national park whose eventual creation Governor Baxter did support was Acadia. In 1901, however, the influential group summoned to Harvard president Charles Eliot's cottage on Mount Desert Island had a modest goal: to organize a "commission to hold reservations at points

12. Common undivided interest works as follows. In a 30,000 acre township, A owns 10,000, B owns the remaining 20,000 acres. Instead of dividing the township, each owns respectively 1/3 and 2/3 of the whole, spreading their liability. The system was probably introduced by maritime merchants like David Pingree who owned shares in other merchants' fleets as a hedge against losing everything if a ship went down.

13. Larry Anderson, *Benton McKaye*, p. 226.

of interest on this Island, for the perpetual use of the public."[14] Their model was the Trustees of Reservations, founded ten years earlier in Massachusetts by Eliot's son. The man who would make it happen was George Bucknam Dorr, "a Boston boy who took a shine to Maine."[15]

The Hancock County Trustees of Public Reservations was chartered in 1903, but aside from a pair of small founding gifts, "the Corporation slept." Then in 1908, a donation of land by "an old friend alike of my family's and President Eliot's" inspired Dorr to set his sights on Cadillac Mountain. "For this the moment was singularly favorable, a great speculative enterprise involving it having come recently to its end."[16]

For the next ten years, Dorr — indefatigable and persistent as Baxter — traveled anywhere at a moment's notice, searching for donors of land and dollars. Summers he spent surveying and investigating deeds on Mount Desert Island, all at his own expense. By 1912, he had created "a splendid holding ... and on it in achievement we could, I thought, fairly afford to rest."[17] Then local realtors asked the legislature to rescind the Trustees' charter. Rushing to Augusta, Dorr proved a most effective lobbyist; the bill was scuttled.

The event convinced him that only a national park could guarantee this "unique and noble tract of land." Within weeks, he was in Washington, staying with Gifford Pinchot, former head of the U.S. Forest Service. He found Congress "loaded up" with national park applications, many of dubious merit; his project, Dorr was warned, would founder with them. National Monuments, on the other hand, required only an executive order.

War had just broken out in Europe, and President Wilson was preoccupied. Dorr enlisted the help of Maine's junior senator, Charles F. Johnson. The first Maine Democrat in years, Johnson had special pull with the president, later confiding that he had told Wilson, "'Mr. President, I don't want you to turn this down!' He wasn't born yesterday; he knew what I meant!"[18] On July 18, 1916, "nearly half the hills of the

14. Charles W. Eliot, letter quoted in Dorr, *The Story of Acadia National Park*, pp. 13–14.

15. Coolidge, *History of the Maine Woods*, p. 641.

16. George B. Dorr, *The Story of Acadia National Park*, pp. 15–17.

17. Ibid., p. 27.

18. Ibid., p. 47.

Another pioneering Maine conservationist was George Dorr, described by Philip T. Coolidge as "a Boston boy who took a shine to Maine." Shown here in a cleft of the rocky shoreline of what is now Acadia National Park, Dorr devoted much of his life to establishing New England's first national park. He did so through persistent action at local, state, regional, and national levels, always faithful to the prototype of a New Englander who would never take no for an answer.
(NATIONAL PARK SERVICE, ACADIA NATINAL PARK'S WILLIAM OTIS SAWTELLE COLLECTIONS AND RESEARCH CENTER)

Island" became Sieur de Monts National Monument, named after the first European master of the land.

Just as Dorr returned to Washington to secure an appropriation for his National Monument, America entered the war. Though not an auspicious time for seeking funds, the appropriation sailed through. It also recommended the land as worthy of a national park, the first in the East. In 1919, Lafayette National Park was established, its name chosen to commemorate "the great events of the period and the war in France."[19]

By now, automobiles had come to Mount Desert Island, despite the opposition of the summer people. The increasing traffic prompted John D. Rockefeller, Jr. to begin a system of carriage trails for his family and

19. Ibid., p. 71.

friends. One of them crossed land owned by the Hancock Trustees, and his involvement with Dorr, as "your silent partner," began.[20]

Realizing how useful carriage roads would be for maintenance, Dorr urged Rockefeller to build an extended system at the core of the park.[21] When some in the summer community were less than enthusiastic about the "disfigurement" of the now-public lands, Dorr rallied the "Governor and State Forestry Service, the political organizations and the Universities"[22] to his side. The governor was Percival Baxter.

In 1928, Dorr acquired Schoodic Point, the park's first parcel off Mount Desert Island. The gift allowed the naval radio station on the island to be relocated, one of Rockefeller's conditions for another major land donation. It also was the occasion of renaming the park Acadia; as the wife of an English lord, one of the donors was averse to its original francophile name.

Two Visions, Two Legacies

The existence of Baxter State Park and Acadia National Park is each due to the unflagging energy of one man. Each was from a prominent, well-to-do family. Neither would take 'no' for an answer, although Dorr's greatest obstacle was often ignorance ("Where is it?" asked a western congressman about Mount Desert Island) as opposed to ridicule (The *Portland Press Herald* called Baxter's project "the silliest proposal ever made to a Legislature.")[23]

They had, however, quite different conceptions of conservation. Dorr was imbued with a European "man in nature" ethos, whereas Baxter wanted wilderness. Baxter bought heavily cut land and let nature take its course. Dorr, through Rockefeller, employed a landscape architect, Beatrix Farrand, to restore the land around Acadia's carriageways, although she used native plants.

Baxter wrestled his park out of corporations whose interest in the land was financial; his deals were with men deeply connected to Maine's business and political life. He was explicit in his desire that the people

20. Judith Goldstein, *Crossing Lines*, p. 216.

21 In 1947, the carriage roads played a vital role in conserving Acadia when they acted as fire breaks and provided access for fire trucks during the great fire.

22. Dorr, op. cit., p. 100.

23. Rolde, *Baxters*, p. 222.

of Maine should be the beneficiaries of his gifts of land. Acquiring Baxter State Park was about Maine's "heritage."

Acadia sprang from a more intimate, not to say selfish, impetus. "Place after place," Eliot complained, "... has been converted to private uses, and resort to it by other than the owner has become impossible."[24] The donors and sellers were private individuals from Boston, New York, and Washington, whose interest was personal recreation. In all Dorr's dealings, being an old friend of the family was what counted.

"THE STAGNATION OF THE PAST NEED NOT MEAN STAGNATION IN THE FUTURE"[25] — 1950–1975

A River Runs Through It

When Walter Wyman — Central Maine Power Company's founder and chief executive, not an elected official — wanted something in the legislature, he sent it in as a bill under his own name.[26] Lobbyists for the Great Northern Paper Company had ensured it got control of the West Branch in 1903, and the company had maintained a strong lobbying base in Augusta ever since. Half way through the century, Maine was still ruled by the state's corporate interests (paper, textile, and shoe manufacturing, and electric power generation) allied with the Republican political power. Industry lobbyists were so powerful that they formed what became known as the Legislature's Third House. [27] A contemporary political study concluded, "In few American states are the reins of government more openly or completely in the hands of a few leaders of economic interest groups than in Maine."[28]

But subterranean forces were fermenting. The upheaval of World War II — and the return of Maine's soldiers less insular than when they left — was followed by the economic growth of the 1950s. Increasing

24. Goldstein, *Crossing Lanes*, p. 182.

25. Curtis campaigning for Governor 1966; *Kenneth Curtis*, Kermit Lipez, p. 20.

26. Neil Rolde, pers. comm.

27. *Natural States, The Environmental Imagination in Maine, Oregon and the Nation*, Richard W. Judd, Christopher S. Beach, p. 33.

28. Ibid.; quote from Duane Lockhard, *New England State Politics*, 1959 p. 79.

numbers of visitors "from away" sought out the places they had seen in *Down East* magazine, first published in 1954. The Interstate System begun in 1956 expanded enormously the area from which tourists could make the trip to Maine,[29] and the Maine Turnpike smoothed their path once they arrived. Tourism quickly became second only to manufacturing as the state's largest industry.

New arrivals — visitors and "in-migrants" alike — had seen 'the way life should be' vanish from their home states, and they wanted to keep their new haven unspoiled. To long-time residents, this looked frivolous next to the need for a job, a skepticism that industry — adept at letting economics trump improvements to conditions along "its" rivers — easily manipulated into a divisive debate over "payroll or pickerel." The rift would grow.

Nothing challenged Maine's claims as "Vacationland" so glaringly as the rivers. Fumes caused people to vomit and the paint to peel off their houses. In *Maine's Life Blood (1958)*, Jerome Daviau, an early environmental activist, painted the state of Maine's rivers as not just a health issue, but as an ecological and economic one — and a political boondoggle.

"The Republican hegemony was starting to wear thin," says Neil Rolde. "The public could see that nothing had been done for years."[30] In 1955, Edmund Muskie became Maine's first Democratic governor since Louis Brann. One of his campaign pledges had been to clean up coastal pollution, and Daviau worked with him on a bill to create a Maine Rivers Authority. To Daviau's frustration: "As time wore on and [Muskie] acquired the 'feel' of his office, the bill became progressively weaker."[31] When it came before the legislature, it died in committee.

Maine's Life Blood concluded with a resounding challenge: "The first order of business is to break [industry] control, restore the legislative function to the people, and rightfully restore our public waters to economic health before the patient dies of septicemia."[32]

29. Federal funding — through a fuel tax — for the interstate seems to have been accepted, even in "fed-averse" Maine. Recalls one engineer, "Our charge from the State Highway Commission was to take that money and do as many miles as you can." "Interstate turns 50 . . . an engineer remembers," Maine DOT.

30. Neil Rolde, pers. comm.

31. Jerome Daviau, *Maine's Life Blood.*

32. Ibid., p. 139.

Making Government More Responsive

To Horace Hildreth, Daviau is "the first environmentalist, unlike Baxter who was a rich guy who bought land ... He was the first one — even before Muskie — who really started raising hell with the paper companies just running their crap into the rivers."[33] Hildreth ran for the state senate as a Republican in 1964 and was defeated.

Nineteen sixty-four also saw Lyndon Johnson beat Barry Goldwater for the presidency in a landslide. Many Mainers voted a straight party ticket — in what was called the "big box" — and Democrats gained leverage in Augusta for the first time in fifty years. A Republican was still governor and his party still controlled the legislature, but the election was "the solid beginning of a competitive two-party system in Maine."[34]

Two years later, Kenneth Curtis, a young Democrat, became governor and presided over a period of astonishing collaboration between pragmatic moderates on both sides of the political aisle. Maine's government expanded dramatically, first through an income tax.

Both parties knew Maine had significant funding needs. When Republicans tried to maneuver the governor into asking for an income tax, says Neil Rolde, who was a Curtis aid, "Curtis asked for a huge one that absolutely shocked them."[35] In response, Republican House Majority Leader Harrison Richardson decided to get "the best income tax bill they could put together ... I am intensely proud of the fact that we passed the income tax the following Friday."[36]

At the beginning of his second term, Curtis persuaded the legislature to consolidate all state agencies into fifteen departments. "When he came in," recalls Rolde, "he found over two hundred separate agencies, departments and so forth reporting directly to him, which was crazy." Support for reorganizing the unwieldy bureaucracy got a boost from Washington when President Johnson made HUD housing and development dollars dependent upon state government reorganization. Maine had to respond to get the funds.[37]

33. Horace Hildreth, pers. comm.
34. Lipez, *op. cit.*, p.17.
35. Neil Rolde, pers. comm.
36. Harrison Richardson, pers. comm.
37. Richard Barringer, pers. comm.

Summing up Curtis's two terms, then-Senator Muskie wrote: "[T]he state government has greatly expanded the breadth of human and social services, has become the leading partner in a productive relationship with federal and local governments, and has responded vigorously to the pressures of growth and development in a rural state."[38]

A Productive Relationship with the Feds

When Muskie spoke of a new relationship with federal and local governments, he was referring to the Allagash Wilderness Waterway, a landmark effort that protects one of Maine's natural treasures. Its creation proved that the state could assert itself at home and in Washington. It was also the first time Maine citizens mobilized over a conservation issue.

As early as 1956, the Maine State Park Commission had suggested public acquisition of the Allagash.[39] A year later, the National Park Service briefly considered a plan for an Allagash National Park.[40] In 1960, U.S. Supreme Court Justice William O. Douglas, recently returned from running the river, tried to enlist the assistance of Percival Baxter. Well aware that Mainers would resist the National Park Service, Douglas suggested that the Allagash be put under the Baxter Park Authority. His idea came to naught.

Then in 1963, a Department of the Interior report found the Allagash to be "a major recreation resource of great potential significance to the Nation." Concluding that private ownership offered "no real assurance" that it would not "eventually be encroached upon by diverse industrial demands," Interior proposed an Allagash National Riverway with "Federal, State, or joint Federal-State administration" under the National Park Service. [41]

Directly affected were Great Northern, International Paper, and the Pingree Heirs. The landowners challenged the federal proposal as "an arrogant disregard for the rights of State Government and private property owners." Hoping to avoid "public control" in favor of "coop-

38. Lipez , op. cit., Preface.

39. Allagash Wilderness Waterway Management Plan.

40. The Natural Resources Council Bulletin, February 1962.

41. Allagash National Riverway report, Bureau of Outdoor Recreation, Department of the Interior, July 1963.

erative regulation,"[42] they seized upon the Allagash River Authority, just established by the Maine legislature, as a more cooperative partner than the feds; but they had already overplayed their hand.

Landowner behavior at a meeting with federal authorities drew a stern rebuke from Interior Secretary Stewart Udall. A number of specific issues "shook my confidence in the good faith of the landowners," he wrote to Governor John Reed. "The main point which was made over and over again was that the federal government need not concern itself with the Allagash River, and that the landowners were perfectly capable of preserving the area." The Secretary's message was clear: either you protect the Allagash, or we will do it for you.

In 1965, the Allagash River Authority recommended establishing the Allagash Wilderness Waterway. Control of all commercial and recreational uses of 145,000 acres of land and surface water between Allagash Lake and the St. John River would reside with the state. It was a "compromise on the part of a great many persons with interests in conservation, timberland management, woods operations, and recreation," reported the *Bangor Daily News*. Added the *Portland Sunday Telegram*, "not to pass it would be to invite the Interior Department to push its Allagash National Riverway plan."

The next year the legislature authorized the Allagash Wilderness Waterway, which would later become the first state-administered project within the National Wild and Scenic Rivers Program. Voters approved a $1,500,000 bond issue to acquire land along the river, to be matched by the federal Land and Water Conservation Fund. In 1970, Senator Muskie, whose bill permitting states to administer wild rivers within the federal system made it possible, dedicated the Allagash Wilderness Waterway. By 1973, acquisition was complete.[43]

Even as protecting the Allagash was being debated, another threat arose. Justice Douglas, whose interest in the river had continued, wrote: "Two power-producing dams are being proposed ...[that] would inundate the entire river basin and make the 100-mile stretch of the Allagash one huge lake."[44]

42. Letter from Lawrence Stuart, Maine State Parks Director, to John Sinclair, Pingree Heirs, February 1, 1965.

43. Allagash Wilderness Waterway Management Plan.

44. "Why We Must Save The Allagash," Justice William O. Douglas, *Field & Stream*, July 1963.

A tidal power station at Quoddy Head, under consideration since the 1930s, required a back-up river dam to generate electricity when the tide was out during peak demand. The first dam that the Army Corps of Engineers proposed was the impetus behind the state's proposal to acquire the Allagash in 1956. Later, a group of businessmen lobbied for a dam below its confluence with the St. John, which would create Grand Allagash Lake and boost tourism in northern Maine. The Maine legislature voted it down.

Opposed to these projects, Secretary Udall supported a dam on the St. John as part and parcel of protecting the Allagash. The Dickey-Lincoln project, he said, "would preserve in its entirety the free flowing nature of the Allagash River and its superb recreational values." Muskie, in his dedication speech in 1970, also linked saving the Allagash with the need "to generate bulk power for Maine and the rest of New England. . . . the Dickey-Lincoln School Project was developed, so that both purposes might be served." With Maine's congressional delegation and Governor Curtis all supporting a dam, the Natural Resources Council of Maine (NRCM) launched a campaign to save the river.

In the spring of 1976, L.L.Bean's senior management team canoed the St. John. Recalls company president Leon Gorman, "I had read John McPhee's piece, 'The Keel of Lake Dickey' in *The New Yorker*, and we were all very much aware of what the dam might do to the river. When we came back from that trip, we decided to get fully engaged with the NRCM." L.L. Bean joined the fight, the first and only time the company used its customer mailing list for fund-raising letters.

Dickey-Lincoln and the 88,000 acres it would flood made famous an obscure endangered plant, the Furbish lousewort; but economics won the day. Although Congress had authorized it, the Dickey-Lincoln project had never been funded. Says Bill Townsend, who was on NRCM's board at the time, "Every year the Maine delegation would try to get some funding for land acquisition and construction. And it just didn't happen. And eventually — it might have been twenty years after — it was de-authorized."

The Allagash Wilderness Waterway continues to be controversial, but it remains a milestone of land protection in Maine. It also spawned an unexpected corollary, a collaboration among private landowners to forestall public management of the growing recreational use of their wildlands. Starting with unmanned gates "as part of an agreement with the State of Maine to keep the number of access points to the Allagash Wilderness Waterway to a minimum,"[45] North Maine Woods is now 3.5 million acres managed for public access and recreation by private landowners in partnership with Maine's natural resource agencies.

Responding to the Prospect of Development

Twenty government bodies had some authority in the Unorganized Territory, notes Esther Lacognata, but that didn't imply "the actual administration or enforcement of these laws. ...Notable by their absence were any statutes or regulations regarding the management or use of the wildlands in the best interest of the people of Maine."[46]

Attorney Horace Hildreth had lobbied for paper companies on some issues, but he always steered clear of their "environmental stuff. Not that they were doing anything illegal. It was just very cynical. They had all these lakes, and they were pretty careful not to destroy their visual beauty." One day, Hildreth knew, they would start selling them to developers.[47]

Elected handlily the second time he ran, Senator Hildreth introduced a bill to zone around the lakes. If the owners wanted to do something besides forestry, they would have to justify it to a board. Unfortunately, says Hildreth, "'zoning' is a poison pill for many people." The Legislature voted to indefinitely postpone the measure.

Hildreth persisted until the Land Use Regulation Commission for the Unorganized Territory was established in 1969. It took another session before the Legislature extended LURC (as the agency soon became known) jurisdiction beyond lands adjacent to lakes to oversee "a modern, comprehensive land-use guidance system for all the unorganized and de-organized territory of Maine."[48] It was a struggle. At one hearing, Majority Leader Richardson stormed, "The paper companies and the big owners, with their usual head-in-the-sand attitude, ... have consistently taken a position against any attempt by the people of this

45. www.northmainewoods.org, *History of North Maine Woods.*

46. *A Legislative History and Analysis of the Land Use Regulation Law in Maine*, Esther Lacognata, 1974.

47. Horace Hildreth, pers. comm.

48. Lacognata, op. cit, p. 26.

state to protect these lakes and these lands, and they have done so even when it was squarely contrary to their own best interest."[49]

Looking back, Richardson has "no idea how it got through. We effectively zoned one half of the landmass of the state."[50] According to Dick Anderson, "There wouldn't be LURC without [Democratic Representative] John Martin of Eagle Lake [later to become Speaker of the House]. Session after session, the landowners tried to get it repealed, and John Martin always protected it."[51] Says Richard Barringer, "LURC exists because in the woods, people can't say no to each other."

Hildreth, Richardson, and another Republican legislator, Jon Lund, were behind another groundbreaking law to regulate development. A coastal town was voting on the approval of a proposed aluminum plant, although the only reason it had a say in the matter was that the developers wanted public financing. Lund raised the question in the House. Should a single town be allowed to vote on something as intrusive as an aluminum plant without involving its neighbors, who would have to deal with the pollution, traffic, and all?[52]

"Getting control out of the hands of either an individual or a single town and spreading it in accordance with the impact was very controversial," Hildreth recalls. Both the Maine Municipal Association and the industry lobby opposed it. When it passed, however, the Site Location of Development law set standards and defined a process to deal with major development projects. It also established the Board of Environmental Protection (BEP), according to Neil Rolde the first such quasi-judicial body in the nation.

"Why did LURC and the Site Location Law pass?" wonders Harrison Richardson. "I have always regarded it as somewhat of a miracle. But there was an impetus, a thrust that said, 'We have to act, and we have to act now.'" According to Richard Barringer, it was also the combination of federal activity at the time to protect the public interest, of greater trust in government than exists today, of effective "advocacy" journalism that aroused the reading public, and Maine's traditional skepticism toward large corporations and their claims.

49. Phyllis Austin, "Maine Land Use Regulation Commission: Past, Present, and Future," MEEPI, April, 2004.

50. Harrison Richardson, pers. comm.

51. Dick Anderson, pers. comm.

52. Jon Lund, pers. comm.

Maine's Green, Bipartisan Golden Age?

"Ken and I were a happy mix," says Rockefeller Republican Richardson of Democratic governor, Kenneth Curtis. "We agreed on virtually everything."

Explaining the bipartisan cooperation on the environment in those years, Richardson says: "In the late sixties, it became obvious that if we didn't learn from the experiences of the megalopolis to the south of us, we were going to lose resources that were very important to what I regard as the Maine way of life. We recognized that if we didn't go forward with some pretty progressive environmental legislation, we were going to — as opposed to learning from history — be condemned to relive it."

Bipartisanship was essential for such an agenda, not least because Republicans controlled both branches of the legislature throughout Curtis's eight years in office, and because Democratic environmental efforts depended on Republican support to counterbalance labor. The unions found it difficult to favor pollution controls that their bosses — true to the "payroll vs. pickerel" playbook — claimed were too costly.

Curtis believed government could make a positive difference in people's lives, and he surrounded himself with a talented inner core of the best and the brightest. "Maine state government had never seen anything like it," says Richard Barringer. But having campaigned on an economic development platform, and supported the Dickey-Lincoln Dam, Curtis was not an obvious environmentalist. When, in 1969, an oil refinery was proposed Down East, could environmentalists "really trust a man with such a history to give full weight to the environmental implications of the Machiasport proposal?"[53]

Afterward Lipez quotes him as saying "I think there's a new element that's been cranked in that everybody is more aware of, and that is that you have to handle environmental matters almost simultaneously with economic development. You can't bring in an industry unknowingly, unaware of the environmental problems. But that doesn't take away from the need for jobs, wages, and a better tax base." [54]

"Curtis was a great governor," says Barringer, "but he would never

53. Lipez, op. cit.
54. Both quotes by Kenneth Curtis from Lipez, op. cit, p.63.

have thought of a statement like that if it hadn't been for *Maine Times*."[55] *Maine Times* was the weekly newspaper started in 1968 by John Cole and Peter Cox. In very consciously bringing advocacy journalism to Maine, their timing was perfect. The paper had an unabashed point of view that was a direct challenge to the powers that controlled the state.

Cox wrote elegant, objective pieces, while Cole's articles were evocative and stirring. The paper prided itself on its reporters' ability to understand the whole picture, not just the immediate event. As Cox wrote of the exemplar of such reporting, Phyllis Austin, "To her, the public's interest in the northern forest was a public trust, and she had a personal mission to keep an eye on it."[56] It also encouraged environmental reporters at other papers. "Bob Cummings," Cox recalled "... said on several occasions that he had used the threat of *Maine Times* to force the Portland papers to let him do the environmental coverage he wanted to do."[57]

During its heyday, *Maine Times* was a must-read for legislators. It appeared, fresh from the printer, on every legislator's desk each Thursday early afternoon. "Legislative business would stop," Barringer says, "as they all read it to find out what was going on. Leaders could not *not* read it. It was indispensable."

In *A Maine Manifest* (1972), Barringer, a trained economist, propounded an alternative to the prevailing "mega-project" approach to Maine economic development. Instead of oil refineries and aluminum smelters along the coast, ski resorts in the mountains, and large hydroelectric dams on the rivers, he argued, the land itself is the key to Maine's sustainable prosperity. It must be carefully conserved and husbanded as a matter of high policy, both public and private. Barringer would go on to serve in the administrations of three Maine governors.

Toward the end of his second term, in 1974, Governor Curtis wrote: "Maine has gone through a long period of relative unattractiveness economically, but now the pendulum is swinging back and Maine is becoming more attractive. The jet plane and modern transportation have narrowed the distances. With the congestion in the cities, Maine is more attractive for industrial development.... And we're in the process of preparing for this growth. Things like the Site Approval Law and our determination to clean up the rivers all point to the fact that Maine

55. Richard Barringer, pers. comm.
56. *Journalism Matters*, Peter Cox, p. 192.
57. Cox, op. cit., p. 188.

is preparing itself very well. The coastal zoning plan, the fact that we zoned the wildlands, our desire to get more land available for public access — it all adds up to the fact that we're preparing ourselves to handle growth probably better than most places."[58]

MAKING UP FOR LOST TIME: LAND FOR MAINE PEOPLE — 1950–1987

State Parks Commission director Laurence Stuart used to "boast that Maine hadn't spent a cent [to purchase] any state park," Bill Townsend remembers. Why acquire public land with hundreds of thousands of private acres open to the public for hunting, fishing and camping? "Which is why," says Jon Lund, "Maine starts out way behind the ball-with land conservation."[59]

When Representative Lund proposed a bond issue for land, he was "stunned" to find that Stuart had no acquisition plans at all. When the commissioner asked him what he should ask for, Lund "grabbed a number out of the air, a couple million dollars or something." Adds Townsend, "And when they got it, they didn't go out and spend the money and come back for more. They hoarded it!"

Created in 1935, the State Park Commission accepted Maine's first state park in 1938, a hundred acres in Aroostook County. It was a gift from the Presque Isle Merchants' Association, which purchased the land for $2,000 through a local fund drive. Parking areas, picnic and camping facilities, and trails were built by the W.P.A. Nor did Maine pay for the state parks added the next year: Bradbury Mountain, Mount Blue, Lake St. George, and Sebago Lake. A cooperative license agreement with the U.S. Soil Conservation Service left custodial duties to SCS's own staff.[60]

"In the thirties, forties, and fifties, land could have been bought real cheap," says Lund, "but we were operating state government on the cheap, and we got nothing."

Even during the '60s, '70s, and '80s, "Maine took minimum

58. Lipez, op. cit, p. 124.

59. Jon Lund, pers. comm.

60. Under Title III of the 1937 Bankhead-Jones Farm Tenant Act, U.S. "Land Utilization" projects bought sub-marginal farmlands incapable of producing sufficient income to support the family of the farm-owner.

advantage of available federal funding to acquire large undeveloped areas of landscape," according to Mason Morfit.[61] His breakdown of the approximately $34 million that Maine received between 1964 and 1986 from the Land and Water Conservation Fund (LWCF)[62] is instructive:

Of $4.4 million (13%) spent by the feds, $4 million went to the Appalachian Trail corridor and $400,000 to Acadia National Park and Petit Manan National Wildlife Refuge. Of $29.5 million (87%) spent by Maine, $17.8 million went mostly for town parks, recreational facilities, boat ramps, etc.; $11.7 million went to the state of Maine, including $1.5 million for the Allagash Wilderness Waterway, and $1.7 million for Bigelow Preserve.

Excluding the last two, the average state LWCF project was less than $50,000, "hardly enough to acquire big pieces of real estate." Morfit also pointed out that acquiring the 34,000 acre Bigelow Preserve was not "the result of a bold vision on the part of the State's political leadership . . . the State was essentially required to do so as a result of a citizen-initiated public referendum in 1976."

Land for Wildlife and Hunters

In the 1940s, the Fish and Game Department bought Swan Island in the Kennebec River. With mowed fields planted to clover and rye, it was a feeding area for migratory waterfowl. If Maine bought land, it was mostly habitat for game species, not the public.

Richard "Dick" Parks started working for the department in 1950. He remembers one commissioner inspecting a property around Brownfield and asking, "What the . . . are you fools buying this stuff for? What a waste of money!" Two thousand acres were already purchased, but the commissioner put a stop to more. Brownfield Bog's 5,700 acres is today one of Maine's favorite birding, hunting, and canoeing spots. They bought the rest, Parks notes wryly, "for several hundred dollars an acre, instead of a dollar or two."[63]

61. Land Conservation in Maine, Mason Morfit, Friends of Acadia Annual Meeting, August 13, 1990. Morfit was formerly executive director of The Nature Conservancy's Maine chapter.

62. LWCF is the primary source of federal funds for land acquisition.

63. Parks's career saw the creation of the Fish and Game Department's Realty Division, which he headed.

A later commissioner, Maynard Marsh, persuaded Senator Harrison Richardson to put in a bill for general fund money to purchase land. "We thought maybe we'd get as much as a million dollars. But Richardson said, 'You can't do anything with a million. We'll put it in for $2 million.' After that we were in the chips, and we started really buying land."

In 1958, Parks embarked on the acquisition of Scarborough Marsh, the largest salt marsh in the state and prime waterfowl habitat. Salt marsh had been filled or drained all along the Atlantic flyway farther south, and the department was looking to preserve it in Maine. The original idea was to purchase just the upper marsh and turn it into a freshwater nesting ground for waterfowl. While negotiating the original purchase, Commissioner Ronald Speers decided to acquire the whole thing from the upland to the sea.

Meanwhile, the U.S. Fish & Wildlife Service was looking for a "big project" around Merrymeeting Bay. Commissioner Speers "didn't like the sound of it. He thought it was taking too much land out of circulation around Merrymeeting Bay."[64] Speers recommended that the governor veto the project. Instead he suggested the feds buy another salt marsh, although he stipulated a setback to allow development along all the roads. The result was the Rachel Carson Wildlife Refuge, which is scattered along fifty miles of coastline between Kittery and Cape Elizabeth.

Restaking a claim

In 1972, reporter Bob Cummings wrote an article about something almost nobody had heard of, Maine's Public Reserved Lands. He had first learned of the so-called Public Lots while working at *The Bath Times*, but the story didn't fit in a small-town daily far from the North Woods. Then he moved to the Portland Newspapers. Casting about for story ideas one day, he remembered the Public Lots. "My editor had never heard of them, as nobody else had. But I did the story."

The Public Lots date back to when Maine was a part of Massachusetts. Whenever its wild lands were sold, the Commonwealth withheld 1,280 acres out of each 24,000-acre township. These "reserved lands,"

64. Dick Parks, pers. comm.

divided equally, would provide for the eventual community's basic needs: church, town government, school, and general municipal purposes. Maine inherited this system in the 1820 Articles of Separation, later rounding down to an even thousand acres to be reserved for the public good.

But instead of building towns in northern Maine, settlers went west. Timber companies and speculators bought up the land, mostly in large tracts. As far as the state was concerned, the Public Lots—some 450,000 acres, scattered, remote, and in many cases existing only as a public right in the deed—were viewed as "an administrative burden."[65] But Maine's constitution prevented their outright sale. Instead, between the 1840s and 1870s, the right "to cut and carry away the timber and grass" growing on the Public Lots was sold to the landowners. For the next hundred years, timber barons first and then paper companies treated them as part of their estate.

Maine's Forestry Department held the view that when the state sold the cutting rights on the Public Lots, the only rights it retained were to the minerals under the soil, nothing on the surface. But Cummings had read the deeds. "They didn't say anything about only owning minerals under the soil. I just speculated that if there was more than a thousand acres, there was a lot more that you could do besides cut trees.... You could camp on it, you could walk on it, you could hike on it, do all kinds of things on that soil."

Cummings's story caused Governor Curtis, who Cummings says had never heard of the Public Lots either, to appoint "a committee to investigate what the hell these things were.[66] And I wanted to keep the story alive long enough for at least the public to take notice. So I wouldn't let the committee cross the street without doing a story on it."

Harrison Richardson, the senate majority leader, argued that if a township were organized, the Public Reserved Lands would revert to the public. He submitted a bill to organize Maine's wild lands into eight Grand Plantations, thus extinguishing the grass and timber rights on

65. Austin Cary, *Maine Forests*, q.v. ante.

66. Interestingly, the public lots are referred to in the summary of the Department of the Interior's 1963 report on the proposed Allagash National Riverway: "With the exception of the great ponds and *school lots* in each township, all of the 1.8 million acres which comprise the Allagash region are privately owned."

the public lots. "That's when I was called a 'Banana Republic dictator,'" he chuckles. "The bill didn't pass."[67]

It didn't pass because the paper companies sued, on the grounds that the Grand Plantation Bill interfered with the right to manage their lands and interests. The legislature would not pass the bill while the issue was before the courts, which Cummings believes was the point.[68] Jon Lund, now attorney general, immediately filed a countersuit arguing that the cutting rights had long since expired. The judge found against the state, which appealed. The case went on for eight years.

In 1974, Robert Hellendale, president of the Great Northern Paper Company, now Maine's largest landowner, quietly approached Governor Curtis with an offer to try and resolve the Public Lots matter, as far as his ownership might allow. Richard Barringer was assigned the negotiations. Earlier that year, he had become the first director of the Bureau of Public Lands, established within the new Department of Conservation to manage 50,000 acres of Public Lots then under state control.

In a series of closed meetings, Hellendale and Barringer agreed to limit their efforts to the 60,000 acres on which GNP claimed "undivided" grass and timber rights; any exchange would be conducted on an equal value-for-value basis in current appraisal. After Barringer and his small staff identified some 58,000 acres in six large blocs of outstanding "multiple-use" value, negotiations quickly led to the historic agreement signed by Hellendale and Governor Curtis, just before he left office. The precedent would ultimately return a half million acres of wildlands to the people of Maine.

The Law Court handed down its decision in 1981. Cummings will remember the call from the State's lawyer to his dying day. "Bob, to the surprise of everyone in the world, including me — we won!" By then, a number of landowners had settled, returning 250,000 acres to Maine. Those that had not faced a new charge: the state wanted compensa-

67. Harrison Richardson, pers. comm.

68. The plaintiffs were Great Northern Paper Company, International Paper Company, and Prentiss & Carlisle. According to an affidavit from John Martin, a lobbyist for Great Northern told him the "overriding reasons [for the suit] was to prevent Senator Harrison Richardson from running for the office of Governor on the Public Lots issue." "Public Lands — Spiteful Suit," *Maine Times*, p. 5, May 25, 1973.

tion for all trees cut on Public Lots except those standing at the time the rights were sold.

Barringer, now director of the State Planning Office in the Brennan administration, again headed the effort to develop a comprehensive proposal that would resolve all outstanding claims to the Public Lots. According to Dick Anderson,[69] who was part of the team, "we convinced Brennan that he shouldn't try to get money, that land was better. We wanted to consolidate the Public Lots, and we wanted a given number of acres of land from each company instead of money. We picked spectacular places!" They were listed in a glossy booklet, by ownership, with an explanation of the analysis behind each selection.

At a summit meeting in the Blaine House, Governor Brennan explained what was in the booklet and handed a copy to each company chief. "They went out of the room in stunned silence," Anderson recalls. "Nobody could believe what we were proposing."

Asking for land instead of cash proved a shrewd move: the companies had a lot of land, it was relatively cheap, and consolidating the state's scattered holdings held substantial benefits for them. Out of the negotiations, which went on for several years, came such treasures as the Mahoosuc Preserve (45,000 acres), the land around Upper Richardson (22,000 acres), Rocky (11,000 acres) and Scraggly (9,000 acres) Lakes, and the entirety of Round Pond (20,000 acres) and Deboullie (24,000 acres) Townships.

Today Maine's Public Reserved Lands — managed for "multiple/dominant-use" including wildlife, recreation, ecology, and sustainable forestry — are a national model. "[O]ne of the ironies of history," writes Cummings, is "that these lots, set up to ensure eventual civilization, may become a major tool in preserving one of the rarest commodities of the final decades of the twentieth century — natural areas and wildlands."[70]

Bonding for Land

Governor Brennan had just months left in his second and final term when a man from Aroostook County wrote to suggest he convene a special commission to assess growing threats to Maine's outdoor her-

69. Then Commissioner, Maine Department of Conservation.
70. "Lots of Confusion," Bob Cummings, *Appalachia*, Winter 198687.

itage. Despite the shortage of time, Brennan responded, "Yes, we are going to do a Special Commission on Outdoor Recreation for the State of Maine." It was the spring of 1986.

The Commission — consisting of some of Maine's most famous outdoor icons — held public hearings all over the state. At one, Anderson recalls, "the guy who wrote the original letter was there. I made a little speech about how he had got the whole thing started." The report made thirty recommendations, one of which was to float a $50 million bond issue for land acquisition.

The bill that became the Land for Maine's Future was written between the gubernatorial election (of Republican John McKernan) and the end of Governor Brennan's term. McKernan's transition team asked for it, and Anderson gave the only other copy to Representative Pat McGowan.[71] The new administration put forward a bond request for $5 million for land acquisition, and McGowan, on the legislature's Appropriations Committee, sponsored a competing request for $50 million to secure "the traditional Maine heritage of public access to Maine's land and water resources or continued quality and availability of natural resources important to the interests and continued heritage of Maine people." In November 1987, the question before the people of Maine was for $35 million. (At Appropriations, $15 million went to an economic development project.)

Maine's nongovernmental conservation community was, meanwhile, moving along a similar path. They invited the Society for the Preservation of New Hampshire Forests' Paul Bofinger to share the results of his sabbatical year, which he had spent investigating bonding for land. The Maine groups were soon working on the idea, but they needed a vehicle. When the governor's commission made its recommendations, Jay Espy remembers, "We said, Wow! There's a government commission discussing outdoor recreation. Get Bofinger's idea to them."

Staff from the Maine chapter of The Nature Conservancy (TNC) and the Maine Coast Heritage Trust (MCHT) had helped draft the bill. Now, they led the campaign to persuade the public to vote for it. A new organization, Citizens to Save Maine's Heritage made up of Maine's environmental and conservation organizations plus a cross section of

71. Pat McGowan is currently Commissioner, Maine Department of Conservation.

business and civic leaders, ran the campaign. An advertising agency developed media ads, and Chris Potholm, a Bowdoin College professor and seasoned political pollster, honed the message.

The idea that the biggest non-transportation bond in state history was for land conservation, says Espy, "seemed a little bit off the charts." None of TNC's state organizations had led a referendum campaign before, and it was the Maine chapter's first foray out of its low-profile role of buying land and into politics. "All of it was brand new to us," says Morfit.[72] According to Potholm, "Hard-core environmentalists were in charge of the campaign and insistent on using posters with no humans in them. ('Our members don't like to see people in the woods!')" Potholm succeeded in convincing them that this attitude could not win. "I insisted that there had to be a coalition of hunters and fishermen, those who wanted to drive their RVs and campers into 'the wilderness' and even the elderly who wanted to look at the foliage from tarred roads."[73]

The Land for Maine's Future bond passed in 1987, two to one. To date, the program has purchased 197,000 acres of Maine land outright, with another 247,000 acres in conservation easements. The fund was renewed with $3 million by the legislature in 1998, and in 1999, the voters approved another bond issue, this time for $50 million. (Voters approved another $12 million in 2005, and $17.5 million in 2007.)

"It has been an extraordinary time for Maine," says Richard Barringer, "one in which the map of Maine has literally been redrawn through the cumulative efforts of many people and many parties, within and without government, all of whom love and are devoted to Maine and its diverse landscape, worked tirelessly and often without compensation, and made a difference for this generation and for generations to come."[74]

MAINE'S FOURTH HOUSE: THE NONGOVERNMENTAL ORGANIZATIONS — 1960–2000

"Billboard, drive-in, billboard, gifts, moccasins, motel, billboard — and each billboard, every splashy motel is followed by bigger billboards and splashier motels in a desperate promotional free-for-all." So John Magee

72. Mason Morfit, pers. comm.

73. *An Insiders Guide to Maine Politics, 1946–1996*, Christian P. Potholm, p. 20.

74. Richard Barringer, pers. comm.

lambasted Maine's roadsides in As Maine Goes, his polemic exhibition of photographs at Bowdoin in 1966. When Maine complied with the Federal Highway Beautification Act three years later, the highways were littered with 2,200 non-conforming billboards. By the end of 1977, they were almost all gone.

By common consent, this was the work of one woman, Marion Fuller Brown. In giving her its 1983 Environmental Award, *Down East* magazine wrote, "Starting in 1959 and working through every group from local garden clubs to the Maine legislature, where she served three terms in the House [of Representatives], Brown worked tirelessly to write, pass, and then defend a law banning the billboards that were blighting Maine's highways." She also mobilized the energetic support of then-Governor James B. Longley. "Working through every group from local garden clubs to the Maine legislature" is a telling phrase. Maine's campaign against billboards (it remains, thirty years later, one of only four states to have banned them) may be the last to rely so completely on the leadership of a single person.

Conflict and Change Bring New and Unimagined Possibilities

"As a rural state with an environmental conservation ethic, artistic and creative traditions, and an increasing demand for social services," wrote Dahlia Lynn at century's end, "Maine has the fourth highest number of nonprofit organizations per capita in the nation."[75] In the conservation arena, we have seen how, by 1987, nonprofit groups were strong enough to lead the public campaign for the Land for Maine's Future. From individuals like Percival Baxter and George Dorr in the early years, to a resurrected state government capable of passing landmark environmental laws, leadership had flowed to the nongovernmental organizations (NGO).

Since the 1960s, nonprofit groups backed by private financial support — new ones like the Maine chapter of The Nature Conservancy (founded in 1956) and the Natural Resources Council of Maine (NRCM) (1959), and existing ones like the Maine Audubon Society

75. "Maine's New Third Sector," Dahlia Bradshaw Lynn, *Changing Maine, 1960–2000*, Edited by Richard Barringer, p. 258, 2004.

(MAS)—had become more and more active. The legislation of the 1970s, plus growing conservation awareness, gave these organizations a further edge as they staked out different (if frequently overlapping) issues on which to focus. NRCM became advocacy oriented, across a broad environmental front. MAS stressed policy and education related to wildlife. TNC acquired reserves for endangered species, while the Maine Coast Heritage Trust (MCHT, founded in 1970) protected open space. (We will consider MCHT in more detail later as an illustration of the growth of a Maine land conservation organization.) Starting in the mid-1980s, the attention of all these organizations began to focus on the forest.

Focus on the Forest. Maine is the most forested state in the nation. Once underutilized, its forest had easily accommodated sportsmen, lumberjacks, and wilderness enthusiasts. Then massive harvesting collided with rising public demand for recreation. A spruce budworm epidemic led to hundreds of thousands of acres being cut for salvage in the 1970s. After river driving was banned, roads to get the logs to the mills started to riddle the forest. The fabled North Woods was now a day's drive away from seventy million people. What they found when they arrived was not pretty: clear-cuts extending for thousands of acres.

In 1983, Great Northern Paper proposed to dam a spectacular stretch of the West Branch of the Penobscot River and sparked the first major clash between NGOs and the forest industry. After a two-year struggle in the media, state agency hearings, and courts, "The Northern" withdrew its proposal; the "Big A campaign," as it had become known, was an environmental victory.

In a more cooperative vein, industry, environmental, and government officials next sought to hammer out a sensible forest policy together, in particular regulating economically efficient, but ecologically wasteful, clear-cutting. The effort failed, and a report by the Department of Conservation, Forests for the Future, left the forest's fate to the industry. MAS drafted legislation to limit clear-cutting and set standards for responsible forestry. The Forest Practices Act passed, but after heavy modification. Rules to implement it were also disappointing. Still, annual clear-cutting dropped from 145,000 acres to 60,000 in three years and ultimately to less than 20,000 acres annually.[76]

76. "Seeing the Forest — and the Trees," *Portland Press Herald,* November 25, 2001.

Land Speculation at a Whole New Scale. In 1987, Sir James Goldsmith sold Diamond International Paper Company to a French conglomerate. Nearly a million acres of northern New England's forest were involved. The buyers financed the deal by selling off so-called highest and best use (HBU) lands, mostly around lakes, for development. Horace Hildreth's fears had come to pass.

The Diamond sale signaled a new threat to the forest. Increased access to inexpensive forestland invited subdivisions for vacation homes, resorts, or investment, while a market glut of salvaged wood made selling off land to developers all the more attractive. Although "forest products firms have not wholly abandoned the idea that timber growing is profitable," warned Perry Hagenstein "...[s]ale or development of separated tracts and of tracts with especially high recreation and development value are increasingly likely."[77]

Of Diamond's 600,000 acres in Maine, LMF funds bought Nahmakanta Lake; traditional logging and forest companies acquired most of the rest. New Hampshire and Vermont paid a steeper price. Warren Rudman and Patrick Leahy, senior U.S. senators from each state, requested the U.S. Forest Service to undertake the Northern Forest Land Study (NFLS). Half the study area, which extended through New Hampshire and Vermont to New York, was in Maine.

The NFLS found increasing value for development "has pushed up the price of forest land to the point where holding prime parcels for traditional uses has put owners at an economic disadvantage."[78] The study was precluded from making recommendations. That was to be the task of the Northern Forest Lands Council (NFLC), a stakeholder group appointed by the four governors. In 1994, it released its vision of "a landscape of interlocking parts and pieces, inseparable, reinforcing each other: local communities, industrial forest land, family and individual ownerships, small woodlots, recreation land, public and private conservation land." To environmentalists, it was short on leadership.

To many in rural Maine, however, a federal study confirmed fears already raised by the aims of several national NGOs: The Wilderness Society's proposed 2.7 million acre reserve around Baxter Park; the

77. *A Challenge for New England: Change in Large Forest Land Holdings*, Perry Hagenstein, p. 10, 1987.

78. *The Northern Forest Lands, A Strategy for their Future: Report of the Governors' Task Force on Northern Forest Lands*, May 1990.

National Parks Conservation Association's search for a Down East national park; a National Audubon vice president who memorably urged "We must take it *all* back!" When U.S. Senators George Mitchell and William Cohen held a public hearing in Bangor, property-rights activists disrupted it, and the senators abandoned the NFLC.

Huge buyouts — land, mills, and all — were leaving ever-fewer timber companies headquartered in Maine. Soon, half the state would be controlled by a dozen multinational paper companies. In 1989, Georgia-Pacific bought Great Northern in a hostile takeover. Two years later, Bowater Inc. acquired it and began the federal relicensing process for its river dams.[79]

The Federal Power Act requires dam owners to mitigate the impact on fish, wildlife, and recreation, and to compensate the public for the private use of the people's resource. Conservation groups and the Penobscot Indian Nation asked Bowater to give up the development rights along the river. The coalition had an uncanny resemblance to the one that prevailed on the Big A. Though Great Northern had changed hands twice, long-time executives remained. They were not going to be defeated a second time.

The new governor, Angus S. King, Jr., considered the environmentalist position on the dams little short of blackmail. However, to pick up where the NFLC had left off, he established the Council on Sustainable Forest Management. Its work had hardly begun when a Green Party activist launched a campaign to ban clear-cutting in Maine.

The Ban Clear-Cutting Referendum. Jonathan Carter's assertion that Maine was not addressing forest practices adequately was true, but his bill raised serious questions.[80] Would landowners sell development lots to offset its economic impact? Would it increase road building and habitat fragmentation? The bill applied to the Unorganized Territory only; how would it effect southern Maine, already threatened by sprawl?

Other considerations heightened the dilemma. The Governor's

79. In 1999, it was split up between a Canadian paper company and an investment holding company.

80. In Maine, a citizen's initiative, once sufficient signatures have been gathered, starts as a legislative bill to be voted up or down; if it is defeated — as is likely with a controversial bill — it goes onto the ballot.

Council needed time. Some foresters were "getting religion;" the Pingree family's forests had just received green certification from the Forest Stewardship Council. The Eco/Eco Forum,[81] a group of environmentalists and businessmen, was encouraging cooperation over confrontation.

MAS and NRCM concluded that Carter's clear-cutting initiative was not the answer. With a group of progressive foresters, they forged an alternative. As it took shape, Governor King embraced it. Rolled out in early summer 1996, the Compact for Maine's Forests strengthened the Forest Practices Act, but its innovative feature was a voluntary forest audit, like a financial audit, to hold landowners accountable for forest health. All the major landowners pledged their support. Voters would have three choices: Carter's clear-cutting ban, the Compact for Maine's Forests, or neither.

Maine's conservation community runs the gamut of ideologies and organizations, and the regional nature of the NFLS had further enlarged the number of NGOs active on forest issues. At the urging of national funders, the Northern Forest Alliance (NFA, eventually twenty-six organizations) had been set up to craft a coherent conservation strategy. Now Maine's environmental community was polarized.

Sierra Club and RESTORE: The North Woods supported the Ban; more mainstream organizations supported the Compact. On Election Day, the Compact fell three percentage points shy of the simple majority required by Maine's Constitution. Taking Ban and Compact together, however, three-quarters of the voters wanted stronger forest regulation. Governor King scheduled an up-or-down vote for the following November.

Its campaign commitment to forest stewardship notwithstanding, in the ensuing year the Maine Forest Products Council withdrew its support for the largely symbolic Northern Forest Stewardship Act. The paper companies dispatched modest restrictions on development in the backcountry proposed by LURC and supported by much of the public. Even a legislative resolve to define acceptable forestry was too much.

Another three-way struggle defined the new campaign. Carter's new organization, the Forest Ecology Network, made common cause with the "property rights" side. When the Compact was defeated by a similar margin as the year before, Richard Barringer commented, "Carter's alliance

81. Standing for Ecology/Economy.

with the right wing to defeat the Compact was a sad, even a tragic day for forestry in Maine." Maine's best chance ever for forest reform was lost.

In the Aftermath of Conflict. With Maine's NGOs divided on forest policy, attention returned to the land. RESTORE: The North Woods supported a 3.2 million acre Maine Woods National Park around Baxter Park.[82] Using geographical information system (GIS) technology, MAS and the Appalachian Mountain Club (AMC) identified five priority "wildland conservation areas,"[83] which became Maine's piece of the NFA's regional strategy.

Shifting its focus to landscape-scale conservation, TNC purchased 185,000 acres on the St. John River (one of the wildland conservation areas) from International Paper, setting a new standard for land deals nationally. Management, also innovative, included a 45,000 acre ecological reserve and sustainable timber harvesting on the rest.

A year later, the New England Forestry Foundation negotiated a conservation easement on an even larger tract, 750,000 acres owned by the Pingree heirs. The price was $28 million, and it removed forever the threat of development.

In barely a decade, conservation groups had upped the ante from buffers measured in hundreds of feet and a few hundred acres, to 30,000 acre townships and entire watersheds. By 2000, three million acres of Maine's forest had been protected by acquisition, easement, or green certification.

Maine Coast Heritage Trust

Peggy Rockefeller and her husband, David, had cruised the Maine coast for years. In the 1960s, they began seeing islands wild one year and developed the next.

Back in the 1950s, "Nobody wanted islands except for fishing rights," recalls Dick Parks, who was already buying land for the state. "When the fishing business went to pot, they just abandoned them." On other islands, says Jay Espy, summer residents "had either modest

82. RESTORE's founder, Michael Kellett, wrote The Wilderness Society's North Woods Reserve report.

83. Androscoggin Headwaters, Western Mountains, Greater Baxter Park, Upper St. John, Down East Lakes.

summer cottages or camps on them, or just used them as picnic islands without building anything. Nobody quite saw the value escalation of Maine's islands coming."[84]

By 1969, coastal prices were increasing "between 10 and 20 percent *annually* (about double the rate for inland property)," according to a University of Maine biologist.[85] Based on an analysis of ecological and historical data, he took the unusual step of prioritizing a list of islands the state should buy before they were built on. "Nobody did that in those days," says Dick Anderson admiringly. "Stuff just came along. Nobody ever sat down and said: Here are the criteria, these are the places to acquire. It was unheard of."[86]

Concerned lest development on islands mar its scenic outlooks, Acadia National Park proposed acquiring seventy privately owned islands in its view shed. "Not surprisingly," recalls Ben Emory, "that idea — government ownership — bombed politically."[87]

Peggy Rockefeller contacted the Park superintendent, John Good, to explore what could be done. She also recruited a fellow sailor, Thomas Cabot, to the cause. Conservationist and businessman, Cabot was a collector of islands who boasted that by the 1940s, he had acquired forty-one islands off the coast of Maine for less than $5,000 total. "I could have made millions," he wrote, "if I had sold them for development. Since 1950 the market value of Maine islands has advanced by a factor of at least one thousand…[B]ut I didn't do badly considering the amount I saved in income taxes by giving nearly all of them to the federal and state governments."[88]

To buy every island they wanted to protect was all but impossible. So Rockefeller, Cabot, and Good investigated a new conservation tool,

84. Pers. comm. Jay Espy was president of the Maine Coast Heritage Trust from 1988–2007.

85. *Classification and Management of Eider Nesting Islands in Maine*, Howard L. Mendel, January 1969.

86. The Department of Inland Fisheries and Wildlife has acquired twelve of Mendel's top twenty islands; TNC has acquired three more; they acquired Great Duck Island together. Brad Allen, DIFW, pers. comm.

87. Emory was executive director of MCHT, then an active board member. His wife Diana is currently on MCHT's board. Unless otherwise noted, quotes through p. 76: Emory, pers. comm.

88. *Avelinda, The Legacy of a Yankee Yachtsman*, Thomas D. Cabot.

the conservation easement. Cabot's idea was to put conservation restrictions on the land by selling it to a third party. Rockefeller's lawyer, however, thought "such a process seemed pretty cumbersome." He suggested they "get a law passed so the restrictions could be put on land directly."[89] The Maine legislature passed the Conservation Easement Statute in 1969.

The next year, Peggy Rockefeller and Tom Cabot founded Maine Coast Heritage Trust.[90] It was to be "an organization exclusively to promote the use of conservation easements and negotiate them for Acadia National Park, primarily on islands." MCHT would not hold land itself, "and Tom Cabot was very firm that we would probably accomplish everything we wanted to accomplish in five years, and then we'd shut it down."

Easements were an ideal tool for protecting undeveloped islands, and in the early days, the park superintendent had full authority to accept them. "You'd just walk in the door, and say, 'We think this is a great easement, do you want it?' And he'd say, 'Yes.' We could get an answer in five minutes. But you sure as hell don't want to waste a lot of time negotiating with some guy who doesn't have title." Dick Parks had already found that "as islands became the thing to own, some of the claims were far from legitimate."

In 1974, at the direction of the legislature, the State Planning Office produced a Registry of Coastal Islands. Without good title, an island reverted to the state: the Department of Inland Fisheries and Wildlife for important seabird islands; the Conservation Department's Bureau of Parks and Recreation for public use; and Bureau of Public Lands for the rest, with Maine Island Trails Association managing the smaller recreational islands.

Congress formerly established Acadia's boundaries in 1986. The fee boundary covered Mount Desert Island and the Schoodic peninsula, but the park could hold easements from Penobscot Bay down to Schoodic Point, including parts of Hancock and Knox Counties. In the public process with towns and the state, certain parcels were legislated into the boundary, others were negotiated out of it.

Emory and his staff "were in touch with a lot of people who owned

89. Ibid.

90. Peggy Rockefeller became its first president and served on the board until her death in 1996.

islands outside the Park's boundary," so MCHT was soon looking for other easement holders. Maine accepted its first in 1972. Until the easement statute was amended, only government agencies could hold them. NGOs could only accept common law "easements appurtenant" that benefited land they already owned.

Emory admits that the early easements were not as protective as they would be now. "We knew they would get better with experience. We were marketing a new product. The first people — in marketing terms — were the early adopters. We needed to get past the early adopters to the people who would take more persuasion. For that we needed a portfolio."

All the early easements were donated, and major fund-raising was not part of MCHT's modus operandi. Its first campaign — in 1981, to purchase a property on a spectacular peninsula in Machiasport — was a modest internal effort from board members and close friends. But it allowed the owner to give a strip of land to the National Audubon Society, which then covered the rest with a very restrictive common law easement appurtenant.

In the late 1980s, MCHT had to raise its fund-raising goals for the Bold Coast program in Washington County. First a headland at one end, Boot Head, then Western Head at the other were threatened with subdivision. MCHT raised the money and bought both. However, when 12,500 acres of land on three and a half miles of shorefront belonging to the Hearst Corporation went up for sale, Emory turned to a Washington DC-based organization experienced in complex corporate real estate transactions.

The Conservation Fund acquired the Hearst property in 1989 with major funding from the Richard King Mellon Foundation. The state purchased the shorefront with LMF funds, and MCHT leased the backland until a permanent owner could be found. After nine years, the state accepted the Hearst property. Though nowhere near the scale of Baxter State Park, it was the biggest land donation to the state since Governor Baxter's.

Accustomed to dealing with federal agencies, the Mellon Foundation had wanted the U.S. Fish and Wildlife to have it, but local politics made a federal entity impractical. Beside the Hearst lands, the town of Cutler was home to the *Downeast Coastal Press*, a conservative weekly whose editorial position was that honest American workers' livelihoods were being threatened by elite liberals hand-in-glove with the feds. Property rights activists "were watching what we were doing very closely in all our work in Eastern Maine."[91]

MCHT had offered assistance to land trusts early on, and Emory helped one or two get started. But land-trust activity in Maine was very limited until the real estate boom of the mid-1980s. Jay Espy, who became MCHT president in 1988, recalls that "anything with water near it was getting bought up and sold through the *Wall Street Journal* and the *New York Times*, sight unseen."[92] As land speculation ran wild across Maine, more and more communities saw land trusts as part of the answer and turned to MCHT for guidance.

Maine's size militated against becoming a statewide land trust (as happened in Vermont.) And, as Espy says, "All conservation is local." In 1986, MCHT started its Community Lands Program to encourage local groups all over the state. (Now the Maine Land Trust Network, it serves some hundred local land trusts.)

When Diamond's forest went on the block the next year, Maine's environmental organizations had no capacity to negotiate deals that might involve hundreds of thousands of acres. Although TNC negotiated the acquisition of Nahmakanta Lake on the state's behalf—at 44,000 acres, it remains the LMF's largest purchase—landscape-scale conservation had yet to become its goal. It was clear that land in the North Woods needed to be protected the way MCHT protected coast.

The MCHT board continued to see the organization negotiating land deals, not holding property, and it debated whether to expand into the North Woods. When it opted to retain its traditional focus on the coast, Espy looked around for another entity to serve the forest. A few years earlier, the Society for the Protection of New Hampshire Forests (SPNHF) had created the Forest Society of Maine to hold a forest easement that no Maine group would accept. After due diligence, a board consisting of three landowners and three conservationists was set up, and SPNHF spun off the Forest Society of Maine.

It was "a turning point" for MCHT, Espy says, and allowed the or-

91. Emory, pers. comm.

92. This and subsequent quotes (unless otherwise noted) through p. 78: Espy, pers. comm.

ganization to focus on protecting whatever islands were still undeveloped. A comprehensive island conservation program, equipped with GIS and a computerized database (based on the state's Coastal Island Registry), brought in all the organizations working on island conservation. As island coordinator, Ben Emory found himself once again reaching out to island owners and working with state and federal agencies to identify priority seabird islands. "The state was there, the feds were there, and the nonprofits were there."

Once more focused on the coast, MCHT was still raising funds project by project. In 1999 the board turned down a proposal to build a $10 million acquisition fund. "But we were losing stuff," sighs Espy. "We just couldn't compete." The next year, three important properties were lost, including Eastern Nubble, a key piece of the Bold Coast.

It was 2000. The Nature Conservancy had just declared victory in its $35 million St. John River campaign. Richard Rockefeller was chairman of MCHT's lands committee and had been closely involved with TNC's success. "Richard had seen what was possible," recalls Espy. At a special board meeting, a major campaign was discussed once again. The staff had the details at hand — what it wanted, where, and the cost. "We started talking about $10 million, we got to $20 million. Somebody said TNC just raised almost $50 million." $100 million, suggested Rockefeller. "And people almost fell out of their chairs."

By the summer, the quiet phase had started, and the Campaign for the Coast was announced a year later, a month before the 9/11 terrorist attack. Despite its economic effects, the campaign finished on schedule in 2006: 194 projects completed; 14,646 acres protected, including 125 miles of shoreline, 56 entire coastal islands, 20 eagle-nesting sites.

Marshall Island was the Campaign for the Coast's largest project and a microcosm of the way MCHT works and the niche it fills. The settled island communities — Swan's, Stonington, and Isle-au-Haut — had excluded Marshall from Acadia's legislative boundaries, and it soon attracted speculators. In the late 1980s, a development consortium planned to subdivide two-thirds of the island, but went bankrupt. When the property went to auction, the bid by an MCHT donor was exceeded by $100,000. The new owners planned on building a house, and Espy thought that was that. "But they started to really understand what the island was about." It took years, but they decided to sell it to MCHT.

The other third of the island was in two parcels. Neither owner indicated an interest in selling, but MCHT let them know its desire to acquire the whole island. "One of them threw a very high number out and said, 'let's go fishing.' And we started talking to them — a long ordeal. We also were talking to the other landowners — that too was a long ordeal. Prices kept going up and up. We bought the first before we knew whether we were going to get the other." In January 2003, MCHT had bought the two-thirds piece; by December they had the rest. "It was the highlight of the campaign, and we got it all."

It is hard not to be struck by how MCHT mirrors George Dorr's activities three quarters of a century earlier; or by how succeeding generations of one family — from John D. Rockefeller, Jr. to his daughter-in-law, Peggy, to his grandson and her son, Richard — have contributed to conservation on Mount Desert Island.

CONCLUSION — BECOMING A PEOPLE OF PROGRESS

In 2000, Roxanne Quimby wrote a check for $2 million, drawn on her very successful company, Burt's Bees. It was her first land purchase in Maine, 10,000 acres in strategic tracts that benefited existing conservation reserves: TNC's Big Reed Pond, MAS's Borestone Mountain, and the Appalachian Trail.

Quimby's activities bring conservation in Maine full circle. Though, as she points out, her wealth was earned rather than inherited, her land purchases, already approaching a hundred thousand acres, are in the mold of Baxter's. Some sixty thousand of them, between Baxter Park and the East Branch of the Penobscot, tie her to the governor's legacy itself.[93]

The twentieth century's last years bequeathed a number of conservation challenges that continue to echo the past. RESTORE:The North Woods' quest for a Maine Woods National Park pits an impressive vision against deep-seated suspicion of federal government that shows no signs of waning. Only time will tell if such an ideal can be achieved, or whether it will remain the enemy of good, if not perfect, land protection.

Plum Creek's proposed development of 425,000 acres around Moosehead Lake is the most massive project LURC has ever seen. It has

93. Roxanne Quimby, pers. comm.

"I visualized a trail that would connect two countries and cultures, link a state and two provinces, and traverse two great watersheds — the Gulfs of Maine and the St. Lawrence," says Dick Anderson, former conservation commissioner and founder of the International Appalachian Trail/Sentier International des Appalaches. Thanks to a 2008 Memorandum of Understanding, executed by Anderson and philanthropist Roxanne Quimby in the photograph above, the trail runs through conservation land purchased by Quimby just east of Baxter State Park.

(Walter Anderson)

rallied conservationists to the barricades as did Dickey-Lincoln and the Big A. And forty years after Horace Hildreth introduced his bill, LURC itself is still targeted by the paper companies and large landowners.

But the twentieth century has also left the conservation enterprise unimaginably richer. Innovative market mechanisms provide tools unknown less than a decade ago, such as TNC's purchase of 200,000 acres next to Baxter Park by first loaning the corporate owner $36 million, then buying it out with New Market Tax Credits.[94]

94. See *Investing in Nature, Case Studies of Land Conservation in Collaboration with Business*, William J. Ginn. pp. 42, 128.

Individuals and communities, NGOs large and small, federal and state governments working together are getting results on a scale and of a complexity never seen before. A model of such collaboration is Mount Agamenticus to the Sea — more casually, MtA2C — saving a natural treasure in the heart of Maine's fastest growing county.

Will "watershed-scale" succeed "landscape-scale" as the next leap forward? With the Penobscot River Restoration Project, a coalition of conservation groups and the Penobscot Indian Nation is working with the owner of three dams to restore salmon runs and revive the river's social, cultural, and economic traditions. Restoration of Maine's signature river should have incalculable benefits for the entire watershed.

Land conservation can be realized with more modest linear efforts as well. Corridors that allow wildlife to move between two established protected areas enhance both. Trails (re)introduce humans to the conservation equation, as will Roxanne Quimby's Thoreau-Wabanaki trail, based on Thoreau's canoe route a hundred and fifty years ago.

Maine's state agencies have opportunities to take conservation beyond traditional acquisition, at less expense. The Forest Service is encouraging forest owners large and small to acquire third-party certification; 41 percent of Maine's forests are already certified, more than any other state.[95] Tying the benefits of tourism to conserving "the Golden Goose" will serve conservation and economic goals.

Future prosperity depends on Maine's natural beauty. How Maine addresses sprawl and the open-space fragmentation it causes will be the twenty-first century's first great conservation challenge. To fight sprawl effectively, conservation priorities must value well-tended lands around urban centers as much as faraway wildlands.

Ironically, Maine's parks, state and national, are seeing attendance drop — by as much as 25 percent at Baxter State Park. A TNC study blames "an explosion of electronic entertainment."[96] The trend casts a shadow over conservation's long-term prospects. If we don't go outdoors, who will vote for LMF or urge their legislators to support LURC?

95. Debate continues as to the value of industry's Sustainable Forestry Initiative (SFI) model for third party certification compared to that of the Forest Stewardship Council (FSC).

96. "Parks: Attendance plummets at Baxter, Acadia," Deirdre Fleming, *Portland Press Herald*, July 28, 2006.

Perhaps the future belongs to Maine's hundred or so land trusts. What group is better positioned to protect a community's backyard? Besides helping towns maintain their identity, land trusts could provide benefits from local food to physical and mental health, to perhaps even breaking the hold of electronic equipment.

A hundred years ago, conservation depended on individual zeal. Over the century, government regulation, then NGO advocacy took their proper place. Continued progress depends on bringing in another partner, industry. Negotiation must replace the all-or-nothing campaign. For just one example, imagine harmonizing clean energy and land conservation goals without it, as sites for wind-power emerge on the coast or in the forest. As Lloyd Irland wrote in 1999, "Political, economic, and technological change has been disorientingly rapid in the past two decades. There is no sign that the pace of change will slacken."[97]

The homily with which Austin Cary finished his paper before the Board of Trade in 1906 still serves as a preamble for conservation in the coming — or any — century. "There is no use in crying over spilled milk, however. What our fathers did we must abide by. On us it is incumbent simply to deal correctly with the conditions of our own time, to think clear and straight in order that in our day and generation no serious mistake be made."[98]

97. Lloyd Irland, *The Northeast's Changing Forests*, p. 114.
98. Austin Cary, *Maine Forests*, p. 182.

ACKNOWLEDGMENTS

It has been an honor to set down the recollections of so many champions of Maine's environment, and a wonderful gift to have had the opportunity to listen to so many of them tell their stories. I trust that the incidents included in this narrative give a sense of their achievements; there were many equally interesting, and not a few amusing, tales that perforce ended up on the "cutting-room floor." The following generously gave up precious hours from the present so that I could interview them and recapture the past: Richard Anderson, Bob Cummings, Ben Emory, Jay Espy, Leon Gorman, Horace Hildreth, Sherry Huber, Jon Lund, Richard Parks, Roxanne Quimby, Harrison Richardson, Neil Rolde, Karin Tilberg, and Clinton Townsend. My heartfelt thanks to you all.

Others kindly answered my questions, often on the spur of the moment, over the phone or at lunch. Thank you: Brad Allen, Department of Inland Fisheries and Wildlife; Rob Bryan, Maine Audubon Society; Thomas Desjardins, Maine State Historian; Tim Glidden, Land for Maine's Future program; Herb Hartman, former director, Bureau of Public Lands; John W. Jensen, The Sharpe Group; Richard Judd, professor of history, University of Maine; Mason Morfit, The Nature Conservancy; Rupert Neily, Maine Coast Heritage Trust; Sandy Neily, Moosehead Futures; Jym St. Pierre, RESTORE: The North Woods; Caroline Pryor, land-use consultant; Richard Spencer, attorney.

I would not have been able to undertake the research and writing of this narrative without the generous support of the Betterment Fund and the Quimby Family Foundation. Personal thanks for their interest go to Carolyn Wollen and William P. Clough, III at the Betterment Fund, and Roxanne and Hannah Quimby at the Quimby Family Foundation.

In Maine, the project was shaped from the start by the combined wisdom of Richard Barringer, Phyllis Austin, Lloyd Irland, and Sam Merrill. The chance to work with fellow collaborators from the other New England states was also a pleasure. Particularly, my appreciation goes to Hank Foster for having the idea in the first place, and subsequently bringing the insights of his many-faceted experience to bear and make it happen, and to Perry Hagenstein who, as president of the New England Natural Resources Center, was ever helpful.

Above all, I am grateful to Richard Barringer for his ever-ready helpfulness on details, administrative and substantial. I called him my "thesis-advisor." His questions and insights, probing beyond the immediate issues into the political and societal context, helped me place conservation in the larger continuum of Maine's progress as a state. Thank you, Dick.

Finally, and as always, I thank my wife, Amy MacDonald, for her never-failing editorial acumen and the time she took away from her own writing to exercise it on this project.

BIBLIOGRAPHY

Anderson, Larry. *Benton McKaye*; Johns Hopkins University Press, Baltimore & London, 2002.

Austin, Phyllis. *The Uncommon Forest*; APF Reporter Vol. 9 #3, 1986.

Barringer, Richard, et al. *Land for Maine's Future: Increasing the Return on a Sound Public Investment*; Muskie School of Public Service and the Margaret Chase Smith Center, University of Maine System, 2004.

Bradford, Peter Amory. *Fragile Structures, The Story of Oil Refineries, National Security and the Coast of Maine*; Harper's Magazine Press, New York, 1975.

Brookings Institution Metropolitan Policy Program. *Charting Maine's Future, An Action Plan for Promoting Sustainable Prosperity and Quality Places*, 2006.

Cabot, Thomas D. *Avelinda, The Legacy of a Yankee Yachtsman*; Island Institute, Rockland, Maine, 1991.

Cary, Austin. *Maine Forests, Their Preservation, Taxation, and Value*; Commissioner of Industrial and Labor Statistics.

Chisholm, Hugh. *Pulp and Paper Industry in Maine*, Paper Industry Hall of Fame, Appleton, WI, www.paperhall.org.

Colby, Forrest H. *Forest Protection and Conservation in Maine*, 1919, Maine Forest Commission.

Coolidge, Philip T. *History of the Maine Woods*; Furbush-Roberts Printing Company, Inc., Bangor, Maine, 1963.

Cox, Peter. *Journalism Matters*; Tilbury House Publishers, Gardiner, Maine, 2005.

Cummings, Bob. "Lots of Confusion," *Appalachia*, Winter 1986–87.

Daviau, Jerome. *Maine's Life Blood*; House of Falmouth, Inc. Portland, Maine, 1958.

Delogu, Orlando E.; Merrill, Sam; Saucier, Philip R. "Some Model Amendments to Maine (and Other States) Land Use Control Legislation," *Maine Law Review*, Vol. 56, No. 2, 2004

Dorr, George. *The Story of Acadia National Park*; Acadia Publishing Company, Bar Harbor, Maine, 1997.

Douglas, Justice William O. "Why We Must Save The Allagash," *Field & Stream*, July 1963.

Fellows, Raymond and Edward J. Conquest. *William R. Pattangall of Maine*, The Maine Hall of Fame.

Ginn, William J. *Investing in Nature, Case Studies of Land Conservation in Collaboration with Business*; Island Press, Washington, Covelo, London, 2005.

Goldstein, Judith S. *Crossing Lines*; William Morrow & Company, New York, 1992.

Governors' Task Force on Northern Forest Lands *The Northern Forest Lands, A Strategy for their Future*, May 1990.

Hagenstein, Perry. *A Challenge for New England: Change in Large Forest Land Holdings*, 1987.

Irland, Lloyd and Theodore E. Howard. "Innovative Forms of Timberland Ownership: What are the Driving Forces?", *The Consultant*, April 1989.

Irland, Lloyd. *The Northeast's Changing Forest*; Harvard Forest, Petersham, Massachusetts, 1999.

Judd, Richard W. and Christopher S. Beach. *Natural States, The Environmental Imagination in Maine, Oregon, and the Nation*; Resources for the Future, Washington, D.C., 2003.

Krohn, William B. *Manly Hardy*; Maine Folklife Center, Orono, Maine, 2005.

Lipez, Kermit. *Kenneth Curtis of Maine: Profile of a Governor*; Harpswell Press, Brunswick, 1974.

Lacognata, Esther. *A Legislative History and Analysis of the Land Use Regulation Law in Maine*, Augusta, LURC, 1974.

Lockhard, Duane. *New England State Politics*; Princeton University Press, Princeton, 1959.

Lynn, Dahlia Bradshaw. Maine's New Third Sector, *Changing Maine, 1960–2000*, edited by Richard Barringer; Tilbury House Publishers, Gardner, Maine, 2004.

Maine Bureeau of Industrial and Labor Statistics, Augusta, 1903.

Maine Department of Transportation. *History of Railroading in Maine*, www.maine.gov/mdot.

Maine Department of Transportation. *Interstate turns 50 . . . an engineer remembers*, www.maine.gov/mdot.

Maine Pulp & Paper Association. *A Brief History of Papermaking in Maine*, www.pulpandpaper.org.

Marsh, George Perkins. *Man and Nature*; originally published 1864; Belknap Press, Harvard University Press, Cambridge, 1965.

McKee, John. *As Maine Goes*; Bowdoin College Museum of Art, Brunswick, Maine, 1966

McKibben, Bill. "An Explosion of Green," *The Atlantic Monthly*, April 1995.

McLeod , John E. *The Northern, The Way I Remember.*

Mendel, Howard L. *Classification and Management of Eider Nesting Islands in Maine*; Maine Cooperative Wildlife Research Unit, January 1969.

Milliken, Jr., Roger. *Eleven Generations of Forest Benefits — Where Do We Go From Here?* Annual Meeting, Society of American Foresters, Portland, Maine 1995.

Morfit, Mason. *Land Conservation in Maine*, Annual Meeting, Friends of Acadia, August 1990.

Natural Resources Council of Maine. *Natural Resources Council Bulletin*, February 1962.

Neily, Rupert. Thoughts about MCHT's origins; internal memo, January 9, 2002.

Osborn, William C. *The Paper Plantation, Ralph Nader's Study Group Report on the Pulp and Paper Industry in Maine*; The Viking Press, New York, 1974.

Pidot, Jeff. *Reinventing Conservation Easements: A Critical Examination and Ideas for Reform*; Lincoln Institute of Land Policy, Cambridge, MA, 2005.

Potholm, Christian P. *An Insiders Guide to Maine Politics, 1946-1996*; Madison Books, Lanham, Maryland, 1998.

Pryor, Caroline. "Women Conservationists in Acadia National Park," *Chebacco, The Magazine of The Mount Desert Island Historical Society* 2005.

Richert, Evan. "Land Use in Maine, From Production to Consumption," In *Changing Maine, 1960–2000*, Edited by Richard Barringer; Tilbury House Publishers, Gardiner, Maine, 2004.

Rolde, Neil. *The Baxters of Maine, Downeast Visionaries*; Tilbury House Publishers, Gardiner, Maine, 1997.

Rolde, Neil. *The Interrupted Forest*; Tilbury House Publishers, Gardiner, Maine, 2001.

St. Pierre, Jym. *Saving the Maine Woods: A Century and a Half of Schemes and Dreams*; Maine Land Conservation Conference, Brunswick, May 6, 2006.

Silliker, Jr., Bill. *Saving Maine*; Down East Books, Camden, Maine 2002.

Sullivan, Mark and R. Alec Giffen. *A Sustainable Vision for Maine's Natural Resources*, Land And Water Associates, 1995.

Thoreau, Henry David. *The Maine Woods*, Ticknor & Fields, Boston, 1864.

United States Department of Agriculture Forest Service and Governors' Task Force on Northern Forest Lands. *Northern Forest Lands Study*, April 1990.

United States Department of the Interior, Bureau of Outdoor Recreation. *Report on the Proposed Allagash National Riverway*, July 1963.

Wilson, Edward M. *The Northern Forest Alliance, An Assessment*; The OMG Center, November 1, 1996.

NEW HAMPSHIRE

COMMON GROUND

JIM COLLINS AND RICHARD OBER

THE FIRES OF CHANGE

When New Hampshire's White Mountains were burning, it was said, laundry hanging on lines in Manchester would turn gray with ash. Wealthy tourists traveling north to the grand hotels found the mountains stripped of timber and smoldering from slash fires.

"Instead of cutting only timber that is matured, everything is cut to the size of five or six inches in diameter, and what remains is cut into firewood or burned at once, leaving a dreary waste," the New Hampshire Forestry Commission reported in 1885. "In Lancaster, the timber and wood are nearly all gone and the mountains are being stripped to their summits. Originally a dense forest covered our state. This magnificent forest has long since disappeared."

In the second half of the nineteenth century, with most of southern and central New England cleared for farming, the timber industry was looking ever farther north to satisfy the growing nation's appetite. The White Mountains were particularly attractive, for they held vast stands of mature softwood and hardwood trees. Timber companies bought large tracts of land in the mountains, cut the forest, and shipped the wood south by train. The bared hillsides eroded, and silt filled the streams. Massive slash piles sparked into conflagrations, often by cinders spewed by passing locomotives.

The forest, the commission declared, was a public resource. Perhaps the state should regulate forestry more stringently and encourage farmers and loggers to reforest their land. These ideas were bold for the times. State government had virtually no control over private woodlands. Indeed, the state had spent the previous seventy years selling off its forests to settlers, including the summits of the White Mountains.

To ignite the public interest, sweeping changes in land-use policy, like the great fires that swept across the White Mountains, needed fuel and the right conditions. In the 1880s those conditions started to develop. First, in 1885, the state of New York set aside a million acres of timberland in the Catskills and the Adirondacks as a public service to protect drinking water downstream. Six years later Congress passed the Forest Reserve Act, setting aside millions of acres of western land the federal government had never sold to private settlers.

In New Hampshire, meanwhile, fires blackened the Zealand Valley at the headwaters of the Pemigewasset River. Crawford Notch was cleared of timber and left barren from the resulting erosion.

In 1889 the Forest Commission warned that deforestation had a direct economic impact on the White Mountain region's 1,100 summer inns and hotels and the $5 million their visitors spent each year; the region was New England's most popular mountain retreat, and the tourism industry relied on the scenery. The New York and Boston press called on the state of New Hampshire to save the forests.

Still the commission remained hesitant: "All the mountain forests in New Hampshire are private property, and we have no more control over their owners' treatment of them than we have over the condition of life on the moons of Mars."

The pressure to change that situation mounted in 1896 when T. Jefferson Coolidge, treasurer of the all-powerful Amoskeag Manufacturing Company in Manchester, produced an alarming annual report. The Merrimack was drying up in summer, Coolidge reported, and flooding in spring. Floods in 1895 and 1896 had shut down the Amoskeag mill, the nation's largest, and thrown 6,000 employees out of work. Coolidge blamed the floods on heavy logging at the headwaters of the Merrimack River in the White Mountains.

Even with this indictment from an industrial giant, public outrage smoldered slowly. Then a small scrap of paper fluttered into the embers. It was a pamphlet titled "The Boa Constrictor of the White Mountains."

The author of the tract, Rev. John E. Johnson, an Episcopal missionary, ignored the statistics on the amount of timber cut, acres burned, and dollars lost to the hotels and mills. He focused instead on what deforestation meant in human terms. And he gave the public a clear and simple villain.

The New Hampshire Land Company had been organized in Hartford, Connecticut, to purchase land in New Hampshire. Speculating on a future demand for timber, the company's 250 shareholders bought cheap farm lots and tax-sale lands, merged them into tracts 10,000 acres and larger, and refused to sell anything smaller to farmers, hotels, and even local towns. Their sole customers were the timber barons—George Van Dyke, J. E. Henry, the Saunders Brothers.

Johnson's pamphlet, which accused the Land Company of systematically deforesting and depopulating the heart of the White Mountains, enflamed key state leaders. In October 1899 Governor Frank West Rollins called on legislators to use state funds to buy the White Mountains as a public park. "The great White Mountains are being denuded and burnt over," he said in a speech, "and the summer tourist turns away in sadness and disgust . . . This must stop. You must arise in your majesty for your own protection and put the heavy hand of authority on these people. Someone will say, 'We can't afford to spend money for these things.' I am tired of hearing that. *You can't afford not to* . . ." The next month the popular weekly *New England Homestead* published three cover photographs of dilapidated wood slab homes in Thornton Gore. The New Hampshire Land Company, the newspaper reported, had squeezed out the residents, and the settlement had become a ghost town. The editors called for a White Mountain Forest Association to address the crisis: "Instant action is imperative. This White Mountain Association should be perfected instantly so as to present the whole matter effectively to the incoming legislature."

Over the next few weeks, letters poured in from farmers, businessmen, outing clubs, women's groups, bankers, and legislators. The Rev. Johnson kept at it, sitting at his writing desk late into the night "for the chance of hitting the devil with an ink bottle."

On February 6, 1901, shortly after finishing his second term, Governor Rollins organized a small meeting at the Board of Agriculture office in Concord. Among the luminaries present were Ellen McRoberts Mason, chairwoman of the Forest Committee of the New Hampshire

Federated Women's Clubs; Nahum Batchelder of Concord, editor, secretary of agriculture, and soon to be governor; George Cruft, legislator, hotel owner, and benefactor of Bethlehem; George Moses, editor of the *Concord Monitor*, secretary of the New Hampshire Forest Commission, and later to be U.S. senator; and Joseph T. Walker, assistant secretary of state. Those in attendance resolved to form a citizen-advocacy group, and Mason suggested a name: the Society for the Protection of New Hampshire Forests (SPNHF). Frank Rollins was elected president of the executive committee.

To keep the fires burning for change, the society hired Philip Ayres, forty-one, a historian who had worked as a reporter and charity organizer before finding his true calling: forestry and political action. Ayres took the job — which would be based in an office in Boston — on the condition that he be allowed to take the issue beyond Concord to Washington if needed.

Ayres spent his first year traveling the White Mountains to see firsthand the devastation of the forests. In the high country he found loggers cutting two-thirds of the spruce trees and leaving them on the ground so the largest trees could roll unimpeded to the logging roads, rail sidings, and rivers.

Initiating a New Hampshire tradition of working with a diversity of interests on forest debates, Ayres hit the road, taking his lantern slides to women's clubs, Grange halls, and teachers' meetings throughout New England. He interviewed timber executives. Orton B. Grown, general manager of the Berlin Mills Company — one of the region's largest landowners and a global pioneer in the use of wood fiber to make paper — explained to him that he hated to cut the timber in the Presidential Range, but could not keep the mills running without more wood. "Our company is interested in preserving the scenic beauty of the state and will contribute toward it, but cannot alone afford to maintain this tract as a park."

Only the national government could do that, Ayres concluded. In 1902 he addressed Congress, arguing that New Hampshire had too small a population and tax base to solve the state's deforestation problem alone. That fall Senator Jacob H. Gallinger of New Hampshire introduced a bill to purchase a national forest reserve in New Hampshire. Reception was cool at best.

"Not one cent for scenery!" thundered House Speaker Joe Cannon.

Rocky Mountain congressional leaders, bridling at the proliferation of Teddy Roosevelt's national parks in the West, were outraged at the idea of using taxpayers' money to buy timberland for federal control in the East. The House Judiciary Committee declared that such a policy would be unconstitutional. Other lawmakers challenged Gallinger to prove a scientific relationship between timber cutting in the White Mountains and the floods and droughts in the valleys downstream. The bill stalled in session after session of Congress. Ayres lobbied from afar, seeking endorsements from politicians, tourism leaders, and journalists. The Appalachian Mountain Club (AMC), founded in 1876, became a major partner in the effort, as did advocates for a bill proposing similar forest protection in the southern Appalachians. In 1906 the two bills were combined into one, but the legislation still stalled.

In 1909 Gallinger again filed his bill. This time, with the support of an influential congressman named John Weeks of Massachusetts, a White Mountains native, the bill called for forest protection "for the purpose of preserving the navigability of navigable streams."

The conservationists had found a common cause with industry: reliable flow of water to power the mills. Weeks filed a companion bill in the House, one that did not mention the Appalachians or the White Mountains at all. It simply said that Congress would appropriate funds to purchase land and forest reserves for "the conservation and improvement of the navigability of a river."

On March 1, 1911, President William Howard Taft signed the Weeks Act into law. The first land purchases established North Carolina's Pisgah National Forest in 1913. Later that year the federal government bought more than 100,000 acres of cutover, burned-out land in the Presidential Range of New Hampshire. The price was $6 an acre. Weeks's broadly worded bill would eventually authorize federal purchases of forestland in forty-eight national forests across the nation. The spark ignited in the White Mountains would roar into a national conservation revolution.

In the 1890s, as the debate over the White Mountains was brewing, the state was given its first public land as a gift — three acres atop Pack Monadnock in Peterborough. Over the next two decades, as the Forest Society, the AMC, and others worked to establish the White Mountain National Forest, a surge of private action would lay the groundwork for what would eventually become New Hampshire's state parks. Land on Mount Monadnock, Mount Sunapee, and Crawford Notch was acquired through citizen campaigns and held by the Forest Society. The primary purpose was not recreation, but protection against overcutting and fires.

But state monies for parks were hotly debated. "Buy the Notch and do it right away," the *Coos County Democrat* urged the state when $100,000 was needed to save Crawford Notch. "The whole people will have to pay the bill," the *Canaan Reporter* countered, "while not one in a hundred, perhaps not one in a thousand, will ever be benefited one cent's worth." In 1915 the state made its first allocation—$5,000—toward the purchase of forestland.

By the 1920s, the touring car was bringing thousands of visitors to the state, many of them flocking to the state parks. Boardinghouses, where a bed and a meal cost a dollar a night, sprouted up in every town. The Appalachian Mountain Club, Randolph Mountain Club, and other outing groups blazed their trails up and around every hill and mountain. Cabins and camping shelters dotted the high pastures and apple orchards.

In 1923, a landmark Franconia Notch hotel called the Profile House burned to the ground. The Profile and Flume Hotel Company announced that it would have to sell some six thousand acres of land to pay its debts. Franconia Notch at the time was drawing more than 100,000 visitors each year, attracted by what one travel guide called a "huge museum of curiosities"—the Old Man of the Mountain, the Flume, the Basin, and the steep walls of the pass itself. The threatened sale drew immediate and vocal response. "King of the Mountains: Save His Kingdom!" cried the editorial page of the *Boston American* on December 27, 1927. "Otherwise, the lumberman's axe will smash, destroy, and lay desolate one of New England's greatest natural attractions."

Once again it was the Forest Society's Phil Ayres and the Federated Women's Clubs who led the campaign. Ayres convinced the organization's executive committee to raise $200,000 to match a state appropriation, which he also secured. With half the private money coming from James J. Storrow of Boston, Ayres set out to raise the rest a dollar at a time by "selling" trees in the notch. Donors received a "Certificate of Purchase" that entitled the bearer to identify and care for one tree. (Seventy years later, residents were still contacting Forest Society

Motorists stop circa 1930 to admire the famous profile of the Old Man of the Mountain in Franconia Notch. The advent of auto tourism redefined the value of scenic, wild lands in New Hampshire, and much of New England, interjecting a powerful new interest into the states' land-use debate.
(New Hampshire Historical Society)

officials wondering if they could find the trees they had bought as children.) When the $400,000 goal was reached and the state bought Franconia Notch in 1928, total state ownership stood at 29,000 acres of land.

But the state's Forestry Commission didn't have the resources to manage the visitors. The Forest Society took care of facilities in the Flume, Mount Sunapee, and Crawford Notch for years. Other private groups maintained trails, supervised beaches, and built shelters at state-owned lands. The pattern would continue for decades. Even when the Forestry Commission was renamed the Forestry and Recreation Commission in 1935, little money flowed in support of its work. On top of modest federal and state funds for work relief projects, administrator Russell B. Tobey was allocated just $11,500 for the parks, which were already welcoming some 300,000 visitors annually. The son of a Massachusetts senator, raised and schooled in New Hampshire, Tobey took his $25-a-week salary and state-provided Model A Ford pickup and went on a mission to educate. He endlessly typed letters and crisscrossed the state promoting parks with a homemade film and stereopticons. He steadily elevated the parks' profiles and attendance, even as money for maintenance and programs remained scarce. In 1942 the legislature declared that the parks would have to pay their own way entirely. (Today, New Hampshire remains one of only two states that don't allow tax-generated revenues to fund their state park systems. The average state provides close to 50 percent.) Undaunted, Tobey focused on expanding the parks' appeal. He added skating rinks and archery ranges to several parks, staged horse shows and concerts, encouraged the use of state land for mass events. He believed in respectful use more than mere preservation. "If our parks are valuable," he said, "we must use them." And New Hampshire citizens and visitors responded.

While public-funding woes and land-use conflicts would challenge both the state park system and the White Mountain National Forest over the coming years, the seminal question of whether public land was appropriate in New Hampshire had been answered with a resounding *yes*. The century began with only three acres in public ownership. Within a few decades some 700,000 acres—more than 10 percent of the state—had been set aside as state or federal land. In this fiercely self-sufficient, Yankee part of New England, a fundamental view of land had shifted. The story of land conservation in New Hampshire starts there.

WISE AND CURRENT USE

By the time Philip Ayres retired from the Society for the Protection of New Hampshire Forests in 1935, the landscape in New Hampshire was changing. The expensive, high-profile campaigns surrounding the state's dramatic peaks and notches had been completed. The wooded slopes of Crawford Notch and the Zealand Valley were recovering.

The state was in the midst of Depression. Slumping demand for paper and wood products had slowed the overcutting in the North Country that had so alarmed and rallied activists years before. Already hemorrhaging jobs to the South, mill towns throughout New Hampshire were crippled by the collapsing economy. In the year that Ayres retired, textile mills in Hillsborough County, alone, lost 110,000 cotton spindles, and the Amoskeag mill in Manchester abruptly shut down, taking 8,000 jobs with it and, ultimately, destroying the economic engine that had driven the state's largest city for more than a century. An 8,000-acre property in nearby Allenstown (now Bear Brook State Park) was opened specifically to give poor and out-of-work families a chance to enjoy camping and the out-of-doors.

Then in the warm, early spring of 1936, a combination of snowmelt, ice jams, and heavy rain created spectacular flooding — the worst riverine flooding the region had seen in 14,000 years — devastating homes, mills, cropland, bridges, roads, and railways. Property damage statewide topped $7,000,000 — an almost unimaginable amount at the time. Two years later, the great Hurricane of September '38 blasted northward up the Connecticut River Valley, flattening 1.5 billion board feet of timber in its path as far inland as the White Mountain National Forest. Three years after that, with the country on the verge of war, with tons of dry slash still littering the ground in the wake of the hurricane, fires burned across the state, including 24,000 acres in the towns of Marlow and Stoddard. It was not the time for ambitious agendas or great land purchases.

Which isn't to say nothing was happening on the ground. One of the first programs of the New Deal — the Civilian Conservation Corps (CCC), signed into law by Franklin D. Roosevelt in 1933 — had rushed thousands of unemployed men to work in the nation's forests and parks, to help preserve and improve natural resources. New Hampshire gratefully put the federal assistance to use. Some 22,000 men, half of them

A woodlot owner near Keene, New Hamprhire, surveys damage in the wake of the great New England hurricane of 1938. Across New Hampshire, some 1.5 billion board feet of timber were flattened by the storm, triggering an unprecedented mobilization of local, state, and federal salvage resources. (U.S. Forest Service)

imported from the cities of southern New England, eventually enrolled in thirty-five separate New Hampshire camps. The majority of them worked in the North Country; a smaller number scattered from West Swanzey and Claremont in the western part of the state to Suncook and Raymond and all the way east to Dover. The CCC camps were charged with a mission of "reforestation, forest protection, and park development." The New Hampshire camps cut trails, built bridges and roads (notably the Evans Notch Road in the rugged Wild River area along the New Hampshire–Maine border and the summit road on Mount Kearsarge), and developed shelters and visitor centers. But while crews in the western states were constructing landmark projects — in Yellowstone, on Mount Rainier, in the Olympic National Park — the CCC here spent an extraordinary amount of time cleaning up after the '36 flood, stabilizing stream banks, battling the spread of gypsy moths and blister rot, fighting fires, and doing the staggering work of clearing brush and salvaging timber left by the hurricane. The Forest Society suspended operations entirely to assist with the salvage.

Overwhelmed by the sheer volume of the storm's destruction and the potential for fire — and with only six or eight months to salvage merchantable wood before pests and fungus set in — New Hampshire turned toward Washington for help. E. W. Tinker, assistant chief of the U.S. Forest Service, opened an emergency office in Boston, called in thirty-two salvage specialists from around the country, and met with all six of the New England state foresters. From that office emerged the

Northeast Timber Salvage Administration (NETSA), funded through the Disaster Loan Corp. Its mission: to use federal, state, and local resources to untangle the colossal fire trap while attempting to reimburse private landowners for some of their timber losses. In New Hampshire, forester Owen Johnson chaired the governor's special commission to mobilize the operation. The epic challenge required the efforts of thousands of WPA workers alongside a small army of laborers, foresters, choppers, scalers, oxen drivers, administrators, and volunteers. All this was before the use of bulldozers and chainsaws became widespread, and twenty years before the skidder; the primary tools were axes and crosscut saws, chains, oxen, horses, and primitive tractors. Ultimately, NETSA salvaged 500 million board feet of timber in the state, of an estimated 620 million that were potentially saleable.

The hurricane and its aftermath left two important legacies. The salvage operation spurred unprecedented cooperation between public agencies and private landowners—a theme that would emerge as a defining aspect of New Hampshire's history of land conservation. And the salvage inspired universal log grading and scaling procedures, accelerated the field-testing of new equipment and practices, and utilized forestry science to a degree not seen before. The embrace of emerging ideas and science would become a hallmark of New Hampshire's state forestry. It would lie at the heart of the new working definition of "land conservation": not only protecting land outright, but managing land responsibly, reducing runoff and erosion, replanting trees, controlling pests. The goal was wise use of the land, long before the "wise use" movement co-opted the term.

Owen Johnson was one of the prominent forward-looking foresters, but he wasn't alone. The Brown Company in Berlin—at one time a world leader in the production of pulp and paper, though ailing since the Depression—had built a reputation for innovation. It had created the industry's first separate research and development facility, employing engineers and technicians but also foresters trained at the University of Michigan and the Yale School of Forestry. Brown Company foresters pioneered the idea of a mill owning enough land to provide all of its own wood fiber, calculated the necessary acreage, and developed harvesting plans that could be sustained over the long haul. One of the company's young foresters, Henry Baldwin, had received master's degrees from Yale in forestry and botany, and had studied at the Swedish

School of Forestry. He'd gone on to become the state's first research forester at the Caroline Fox State Forest in Hillsborough, the state's official research forest. Along with other leading foresters such as Will Brown and Howard Mendenhall, Baldwin believed passionately in science, ecology, and education. Their breed would serve as a bridge between classically trained foresters and the environmentalists who would follow after them.

Under the new leadership of Lawrence Rathbun, meanwhile, the Society for the Protection of New Hampshire Forests continued promoting its view of the state's woodlands as a precious renewable resource, with room for industry *and* tourism, conservation *and* development. Rathbun, erudite and well-connected, partial to horn-rimmed glasses and bow ties, schooled at Harvard and Yale, was especially interested in educating landowners, both large and small. He was determined to help spread progressive techniques across the state. He started up a magazine for society members, *Forest Notes*, which resembled a technical forestry journal. Working with the New England Wildflower Preservation Society, he pushed the society's New Hampshire Nature Camp — an idea modeled on teacher-education programs that several state universities had adopted following the devastating years of the Dust Bowl. The Nature Camp brought 30 teachers from around New Hampshire to the society's Lost River property for intensive, week-long workshops with natural resource experts. Several of the experts brought in by Rathbun would go on to play important leadership roles in the state's history of conservation, including Tudor Richards with the Audubon Society and Ted Natti, the long-time state forester. Rathbun expanded the program to reach high school students directly, through the Youth Conservation Camp at Bear Brook State Park. Some of the students who came through the youth camp would later make contributions of their own, including future state forester Jack Sargent. The curriculum in both programs — covering the "study of the interrelationships between all species of plants and animals and the soil and water on which their habitat depends" — was novel in the 1940s. It picked up the urgent call of Aldo Leopold, who was championing such interrelated awareness from his lonely perch in Sand County, Wisconsin. Rathbun helped launch New Hampshire's Tree Farm program soon after the original program made its 1941 debut in Washington State. In its first year, twenty-four New Hampshire landowners agreed to meet requirements that demonstrated a commitment to long-term stewardship of their land.

A major political victory helped enable such stewardship. Since 1912, the Forest Society had continually lobbied to change the state's timber tax. The tax—an annual 2 percent levy based on the value of a parcel's standing timber—essentially penalized people who left their trees to grow and rewarded those who cut them down. If the same logic were applied to agriculture, state forester Edgar Hirst noted, "The inevitable result would be the picking of apples while still green, the gathering of corn in the silk, the cutting of hay before it is ripe." In many cases, struggling landowners had to liquidate their woodlots just to pay the tax bills.

To change the law, though, the legislature needed to amend the state constitution to allow different classes of land to be taxed differently—a radical change in the state's traditional view of property. Finally, the legislature bowed to logic and relentless pressure by Rathbun and others, the voters approved the amendment, and in 1949 Governor Sherman Adams signed the "yield tax" into law. From then on, the tax was levied only when timber was actually harvested. That single change in the tax law had immediate and far-reaching effects on the way owners managed their land. It encouraged forestry with a longer view.

Alongside this important gain in public policy, education remained the story for much of the 1950s and '60s. The Tree Farm program expanded. The Forest Society hired the first full-time educator of any conservation group in the sate, a school-camp pioneer named Les Clark who would influence generations of young people over a twenty-two-year career. The society cosponsored education projects with garden clubs, with industry-minded organizations such as the Granite State Division of the American Forestry Society, and with the University of New Hampshire Cooperative Extension Service, which was broadening its mission in response to the region's dwindling agriculture. The society's approach very much fit the state's history and ethos, which left so much up to private citizens and private initiative. New Hampshire didn't share the public land-grant tradition of the upper Midwest, or have the activist government of neighboring Vermont. It had strong industry interests, but nothing like the singular dominating influence of Maine's paper industry. Unlike Massachusetts and Connecticut, the state's citizen legislature tended to mistrust regulation of all kinds, even the public funding of public land. Years later, Henry Tepper of The Nature Conservancy would say, "You can do great land conservation in New Hampshire. You just won't have state money to pay for it." An educated citizenry was key.

In the fall of 1961, Larry Rathbun hired a young forester named Paul Bofinger to help work through a backlog of Tree Farm applications. Bofinger had grown up in a 1930s subdivision on Long Island, next to potato farmland that was bulldozed when he was still a kid. He'd studied fisheries biology and woodlot management at Cornell, then forestry in earnest at the University of Michigan, before landing in New Hampshire with a small lumbering outfit that put him to work for five years "from stump to finished product." He was an "Eisenhower Republican," down to earth, savvy, curious, with persuasive and infectious energy. He befriended state forester Edgar Hirst, and worked closely with the veteran woodsman Owen Johnson, and came to consider both men mentors. During the three months he inspected Tree Farm applications, he met dozens of forestry consultants and all ten of the state's county foresters. (New Hampshire had more professional foresters per acre than any other state in the country then, and still does today.) He worked through the backlog, and Larry Rathbun kept finding more work for him, and Bofinger kept learning about the state and building relationships.

At that time, the Society for the Protection of New Hampshire Forests had one full-time education director, one secretary, Bofinger, and Rathbun, who was drawing a half-salary and spending more and more of his time traveling. They worked out of three rooms at 5 South State Street in Concord. Bofinger began a close relationship with Eugene Struckhoff, executive director of the Spaulding-Potter Charitable Trust, which shared office space in the building. The trust (which later morphed into the New Hampshire Charitable Foundation) became one of the Forest Society's most reliable sources of funding. Bofinger got in the habit of walking down to the State House early each morning, to read the House and Senate journals and to bump into legislators. He made a point of going out to breakfast or lunch with various agency heads. Concord was a small town, and New Hampshire a small state, and Bofinger learned who the players were. "Back in those days," he recalls, "there weren't separate funders and separate doers. They were the same people."

But even as he took on an increasingly large role with the Forest Society, the pressure on the state's land was ratcheting up. Following the war, abandoned farmland had been snapped up for $10 an acre. Field pines and scrubby woods were inexorably reclaiming open space — a defining story of much of the state, ever since the "great merino craze" of the mid-1800s when hundreds of thousands of sheep dotted pastures and hillsides everywhere, and forests covered just 48 percent of New Hampshire's total land area. Now, after decades of relatively slow growth and a wave of farm closings, demographic shifts were altering the face of southern and central New Hampshire. The price of farmland shot up to $100, $200, and in some areas — incredibly — $1,000 an acre. Growth spreading out of northern Massachusetts was making land in New Hampshire more profitable for development than for growing trees or raising crops or raising cows. Real estate developers boomed with the interstates, bought large tracts relatively cheaply, put in roads, and built neighborhoods farther and farther from town centers. The new households demanded new schools and other services, and towns raised property taxes to meet the need. Unable to keep up with the taxes, many landowners sold properties that had been in their families for generations.

The state government — in a rare moment of largesse — took notice. In 1963 the legislature, at the urging of State Parks Director Russell Tobey; Mary Louise Hancock, Director, State Planning Office; and Democractic Governor John King; ended decades of meager appropriations and reliance on private land donations. It authorized a $9 million bond issue — a huge figure by New Hampshire standards — to expand the state park system. Just after the legislative session ended, Governor King called Ted Natti, the state's chief of lands and forest management into his office. "Ted," he said, "I want you to start buying land before the word gets out that we have all this money to spend." Natti, a strong advocate of conservation who would become state forester in 1967, immediately began acquiring park-quality lands, some where prices were still as low as $50 an acre. The state added thousands of acres to Bear Brook and Pawtuckaway parks in southern New Hampshire; the only public beach on the state's largest lake, Winnipesaukee; a priceless white sand beach in Rye; the summit of Mount Washington; critical access to Lake Umbagog in the North Country.

Yet more effort was needed. In the face of relentless growth and rising property taxes, a broad coalition calling itself SPACE— Statewide Program of Action to Conserve our Environment—set out to amend the state constitution to allow underdeveloped land to be taxed at its much lower "current use" value rather than its potential development value. The effort involved thousands of citizens in dozens of organizations, including the Forest Society, the Audubon Society, the Farm Bureau, and the New Hampshire Timberland Owners Association. Two other states, Maryland and Connecticut, had passed versions of current-use tax law, but the culture in New Hampshire promised a harder sell. The seasoned public relations firm, Jackson,

Voters in 1968 said "yes" to an historic ballot initiative that enabled New Hampshire land owners to be taxed according to their land's current use rather than its potential development value. Key supporters of "Question #7" included Dick Kelly, Paul Bofinger, Miriam Jackson, and Tudor Richards.
(Society for the Protection of NH Forests archives)

Jackson & Wagner, warned the Forest Society it would need to raise at least $30,000 for the campaign.

The issue was put to the voters in November 1968. On the Sunday before the vote, the conservative statewide newspaper, the *New Hampshire Sunday News*, came out with an editorial opposing the amendment. On the night before the vote, Bofinger—who had officially taken over as head of the Forest Society in 1965—got a call from Miriam Jackson of the public relations firm, who told him to get out and distribute more leaflets. Tired from months of heavy campaigning, Bofinger didn't jump at the suggestion, but he pulled on his boots and went out to leaflet cars in a Concord shopping center. The next day, voters approved the current use amendment by just 126 votes. Governor John King, who supported the program, said there would be no recount. Years afterward, Bofinger would still wonder if those last-minute flyers had made a difference.

The coalition immediately started lobbying the legislature to write the enabling laws. One tactic involved taking a busload of lawmakers through the southern tier to see recent development for themselves. "Some couldn't believe their eyes," recalled one of the society's consultants, Isobel Parke. The momentum grew. Working with key legislators, the coalition crafted a system of differential taxation that addressed core concerns and garnered broad bipartisan support. The program's ten-acre minimum allowed smaller landowners to participate. A penalty charged for land taken out of the program gave towns a chance to recoup potentially lost revenues. The influential farm and forest industries backed the plan. Municipalities without planning or zoning ordinances saw the possibility of some breathing room. The program earned the support of Republican Governor Walter Peterson, and even that of his successor, the archconservative Meldrim Thomson, who signed the bill into law in 1973. A quarter of a century later, studies would show that open space contributed one-fourth of the state's overall economy, some $8.2 billion annually, and that towns with more open space had, in fact, lower property taxes than more developed towns. Long-time agriculture commissioner Steve Taylor believed the law was even more important for protecting working and open farmland than it was for forestland. "Open land is often more valuable and more easily developed," he noted.

The current-use victory—daunting and difficult as it was—may have become the single most important piece of New Hampshire's land-conservation history: by the year 2000, almost half of the land in the state would be enrolled in the program, more than the White Mountain National Forest, all of the state parks, refuges, sanctuaries, and easement-protected acreage, combined. As Steve Taylor put it, "Current use was the frontline tool."

WATER'S EDGE

In the early1960s, Merrimack River fishermen sometimes found themselves thigh deep in water coated with reddish scum — raw organic waste from toilets upriver. Hydrogen sulfide fumes from the paper mill in Lincoln peeled paint off buildings in Bristol, thirty-odd miles downstream. Dyes dumped by textile mills left their stain on the river — some locals called it the "Merrimuck" — and fish that survived smelled as appetizing as their environment. "It was so bad," said fisherman and lifelong water-quality professional Terry Frost, "you could plunge your hand into the river up to your wrist, and not be able to see your fist."

The Connecticut River had become so polluted that Dartmouth College suspended its annual downriver "Trip to the Sea" for more than two decades. During the spread of Dartmouth's national reputation for outdoor programs, the Connecticut was called "the best landscaped sewer in the nation," and many students went four years barely knowing the river even existed. The discharge pipe from Mary Hitchcock Hospital in Hanover emptied into the river just above the docks of the college's canoe club.

The Forest Society — as it had a half-century earlier when it helped put rivers at the center of the Weeks Act — understood the inseparable connection between clean water and properly managed land. The society and State Forester Ted Natti got the support of the New England Lumberman's Association, and pushed New Hampshire to become an early adopter of "Best Management Practices" — specific guidelines to prevent logging-related water problems such as runoff and erosion. Importantly, New Hampshire became one of the first states in the country to regulate cutting along stream banks and other waterways, and to require soil and vegetative buffer zones.

When developers started eyeing salt marshes along the Hampton and Seabrook coastline, New Hampshire Audubon and, soon after, the Forest Society set out to acquire key parcels to secure their water-cleansing qualities. The two organizations helped form the New Hampshire Better Water Committee (BWC), which worked behind the scenes and on the ground to support the work of the New Hampshire Water Supply and Pollution Control Commission. "We realized we'd never be able to afford to buy all the land necessary to help protect water supplies," Bofinger recalled. "So we took a clue from Massachusetts and focused on legislation." New Hampshire's resulting wetlands protection law was the second in the nation. A few years later, the Better Water Committee helped push through the country's first law requiring a statewide review of septic systems. With the passage of the Water Quality Act in 1965, federal dollars started flowing into state sewage treatment plants, and the BWC soon focused on persuading the legislature to increase matching funds for these projects. It was successful: at one point, New Hampshire had the highest percentage of state and federal aid of any state in the country.

Along with the landmark state and federal water policies, an equally crucial force was emerging at the local level—conservation commissions.

In 1960, the Society had sent Les Clark to the first New England Conservation Commission Conference, held at Harvard University. Paul Bofinger had written letters to society members and traveled the state to educate people about the potential of this kind of local oversight and control. With society encouragement, the state had passed the appropriate legislation, and by 1965 eighteen New Hampshire towns had formed conservation commissions. It wasn't until 1973, though, when the legislature extended the authority of local commissions to wetlands protection, that the true importance of conservation commissions became clear. Systematically, wetlands began to be saved by people who knew where they were and who they were up against. For the first time, local developers and selectmen had local watchdogs to answer to. In New Hampshire, that amounted to oversight of the first degree.

With his seminal leadership of the current-use campaign, the establishment of the Better Water Committee, best management practices, conservation commissions, a high-profile fight to protect the Androscoggin River from a massive dam south of Errol, and other conservation successes, Paul Bofinger showed how much influence a single citizen or small group could have. His approach, he said, followed "the New Hampshire Way." The compromises and solutions came about as they would at a town meeting: people from different backgrounds, face to face with each other, debate and differ and arrive at solutions, and then leave as neighbors who still live with and rely on one another. The proximity and personal relationships in Concord fostered trust and continuity, as well as strategies.

Of course, there were natural rivalries and resentments over such cozy arrangements. There was an image, unfairly simplistic, that "Paul and Walter Peterson and a couple of agency heads could get together and work out current use over a plate of eggs at the Highway Hotel." This wasn't small-town democracy; it was a big-city, smoke-filled back room. The truth in it was this: through the 1960s, the Forest Society was the only conservation organization playing at the state level — and when a more local approach was necessary, it was often the Forest Society, alone, that understood how to help or create smaller organizations to get the work done. The New Hampshire Audubon Society — unlike the powerful and well-established Massachusetts Audubon, and unlike states that were fully functioning chapters of National Audubon — was for a long time a tiny, strictly volunteer organization until Tudor Richards became executive director in 1968, and even then the group struggled for funding. National groups that were making inroads elsewhere in New England — The Nature Conservancy, The Sierra Club, The Wilderness Society — had little credibility with legislators or locals, and almost no place at the table in New Hampshire. There was no spotted owl to inspire national controversy, no snail darter to bring in the national media and the Supreme Court. Despite the tremendous work of long-time directors such as Fran Belcher and Tom Deans, local residents were even suspicious of the Appalachian Mountain Club, whose offices were on Joy Street in Boston. Eve Endicott recalled the frustrating years she tried to work in New Hampshire out of the Eastern Regional Office of The Nature Conservancy: "I would drive up to Concord from Boston, so I was a carpetbagger by definition. The Forest Society was Big Brother then. They didn't have any interest in a little sister, let alone another big brother! It wasn't easy. But I'd have felt the same way — 'Who are these guys? They don't know what's going on up here.' But The Nature Conservancy was doing eco-regional planning, looking at endangered species in New Hampshire as part of the eastern coastal region or the northern Appalachian region. Unfortunately, species don't recognize state lines."

The White Mountain Advisory Committee illustrated the problem some national groups had working in New Hampshire. The National Forest Service, charged with drafting and implementing ten-year management plans for the White Mountain National Forest, asked the state to formally convene a committee to work out issues and compromises ahead of time, to ensure the support of the management plan once it was announced. The advisory committee included a wide range of perspectives: Paul Bofinger of the Forest Society; Francis Belcher of the Appalachian Mountain Club; Rachel Terrill of the Federated Sportsman's Clubs; Dick Hamilton, president of White Mountain Attractions, representing the tourism industry; the Saunders Brothers' Berm Garland, representing the wood industry; Abigail Avery of the Sierra Club. ("Once you had Garland and Avery," recalled Bofinger, "you had it. Once you had each end, the middle followed.") There was a spirit of cooperation and trust on the committee, including from Abigail Avery, who went along with what her local grassroots people told her. Only the Wilderness Society—which sent a representative to the meetings only after most of the compromises and agreements had been worked out, with a mandate from its national constituency—opposed the committee's final recommendations. Its challenge failed.

The Forest Service eventually abolished the formal advisory structure, which it had relied on for the planning in several national forests around the country. But New Hampshire decided to keep its advisory committee anyway, to serve the same function in an unofficial role. Many of them would later meet quarterly in something called the New Hampshire Natural Resources Forum, created by Sharon Francis, a recent transplant from executive-level conservation work in Washington, D.C. The forum encouraged conservationists from around the state to explore the interrelatedness of woodlands, pollution, incinerators, traffic, agricultural land, and water.

Increasingly though, with the broadening of the environmental movement, "the New Hampshire way"— not resorting to lawsuits, not relying on government pressure, inviting industry and tourism into the conservation debates, searching for common ground rather than clinging to higher ground—was seen by some as too soft and accommodating. At the same time, New Hampshire was becoming a more complicated place, and common ground harder to find. When local residents took on Aristotle Onassis over his plans to build an oil refinery on Durham Point, or when the Clamshell Alliance protested the construction of the nuclear power plant at Seabrook in the early 1970s, the implications rippled out far beyond New Hampshire's borders, and more people wanted a say. Consensus and democracy are rarely neat. New Hampshire increasingly found itself in the midst of regional and national debates over politics and energy, the economy and the environment.

Meanwhile, environmental consciousness was rising, the federal Clean Water Act was working, pollution and discharge standards were tightening. By the 1980s Terry Frost was fishing and paddling in a clear-running Merrimack River. Dartmouth students had long since resumed their annual canoe trip to Long Island Sound

Lake associations blossomed throughout the state, and began an impressive state-coordinated water-quality self-monitoring system that became the envy of the region. When the Federal Energy Regulatory Commission received an application for a small-scale hydro dam on the scenic Wildcat Brook in Jackson, a young congressional aide (and Jackson native) named Tom Burack realized that only another federal program would have the power to stop the ill-advised project from moving forward, and helped get the river designated as a "study river" in the federal Wild and Scenic River program. Through the leadership of Senator Gordon Humphrey, the river later became the first fully protected river in New Hampshire under the federal law, and remains wild and scenic to this day. (Burack eventually became the state's commissioner of environmental services.) There were wetland mitigation victories at ski areas and highway construction projects. There were legislative advances — many of which increased buffer zones along water edges — perhaps none as important as the 1991 passage of the Comprehensive Shoreland Protection Act, which set statewide minimum standards for setbacks along rivers and lakes.

By century's end, a new generation of spokespeople reminded us of our place, and also that our place was inextricably tied to a larger world where actions of all kinds had impacts downstream. Their message found a metaphor in the establishment of the Connecticut River Valley's Silvio O. Conte National Fish and Wildlife Refuge. For the first time, a national refuge was established that did not rely primarily on federal land acquisition. Instead, the refuge would depend on unprecedented public and private cooperation: on outreach and the coordinated efforts of dozens of stakeholders within a four-state, 7.2-million-acre watershed, including private landowners, farmers, fishing and hunting clubs, local and national conservation groups, state and local governments, national and state agencies, and local businesses. The mechanisms for protection included education, incentives, sustainable land management, conservation easements, and the implicit sense of the common good shared by more than two million residents in the watershed.

The multipartner approach represented a wholly new model for national refuges. In 2006, it would be put to the test when the U.S. Fish and Wildlife Service proposed creating a new federal land acquisition unit in southwest New Hampshire within the boundaries of the refuge. In a widely circulated letter to the service, the Connecticut River Joint Commissions of New Hampshire and Vermont opposed the plan. "Large areas of land under federal ownership and management runs counter to the carefully considered Action Plan and Environmental Impact Statement of 1995 which favored habitat protection through partnerships with communities and organizations, and rejected the option of federal land ownership," the commissions' cochairs wrote. "While we recognize that these views may run counter to recent thinking within the U.S. Fish and Wildlife Service, we are certain they reflect the best interest of our part of the watershed. They are faithful to the legislation that established the refuge."

In other words, throughout the multistate watershed, the New Hampshire way was still alive.

TRUST IN NEW HAMPSHIRE LAND

By 1987, New Hampshire was the second fastest growing state east of the Mississippi. Its population had passed a million and was increasing at 20,000 annually—twice the rate of neighboring New England states.

Although the growth brought much prosperity, the impact on the landscape was inescapable; forests and farms were disappearing at a rate of 20,000 acres a year. New Hampshire was eating its seed corn—its scenery, its natural resources, its way of life. "There was a lot of concern about what was happening to New Hampshire," said Stephen Taylor, the state's commissioner of agriculture. "A lot of people felt that the state was like a big truck rolling downhill with no brakes. You could steer the damn thing a little but you couldn't stop it."

Planners at all levels were steering with all their might. State legislators authorized locally approved zoning ordinances, which allowed Charlestown to classify 25 percent of its land as a "watershed zone," for instance, and Lyme to set minimum lot sizes of fifty acres in a remote part of town. Extension forester Phil Auger reached out to Seacoast area developers to encourage more environmentally friendly practices. Auger, Frank Mitchell, and Jeanne McIntyre worked with the town of Deerfield to create a natural resource inventory — a critical tool for managing and setting priorities that became a model for towns and conservation commissions across the state. But applications to subdivide, build, dredge, and fill buried local and state regulators. Towns in the southern tier saw building permit applications double from one year to the next and grappled with subdivisions of 100 lots when they had previously seen no more than a dozen. Local land trusts were starting to coalesce around the growing interest in conservation easements and other voluntary land protection mechanisms. But money was scarce and many owners of high priority open space could not afford to make donations.

Paul Bofinger took a sabbatical from the Forest Society to find a new solution. "I set out thinking that somebody else had the magic wand — the experience or technique that I could bring back to New Hampshire," Bofinger said. "But in the final analysis I realized that wasn't going to work. Whatever would work would have to be something that fit the times and the New Hampshire scheme of things."

Bofinger figured that the only suitable response was to offer cash to owners of important conservation land. The concept wasn't new, but the scope was. To make a dent, Bofinger calculated it would take 100,000 acres and $50 million from the legislature.

Society trustees and staff approached leaders from all arenas — conservation, business, finance, tourism government, philanthropy, and civic groups — with questions as well as answers. The New Hampshire Wildlife Federation and the Homebuilders Association alike were asked for support and advice.

Several elements quickly became clear. First, the times called for an all-out effort. No token project would meet the conservation needs of a burgeoning state that had not made a major commitment to land acquisition since 1963. Second, no single entity was big enough to pull it off. Despite missions that might seem at odds, groups would have to work in unprecedented harmony. And third, public money could do only part of the job.

"Even collectively there aren't enough public dollars," said Lewis Feldstein, president of the New Hampshire Charitable Foundation. A new nonprofit organization would need to be created. In 1986, 150 state leaders became incorporators of the Trust for New Hampshire Lands (TNHL).

With seed money from the charitable foundation, the society set out to inventory the best of what was left of New Hampshire. A team of interns ended up with an inventory of more than 300,000 acres of priority lands. Bofinger took a map and a proposal to Governor John Sununu, who promptly endorsed the plan.

As nature abhors a vacuum, New Hampshire abhors bureaucracy. So the Trust for New Hampshire Lands didn't ask for one. Private donations would fund administrative costs; public money would fund acquisitions; the effort would have a six-year life span; and as much land as possible would be protected with conservation easements to keep land productive and on local tax rolls.

Led by Rob Trowbridge, publisher of *Yankee* magazine, the trust's team of high-end volunteers raised $700,000 in private donations almost overnight; in two years an astounding $3.2 million came in from nine foundations, nearly 100 businesses, and some 2,000 individuals. Meanwhile, staff and volunteers from conservation organizations presented more than 100 slide shows wherever people were willing to gather: garden and gun clubs, churches, schools, community and service groups. Those who liked the idea were asked to sign a simple pledge saying so. Editorial support by the state's newspapers, including the conservative *Union Leader*, was unflagging.

When the legislation seeking public funds was drafted and filed in 1987, Senate president Bill Bartlett assigned it the prestigious ranking of Senate Bill 1. The leadoff witness at the first hearing was Congressman Judd Gregg, who would succeed Sununu as governor and become a United States senator. Gregg was followed by bankers, loggers, hunters, agency heads, taxpayers, and farmers. Nearly 8,000 pledge cards were delivered to the committee.

The bill overwhelmingly passed, creating the New Hampshire Land Conservation Investment Program (LCIP) and appropriating an initial $20 million. Subsequent appropriations of $18 million and $7 million in 1989 and 1991, with interest, made a total state investment of $48.76 million.

The legislation called for a state-appointed, fifteen-member board of directors to decide how to spend the money. The private Trust for New Hampshire Lands would work with towns and landowners to bring projects to the board. The legislature charged the program with focusing on land offering "aquifer recharge areas, forested wetlands, recreation lands, areas of special scenic beauty, plant and wildlife habitats, critical farmlands, undeveloped shorelines, wetlands, flood storage areas . . . and parcels contiguous to or enhancing land already protected from development." Virtually all protected lands would have public access.

One-third of the money would be available to municipalities for locally significant lands and two-thirds for land to be acquired or overseen by state agencies. Municipalities had to contribute at least half the value of any project, but the match could be made in land rather than cash. This assured leverage and access to funds for less wealthy towns. The Audubon Society of New Hampshire helped train local community volunteers to develop and advocate for applications, assisted by professional land agents from the trust.

From 1987 to 1993, the TNHL/LCIP partnership protected 379 parcels covering 100,876 acres from Portsmouth to Pittsburg. Preserved for good were Canterbury's Shaker Village, the back 40,000 of Nash Stream Forest, Lake Umbagog's wild shoreline, dairy farms on the Connecticut River, and stretches of Great Bay. Community activists secured scores of lesser-known special places — riverbanks and sugar bushes and farms, quaking bogs and mountaintops, vernal pools, intervales, and open vistas.

Private contributions, landowner donations, and tough negotiating by the private Trust stretched the state's investment of $46.4 million into $83.3 million worth of protected land. Other states, including Maine, Vermont, and Massachusetts, had similar programs in the 1980s. But in terms of efficiency and reach, only one other state in the country — Michigan — even came close.

The partnership sunsetted as promised, but hundreds of New Hampshire residents were trained in the ways of stewardship. "Behind every piece of land in the matching programs might have been fifty or a hundred local people who had a feeling of sharing and accomplishment," said Charlton MacVeagh of Marlborough, who sat on the board of both the LCIP and the TNHL. MacVeagh and others pointed out that the partnership helped renew commitment to conservation at the municipal level and spawned or nurtured several very successful regional private land trusts. Between 1994 and 2003, the Squam Lakes Association, Lakes Region Conservation Trust, and Squam Lakes Conservation Society protected close to 20,000 acres of forests and shore frontage by raising more than $12 million dollars in voluntary donations. The Ausbon Sargent Land Preservation Trust has steadily, systematically protected land around New London and Mount Sunapee. The Upper Valley Land Trust, protecting open and wooded land along both sides of the Connecticut River, would be cited by the Land Trust Alliance as one of the most effective land trusts in the nation.

Before 1986, only a few organizations, chiefly the Forest Society, were doing significant levels of land protection work. By 2000 a dozen land trusts were collectively protecting many thousands of acres every year, and municipalities were starting to pass significant bond acts and appropriations for local projects. The successor to the TNHL/LCIP partnership, the New Hampshire Land and Community Heritage Investment Program (LCHIP), was established by the legislature in 1999 with the ultimate goal of investing $12 million a year in conservation. It is no stretch to find in all those efforts genetic material from the Trust for New Hampshire Lands.

In February 1988, with the Trust for New Hampshire Lands and the Land Conservation Investment Program in full swing, an announcement rocked the North Country: More than a million acres of land once owned by the match company Diamond International in northern New England and New York was going on the market.

Larger parts of the industrial northern forest had changed hands with increasing frequency in recent years as paper companies bought and sold assets—that was nothing new. But the paper industry had grown volatile, and new, unfamiliar bidders were crowding the table, and the scale of the announced sale was stunning.

Forest economist Perry Hagenstein and others had been warning for some time that changes were in the wind. For generations, the vast forested tracts had been held by companies and families in an uneasy, unspoken trust. Conservationists, paddlers, snowmobilers, hunters, fishermen, loggers, anyone with a stake in the land — in fact anyone who took comfort simply knowing all that unbroken green space was up there — could accept the temporary state of the cutting or the spraying or the roads or the gates or whatever, because the land, itself, would always be there, and given enough time could always recover. The actual ownership of the land didn't fundamentally matter. As long as the camp leases were renewed. As long there was access. As long as there were trees. For the first time in northern New England's history, though, large tracts of industrial land were becoming more valuable for speculative real estate than they were for wood and recreation. Once development became part of the equation, all bets were off. Especially valuable — and vulnerable — were lake and pond shorelines, river corridors, ridgetops offering distant views. When a real estate developer promptly bought ninety thousand of the Diamond acres in New Hampshire and Vermont, you could almost feel the crack of a shattered promise shake an entire region.

With visions of runaway development parting the "granite curtain" of the White Mountains, private conservation groups — both local and national — rushed to join public agencies to buy most of the New Hampshire acreage. Included was the largest single outright conservation purchase in the state's history: the 40,000-acre Nash Stream watershed. The N.H. Division of Forests and Lands bought the fee interest, using a grant from the state's Land Conservation Investment Program, and the US Forest Service bought a conservation easement to help fund the deal. After years of struggling to get a foothold in New Hampshire, The Nature Conservancy played a key role in securing the easement. As Eve Endicott recalls, "We were just trying to afford the rent in Concord. But once the big timber parcels went on sale, it was, 'All hands on deck!'"

In the Diamond transactions were the seeds of the Forest Legacy program — a critical joint partnership between states and the U.S. Forest Service — which would grow into the most robust source for funding of conservation land in the northeastern states. Vermont, New York, and Maine scrambled to find conservation solutions of their own, as well, even as groups across the entire Northern Forest recognized now that, on some new fundamental level, they faced a common threat, and a common future.

In 1989 Congress, led by Senators Warren Rudman of New Hampshire and Patrick Leahy of Vermont, funded the Northern Forest Lands Study, a close examination of the trends that drove the Diamond sale. The study concluded that a wide range of economic and social factors would compel additional conversions of wild and working forests for development. An accompanying task force appointed by the region's governors suggested five conservation strategies: incentives for landowners to keep their property, public acquisition of land and conservation easements, better land use planning, closer community involvement, and diversified economic development. Both the study and the task force called for a high-level Northern Forest Lands Council to recommend specific actions. New Hampshire's appointed members were the Forest Society's Paul Bofinger, Henry Swan of Wagner Woodlands, and journalist John Harrigan.

As the Northern Forest became the region's hottest environmental issue, early hopes for a deliberative and consensus-based solution seemed remote. At one far end of the debate were people calling for huge new national parks and draconian forestry laws; at the other end were property rights advocates from the growing "wise use" movement who complained of a massive government land grab and takings. Shouting matches and animal costumes were not uncommon at public meetings; in upstate New York, property rights extremists threatened public conservation agents and vandalized their property. Across the region, the stakes couldn't have been higher: if this deal got botched, it would jeopardize any conservation plans in a suddenly uncertain future.

After an enormous amount of work by the Northern Forest Lands Council, which was directed by respected forester Charles Levesque of New Hampshire, a consensus gradually emerged that three building blocks were needed to sustain both human and natural communities in the Northern Forest: healthy and productive forests, strong and diverse local economies, and protection of special places. When the council released its final report in 1994, it called for thirty-seven actions at the private, local, state, and federal levels. Recommendations ranged from regulatory reform to bioreserves, from tax relief to logger education. "We believe urgent and sustained action is necessary for the long-term integrity, character, and productivity of the Northern Forest," the council reported. "Accomplishing this goal will demand new, imaginative thinking and doing."

The council left this menu of action for others to enact. For several years a coalition of groups attempted to codify much of the work in an omnibus congressional bill called the Northern Forest Stewardship Act. When that legislation met with resistance from western senators, advocates shifted strategies to push individual pieces, especially at the statewide levels. Predictably, some of these recommendations languished and the experience in the four states varied.

Among the most successful advances in New Hampshire were the development of a comprehensive set of sustainable forestry practices by a wide spectrum of stakeholders, ranging from ardent preservationists to timber company executives, entitled "Good Forestry in the Granite State"; federal and state land-protection successes at Lake Umbagog and the White Mountains, many of which were aided tremendously by Judd Gregg as both governor and U.S. senator; and the expanded use of conservation easements to protect working timberlands and the embedding of the forest legacy program in federal land-protection policy.

In the years after the Northern Forest Lands Council report, change continued to affect New Hampshire's North Country. Pension funds and other nontraditional timberland investment organizations stepped up their purchases of land. Some owners continued to cut too hard, too often, or both. In 2000, International Paper (IP) bought Champion International, the state's largest landowner. IP promptly entered into negotiations with Lyme Timber Company and the state on a huge transaction that would end up with close to 150,000 acres of working forest under easement with the New Hampshire Division of Forests and Lands and 25,000 acres owned in fee by the state's Department of Fish and Game—a natural, forever-wild tract secured by The Nature Conservancy and the Trust for Public Land. The landscape was shifting on several levels. The massive paper mill complex in Berlin changed owners twice and, in 2006, finally closed for good.

Economic, and even global, challenges loomed. But a new land ethic—informed by Rachel Carson and Aldo Leopold, but also by foresters and loggers and environmental activists and writers and legislators—had taken root.

The century had begun and ended with fierce, innovative, and ultimately successful public debates over uses of New Hampshire's most important natural resource: its forested lands. In 2001, the Society for the Protection of New Hampshire Forests marked its hundred-year anniversary at a renovated White Mountain grand hotel, surrounded by a lush and once again living landscape. There was no better place to celebrate.

ALL IN

Eleanor Briggs liked to tell the story of how she came back to her family's Hancock property after college one year, and hiked with her father to the top of nearby 2,000-foot-high Mount Skatutakee—how beautiful and wild the land seemed to her. Her father told her, "You know, I own that mountain." Over the remaining years of her father's life, the two would argue often over the fate of Skatutakee and the two thousand acres surrounding it. Her father saw the potential for home sites, even a possible ski area. But Eleanor felt a deep connection with the nature of the land, and wanted to preserve it for its own sake and so that others might also be moved by it.

After years of sometimes tense lobbying, Eleanor finally persuaded the lawyers holding the land in trust to permit her to put a conservation easement on the whole thing. The easement would be held by the Forest Society, but the land would be managed by a small nonprofit Eleanor had started, the Harris Center, which she'd created in 1970 as a place for nature workshops and educational programs. Fifteen years later, the Harris Center found itself in the nascent movement of local land trusts.

Meade Cadot, the director of the center, made the case to the Harris Center board that doing local land-trust work could show residents of the Monadnock region how to protect land, which fit neatly into the center's educational mission. Cadot was persuasive. Originally from Delaware, he had a passion for birding and a broad perspective. He had received master's and Ph.D. degrees in geology from the University of Kansas, had worked summers at the Smithsonian, and in the field in the southern Indian Ocean, Tierra del Fuego, Bermuda, and parts of the U.S. He had settled in New Hampshire partly because, years earlier, Tudor Richards had offered him a soft-money position at the Audubon sanctuary at Willard Pond in Hancock. And partly because he was drawn to the place. In all his travels, he'd seen only one other region — America's northern Great Lakes states — with a comparable history of private land protection and strong land ethic.

In the towns around Monadnock, that land ethic came from a mix of year-round natives and educated summer people, of old money and old families and newer residents who had fled suburbia and moved back to the land and into the country. A surprising amount of land in the region already had been set aside by individual families, small-scale organic and specialty farms were sprouting up, vibrant, volunteer lake associations monitored water quality and shoreline development, activists fought the state's plans to turn Route 101 into a major east-west superhighway. But land values were jumping as the region felt the pressures of the widening growth rings around Boston and southeastern New Hampshire. Neither the state nor the Forest Society, Cadot argued, was going to conserve enough of the Monadnock Region. He drew a rough circle, centered on Eleanor's land, including the nearby Willard Pond Audubon Sanctuary, and proposed that the Harris Center take on as part of its mission the goal of connecting and adding to these nodes of contiguous protected parcels.

The board was persuaded. In the coming years, working with the Forest Society, the Monadnock Conservancy, and other groups, the Harris Center added parcel after parcel to its "Supersanctuary." It created conservation easements, brokered easements with third parties, accepted land donations. It took developed properties, deed-restricted them, and resold them at a profit, and used the proceeds to create a land-protection loan fund. When the owners of the Andorra Forest decided to place the land under easement with the Forest Society, the Harris Center helped delineate a 2,650-acre part of the forest in Stoddard that would remain "forever wild." It combined resources with the New Hampshire Audubon Society, Forest Society, and the Trust for Public Land to purchase protection for strategic parcels, and encouraged The Nature Conservancy to protect a noteworthy Atlantic white cedar swamp north of Route 9 — covering some 1,200 acres in Antrim and Windsor. Over two decades, the Harris Center would grow to manage some 7,000 conserved acres, and its Supersanctuary greenway would stitch together 15,000 acres within a 100-square-mile area.

The work represented an emerging trend across the state and around the country: at the turn of the new century nearly a thousand additional land trusts nationwide were working to conserve land at the local and regional levels. In New Hampshire, alone, twenty-five land trusts were members of the national Land Trust Alliance, from the Ammonoosuc Conservation Trust to the Strafford Rivers Conservancy. Many of them disregarded political boundaries, pursuing, instead, larger "landscape scale" projects, trying to protect entire watersheds, ridgelines, and wildlife corridors. Working across town lines and between organizations fit the historic pattern of cooperation in New Hampshire. As the new century began, local municipalities and local and regional land trusts were spending ten times the amount on land conservation that the state was. And the Forest Society, under the direction of Jane Difley—an effective facilitator and consensus builder—had, in a sense, reached one its highest goals: educating and empowering citizen groups everywhere to do the work that for so many decades the Society had been forced to do alone.

There was yet one more story of collaboration.

Counting the tidal shores of Great Bay Estuary, New Hampshire's modest coastline covers some 150 miles. The estuary is one of the largest on the Atlantic coast and, stretching ten miles inland, one of the most recessed. Geologically, it is a drowned river valley of tidal waters, deep channels, and fringing mudflats. The estuary receives its fresh water from seven major rivers. Three of these rivers – the Lamprey, Squamscott, and Winnicut—flow directly into Great Bay. The others—the Bellamy, Cocheco, Oyster, and Salmon Falls—flow into Little Bay and the Piscataqua River. With Great Bay's average depth of less than nine feet, the changing tides make a dramatic difference. Its 8.9 square miles shrink to less than half that area during low tide, leaving much of the bay exposed as mud flats.

By the mid-1940s, the productive, species-rich estuary had become a quiet backwater, polluted with decades of sawdust, industrial discharge, and raw sewage from upstream mills and towns. A state government panel offered many recommendations, including pollution cleanup, improved fisheries, better access, and a Great Bay Authority to help develop the area. It also suggested setting aside thousands of acres as a state park and game preserve. While improvements to water quality eventually came about through sewage treatment facilities, most of the commission's other recommendations were ignored or forgotten.

The area surrounding the estuary, however, didn't remain a quiet backwater. From the 1970s through the 1990s, the population of Rockingham and Strafford counties jumped 86 percent. The two counties lost an average of 2,230 acres of natural landscape per year to development. Loss of open space was eroding the traditional small-town atmosphere and destroying diverse upland and estuarine habitats throughout the region. The complexity of the land-use and ownership issues, and the sheer size of the challenge, required a mobilization of a kind not seen before along the coast.

While the fight over Aristotle Onasiss's proposed oil refinery in 1973 galvanized local appreciation for Great Bay and its resources, scientists working on a larger scale would later put the estuary on the map for conservation. In 1986, the North American Waterfowl Management Plan – ratified by the United States, Canada, and Mexico – identified Great Bay as one of only a handful of focus areas for waterfowl conservation. Great Bay's abundant wetlands and associated uplands provided critical waterfowl wintering, migration, and production habitat, according to the plan.

Three years later, some of the leading opponents of the oil refinery successfully petitioned state and federal agencies to designate Great Bay as a National Estuarine Research Reserve. The designation in this program of the National Oceanic and Atmospheric Administration (NOAA) helped secure funding and resources for land protection, research, stewardship, and conservation measures both in the water and its surrounding lands. Also in 1989, the North American Wetland Conservation Act (NAWCA) was established by Congress. NAWCA's intent was to help conserve wetland ecosystems that are critical to waterfowl and other migratory birds, fish, and wildlife. The act provided funds and encouraged partnerships in priority areas identified in the 1986 waterfowl plan. Great Bay was one of those priority areas.

Around the same time, big changes were afoot with the largest landowner on Great Bay: Pease Air Force Base. The former Strategic Air Command site was slated for closure by the military, and a myriad of questions arose about the future use of the base's 4,000-plus acres, including prime shorefront on Great Bay. Judd Gregg, governor at the time, helped lead transition efforts that culminated in establishment of a commercial and industrial park, an international airport, and a federal wildlife refuge. In 1992, Great Bay National Wildlife Refuge was established, consisting of about 1,100 acres of the former base — including seven miles of shoreline on the bay.

By 1994, the opportunities presented by the research reserve, the North American Waterfowl Management Plan and NAWCA had caught the attention of what would become an extraordinary partnership for land protection around Great Bay. Spearheaded by The Nature Conservancy, the Great Bay Resource Protection Partnership — an affiliation of nine organizations and agencies — secured funding for its first two properties: 176 acres in all on Durham Point, right in the area once eyed for the oil refinery.

From the fall of 1994 through April 1997, this consortium embarked on a long-term strategic planning effort, and defined its project area as 272,000 acres in 24 towns around the estuary. The resulting Habitat Protection Plan identified 14,200 acres of high-value habitats and has served over the years as a blueprint for the partnership's progress.

Following completion of the plan, the group of conservation entities formalized the collaboration. A key decision at the outset was to avoid creating a new institution with its own legal standing. The Partnership's principal partners would consist of Ducks Unlimited, Great Bay National Estuarine Research Reserve, New Hampshire Fish and Game Department, The Nature Conservancy (N.H. Chapter), Society for the Protection of New Hampshire Forests, U.S. Environmental Protection Agency, U.S. Fish and Wildlife Service (Great Bay National Wildlife Refuge) and the U.S.D.A. Natural Resources Conservation Service. A crucial component was the inclusion of associate partners (local land trusts and county conservation districts) and community partners (twenty-four municipalities and nine additional conservation organizations). From the beginning, the group had done tremendous work in the grass roots.

The Nature Conservancy, the group decided, would manage the federal grants and negotiate land acquisitions on behalf of the partnership. But central to its success would be a culture of making decisions by consensus, a model that reinforced a focus on common goals and broad-based solutions. Robert Miller, who led the conservancy's land acquisitions for the Partnership's first ten years, urged partners to "leave the bowling shirts at the door" — referring to shirts with affiliation logos — and fostered a spirit of collaboration.

In 1997, the Great Bay Resource Protection Partnership was awarded a prestigious award from the EPA's New England region, the Environmental Merit Award. The partnership was recognized as an effective and energetic start-up conservation entity using science to guide land protection efforts with willing sellers.

In the first thirteen years of its existence the partnership secured $54.4 million in federal funding — a staggering amount, most of it from NOAA, with the support of Senator Judd Gregg — which it leveraged to generate even more funding sources, including other federal, state, and municipal programs, and private donors. To date, the partnership has protected more than 4,800 acres in the Great Bay region, including nearly 1,000 by conservation easement, a total of seventy-six deals.

"We couldn't wait for institutions to do it. There were no other similar partnerships when we started," said Dea Brickner-Wood, the partnership's coordinator, and she could have been speaking more broadly, for the Forest Society at the turn of the century, for the Extension Service, for the Silvio Conte Refuge, for the Harris Center, for the history of land conservation in New Hampshire. "We had no manual on how to do this."

"There are a lot of ideas out there that go nowhere — ideas every bit as smart as the Trust for New Hampshire Lands," Lewis Feldstein had said a decade earlier. "This idea could have been nibbled to death if each person took a chunk out of it. But people worked together and in the end, we were all reminded that our collective action can accomplish great things."

NOTES

The Fires of Change: Excerpted and adapted from articles by Martha Carlson and Richard Ober in *Forest Notes* and Ober's *People and Place: The First Hundred Years*. Society for the Protection of New Hampshire Forests, 2001.

Wise and Current Use: Includes adapted material from "Current Use" by Richard Ober, Liz Lorvig, and Isobel Parke, originally published in *People and Place*.

Water's Edge: Includes excerpted and adapted material from "Reclaiming New Hampshire's Waters" by Suki Casanave, originally published in *People and Place*.

Trust in New Hampshire Land: Excerpted and adapted from an essay by

Ralph Jimenez and originally published in "Land for New Hampshire," the final report of the Trust for New Hampshire Lands and the Land Conservation Investment Program. Includes adapted material previously published by Richard Ober.

All In: Includes excerpted material from "The Great Bay Resource Protection Partnership" by Eric Aldrich of The Nature Conservancy.

UNSPOILED VERMONT

THE NATURE OF CONSERVATION IN THE GREEN MOUNTAIN STATE

ROBERT MCCULLOUGH, CLARE GINGER, AND MICHELLE BAUMFLEK

When novelist Dorothy Canfield Fisher adopted Arlington, Vermont, as her home in 1915, she became part of a growing body of artists, writers, poets, and musicians who had discovered in Vermont's Arcadian landscapes a sense of time standing still; a place where one could live simply and in quiet harmony with community and surrounding environments. Two decades later, in an essay for the Works Progress Administration publication *Vermont: A Guide to the Green Mountain State*, Fisher recalled her godfather's thoughtful counsel:

> What ought to be done with the old state is to turn it into a National Park of a new kind — keep it just as it is, with Vermonters managing just as they do — so the rest of the country could come in to see how their grandparents lived.[1]

In many ways, the history of conservation in Vermont during the twentieth century has been defined by efforts to heed that advice — to keep Vermont just as it is. Beginning in 1933, the Bureau of Publicity — soon to be the Publicity Service of the aptly titled Vermont Department of Conservation and Development — captured this spirit in a series of promotional pamphlets titled *Unspoiled Vermont*, which the service printed using variations of the phrase into the 1950s.[2]

Paradoxically, the image of an unspoiled Vermont so often called to mind today is an image that has been crafted for much of the twentieth century by those seeking to capitalize on the state's scenic rural qualities in the name of economics, principally tourism. That image is not always shared by those whose families have worked the difficult land in Vermont for generations, nor by those who understand how harsh its rural environments can be.

The jostling caused by these two perceptions of Vermont, the unspoiled ideal and the reality of economic imperative, continues to frame the conservation dialogue today. Most of the tangible results that have emerged from that debate have tilled workable ground between the two. Yet unlike more populated regions of New England, where farsighted conservation initiatives took root late in the nineteenth century or early in the twentieth century, Vermont's contributions to conservation in the New England region developed slowly.

Isolated landmark events did occur, notably George Perkins Marsh's seminal 1864 treatise, *Man and Nature*. Other noteworthy initiatives, if less dramatic, developed during the nineteenth century, as well. For instance, concern about depleted forests and silted streams, and the impacts on fish and game populations, led to nascent government efforts to manage those resources more carefully following the Civil War. Experimental forestry practices also surfaced early in Vermont on private estates in Woodstock and Shelburne. And, wealthy benefactors donated extensive tracts of woodland to the state for conservation purposes during the early twentieth century, providing a foundation for a system of public forests and parks. More recently, beginning in 1965, investigations of declining spruce forests on Camel's Hump identified acid deposition as the probable cause, prompting national legislative initiatives. Developments such as these confirm a special insight that advances human understanding about the role of ecology in society, and represent one channel of conservation successes in Vermont.

For the most part, however, a concerted public voice favoring conservation, evident in legislative and administrative policy making, evolved only gradually. In part, growing recognition about the value of scenery to the state's economy began to play a forceful role — an objective stretched evenly across an entire century, a second channel marked principally by economic concerns. Indeed, the tourism industry in Vermont has done its job well. In some instances, too, awareness of these values has translated easily into ecological understanding. Yet the human pressures that often pit ecology against economic growth remained a distant problem in most of Vermont during the first several decades of the twentieth century. Not until proposals for the Green Mountain Parkway surfaced, and were rejected in a town meeting referendum in 1936, did an intense dialogue about how best to keep Vermont unspoiled begin to take shape.

Arrival of the interstate highway system during the 1960s, however, brought fresh legions of out of staters, some to visit, but many to live—at least part of the year. The onslaught of change that traveled these express corridors tested the depth of commitment among Vermonters to keeping the state unspoiled, sharpening debate in the process. From that complex and conflicted mix of social, economic, and political concerns or fears, Vermont's most important decades of conservation policy making emerged. Many of the initiatives adopted were innovative, indeed groundbreaking, not just for the New England region but for the entire country. These include the state's land-use permit law, anti-land speculation tax, restrictions on billboards, bottle tariff, a political alliance between affordable housing and conservation, and flexible design standards for the state's highways and bridges, to name a few. Others have borrowed and strengthened tools in use elsewhere, including town and regional planning initiatives, financial and other incentives to encourage downtown development or strengthen village centers, nonprofit land trusts that seek to preserve open lands, legislation that controls clearcuts, and current use tax incentives. Nearly all, too, reinforce the vision of Vermont's—and the country's—founding conservationist, George Perkins Marsh.

MAN AND NATURE

Biographer David Lowenthal has described *Man and Nature* as one of the nineteenth century's two seminal books on the subject of humankind and its relationships with the natural world, the other being Charles Darwin's *On the Origin of Species*. Marsh's 1864 treatise provided the foundation upon which the modern field of ecology has been built, principally that nature's systems are intricately linked through a labyrinthine structure with nearly invisible bonds. Equally important, humanity is part of — not separate from — that overarching ecology and, through misstep, is capable of irreversibly altering nature's complex dynamics to the detriment of all forms of life. Drawing from these observations, Marsh warned that earth's resources are not inexhaustible and that without perpetual stewardship by humans, nature will not regain its equilibrium. In turn, these maxims became cornerstones of the conservation movement during the twentieth century.[3] The import of Marsh's study is global in reach, and the book immediately found an international audience, helping to instill broad public and professional awareness about the myriad ways in which humans transform their environments. In part, Marsh's stature as a linguistic scholar, diplomat, congressman, and lawyer helped to place the book in its fitting international context. He also completed much of the writing while living in Italy, compiling lengthy segments at a small coastal village, Pegli, near Genoa during the winter of 1862–63. On weekdays he remained in Turin, fulfilling his responsibilities as American minister to Italy.[4]

Yet Marsh had begun writing the book while living in Burlington, Vermont. His investigations there, and in many parts of Europe, may have been inspired by impacts to environments he observed much earlier, initially as a boy in Woodstock. Thus, the work should not be separated from its Vermont sources. In particular, Marsh had ample opportunity to observe the state's depleted forests and eroded, overfarmed topsoil. Decades later, as Vermont's fish commissioner, Marsh studied the effects of deforestation on the state's streams and rivers — dry in summer but swollen and filled with sediment following heavy rains or rapid thaws. Drastically altered stream beds destroyed spawning grounds for fish, decimating their populations and providing further evidence of the important connections between plant and animal habitats. Ultimately, the relationships between wasteful forestry practices and heavily eroded and altered watershed systems would become integral to his conclusions regarding nature's complex linkages.[5]

In less than a decade, *Man and Nature* had become instrumental in shaping national forestry policies, especially the matter of forest influences and watershed protection. In 1873, three years before he became the inaugural head of the Department of Agriculture's Forestry Bureau, physician Franklin Hough presented a paper to the American Association for the Advancement of Science at its meeting in Portland,

Maine. Titled "On the Duty of Governments in the Preservation of Forests," Hough's appeal drew from Marsh's study, emphasizing ties between deforestation and floods, drought, inadequate water supplies for cities, and waterways no longer commercially navigable. The association's report to Congress included a copy of Hough's paper and helped to secure passage of the Timber Culture Act that year, encouraging the planting of trees by those who were settling the Great Plains, and also leading to the appointment of a national forestry commission.[6]

By 1891, Congress had passed a second forestry law, what later became known as the Forest Reserve Act, authorizing American presidents to set aside designated areas of forested public land and leading to the creation of America's first national forests. Six years later, the law was amended to clarify the role that forests play in securing favorable conditions for water flows and furnishing a continuous supply of timber.[7] The theme of keeping navigable waterways clear, as well as preventing forest fires, would reappear in the justification for the Weeks Act. Passed in 1911, this law mandated efforts to acquire land for national forests in the East.

The magnitude of Marsh's work, and its widespread influence—global, national, and regional—should not obscure the book's contributions closer to home. Apart from the tangible relationship between Vermont's landscapes as they appear today and public awareness of ecology stemming from *Man and Nature*, Marsh's influence comes to ground in at least two profoundly symbolic places. Each is a monument to Marsh and to Frederick Billings, who became Marsh's principal benefactor in Vermont. One is the Marsh-Billings-Rockefeller National Historical Park, the site of Marsh's boyhood home in Woodstock. Billings acquired the property in 1869 and began enlarging the site, developing it into a model of forest management and agricultural practice. Today, it is the nation's only National Park devoted to the interpretation of conservation history.

The second is Billings Library at the University of Vermont in Burlington, a monument no less significant, potentially, than the spiritual home the two men shared in Woodstock. Throughout his life, Marsh had been a habitual buyer of books. Late in life he began negotiating for the sale of his library to the University of Vermont, principally to provide an income for his wife but also because of his ties to the university. Those negotiations remained at a standstill when Marsh died in 1882. Billings promptly purchased the collection, paid for its transport from Italy, and donated the books to the university. Within a year, he added a second donation for the cost of constructing a fireproof library to house the collection, and he urged the university to retain America's leading architect, Henry Hobson Richardson, to design the building. As the true cost of Richardson's architectural genius became apparent, Billings more than doubled his original pledge of $75,000.[8]

In its completed form, Billings Library is neither what Richardson truly wanted nor what its benefactor had anticipated, the consequence of two men engaged in the tortuous struggle that pits artistic expression against practical-minded restraint. Although university president Matthew Buckham served as an intermediary, the power of Richardson's architecture ultimately prevailed. To Billings credit though, the symbolic value of a building designed by one of America's heroic

The Marsh-Billings-Rockefeller National Historical Park in Woodstock, the home of pioneer political, business, and philanthropic leaders George Perkins Marsh, Frederick Billings, and Laurance Rockefeller is where the concept of individual land conservation stewardship first evolved and was practiced. Nearby Mount Tom is the site of one of the earliest efforts at sustainable forest, farm, and recreational management in New England. The park itself is a national center for studies relating to conservation stewardship.
(National Park Service)

architects, to house the library that contributed to one of the country's most important works in the field of conservation, is a profound legacy. Here, a union of cultural and natural resource protection, the full potential of which remains largely ignored in America, can find a much-needed home. Ironically, the building's latent ability to serve that purpose seems to remain unrecognized by the university, as well. As Marsh's pleas for ecological stewardship gain urgency with each passing year, such public symbols can play an increasingly valuable role in shaping public attitudes.

MARKETING UNSPOILED VERMONT

Although *Man and Nature* offered a lofty point of beginning for ecological perspective in Vermont, the actual practice of conservation in the Green Mountain State has long been tied, either directly or indirectly, to more mundane economic considerations: tourism and recreation. Vermont's natural resources have been attracting visitors from afar for more than two hundred years, and not surprisingly, awareness of the economic value of these resources has often produced an impulse to protect. Unlike other aspects of conservation history in Vermont, though, few single events (arrival of the automobile excepted) can be isolated to signal dramatic advances in the state's tourism industry. Instead, the process has been one of gradual maturity over time.

Moreover, the complex and sometimes divergent paths of conservation and tourism have hardly been trouble free. More than occasionally the goal of marketing Vermont's unspoiled rural countryside has stumbled into concerns about the underlying consequences of those efforts, social as well as ecological. The proposal for the Green Mountain Parkway marks just one example of that debate taking early shape, but it continues to surface in many other sectors, including the ski industry and, very recently, the development of wind power.

Despite the potential for conflict, recognition that the long-term goals of tourism and conservation are inextricably linked has shaped the course of both. Simply stated, each is dependent on the other. Disregard the dictates of ecology, and opportunities to convert a rural idyll into economic gain will inevitably decline. Ignore a marketable value, and opportunities to build a broad constituency of support for protecting resources will be lost. As in many areas touching matters of conservation, the challenge lies in resisting short-term actions that undermine long-term vision.

If the imperative of balancing these often competing goals has not always been a part of tourism in Vermont, the type of activities occurring today have ample precedents that reinforce the need for striking such a balance. Railroad companies, for example, became skilled at marketing tourism, helping to promote leisure as a type of middle-class product and selling landscape as a backdrop for these recreational activities. For example, the Rutland Railroad's 1897 tourist publication, *Heart of the Green Mountains*, claimed that the line offered the most picturesque views in New England. The company supported that contention with a special art supplement to its booklet, containing numerous photographs of Vermont scenery. The Central Vermont Railroad, the Rutland's principal competitor, offered similar literature during the same period, including one publication called *Summer Homes Among the Green Hills of Vermont and Along the Shores of Lake Champlain*, which advertised vacation properties — many of them abandoned farms — for sale or rent.[9]

By the 1890s the marketing of unspoiled Vermont had become institutionalized, both in the private and public sectors. The Vermont Development Association, chartered in 1897 and comprised of 150 representative business and professional men, continued the campaign to entice tourists that railroads had begun. By then, however, state government had already taken similar steps, partly to reclaim abandoned farms and partly to recapture travelers who had drifted to other regions, notably resorts in the more rugged White and Adirondack mountains, causing a decline in the state's mineral spring spas and summit hotels.[10]

In 1891, the Vermont Board of Agriculture began publishing a series of pamphlets, the first of which was titled *Resources and Attractions of Vermont with a List of Desirable Homes for Sale*. Part of a broad, multifaceted mission to preserve farming, the board's alliance with tourism evolved into a two-pronged strategy. One involved a campaign to solicit the purchase of abandoned farms as seasonal vacation homes, and in 1893 the board issued a second publication, *A List of Desirable Vermont Farms at Low Prices*. The other tactic encouraged farmers to take in summer boarders, a growing business aimed at city dwellers in search of a farm holiday — clean air, fresh food, and wholesome living. Both strategies, however, depended on the marketing of Vermont's landscape as a pastoral ideal, an important shift from earlier emphasis on mountainous scenery. Another of the board's 1893 publications, *Vermont: A Glimpse of its Scenery and Industries*, reflected that tactic. Focus on this nostalgic view of a simple, rural life in New England took formal shape with adoption of Old Home Week, first celebrated in New Hampshire in 1899, but quickly adopted by Vermont and other New England states.[11]

In 1911, Vermont's newly established Bureau of Publicity assumed responsibility for promoting tourism, just as the prospect of automobile travel loomed on the horizon. A 1915 booklet by the publicity agency, written by Mortimer Proctor and Roderic Olzendam and titled *Vermont: The Unspoiled Land*, became one of the state's first publications aimed at automobile tourism, and many of its photographs were views from roads. The increasing mobility of automobile travelers changed the pace of tourism in Vermont dramatically, and with it the breadth and reach of promotional imagery. Photography accounted for many of those images, whether in the form of government-sponsored publications or in countless postcards available for purchase at general stores and corner markets alike. Most of these images continued to portray Vermont as a place of unspoiled pastoral beauty linked to its history of farming, and ties between the Department of Agriculture and Bureau of Publicity remained strong.[12]

Although the promotion of tourism steadily gained a broader base, initiatives to preserve the aesthetic qualities of Vermont's landscapes developed only slowly in the context of tourism. During the late 1920s and early 1930s, efforts to keep farms tidy and roadsides attractive became part of campaigns sponsored by official bodies such as the Vermont Commission on Country Life and its subcommittee on Tourist and Recreation Facilities. The State Chamber of Commerce urged local roadside improvement committees to remove junked automobiles, straighten tipsy fence posts, and remove unnecessary signs. As in other parts of New England during this period, concern about billboards and other roadside clutter gathered voice, even to the extent of criticizing the tourism that led to much of that clutter.[13]

During the Depression years, the Bureau of Publicity continued to send invitations to out of staters, urging them to buy Vermont's marginal farms as seasonal retreats. Dorothy Canfield Fisher's 1932 booklet published by that bureau, *Vermont Summer Homes: An Open Letter*, even hinted at the right type of people who would find such homes appealing. With publication of the first issue of *Vermont Life* in 1946, state government's official magazine, promotion of Vermont's beauty to travelers acquired its most polished format.[14]

All the while, another more subtle shift in attitude toward Vermont's landscapes began to occur, one that turned away from the Arcadian pastoral and became more expressive of landscapes as representative of human activity. Formation of the Southern Vermont Artists association in 1933 contributed to this shift, helping to unify a group of artists who were diverse in technique and approach but who, in many instances, were also bound by strong ties to landscape imagery. Origins of the association are traced to five artists, the Dorset Painters, who first exhibited together in 1922 at the Town Hall in Dorset. That coterie included Edwin Child, Francis Dixon, Wallace Fahnestock, John Lillie, and Herbert Meyer, but annual exhibitions introduced new artists and led to incorporation in 1933. The association ultimately embraced a number of well-known regionalists, among them Luigi Lucioni and Paul Sample. Sample, in particular, acknowledged the important and complex social and economic relationships between humans and the land and confronted the mythical impressions that had become attached to Vermont's landscapes as the result of promotion in the name of tourism. His paintings often point specifically to the underlying social, economic, and class divisions brought into sharp and sometimes harsh focus by the arrival of visitors from afar. Such divisions, too, have ultimately influenced the course of conservation in Vermont, the beginnings of a dialogue about the manner in which Vermont's rural landscapes should be interpreted.[15]

Today, forums for the debate about aesthetic qualities in landscape conservation are found principally in Act 250, the state's land-use permit law that measures proposed projects against various environmental criteria, including aesthetics. Yet similar debates also take place in the context of local design review ordinances, enabled by planning and zoning legislation enacted in 1968. The debate can also surface in the context of conservation efforts by the state's nonprofit land trusts, and the ways in which segments of protected lands are developed. Often, though, such debates about aesthetics seem to occur separately from the broader dialogue that engages matters of environmental ecology.[16]

RECREATIONAL FOREST USE

Despite the emphasis on a pastoral, picturesque landscape that influenced tourism in Vermont during much of the twentieth century, the state's Green Mountains have also provided more than just a scenic or picturesque backdrop for those who visit the state. In particular, recreational use of forests has played a vital role in tourism and has appealed to individuals who are, in some cases, more intimately connected to the land than those travelers who seek an imaginary rural idyll to escape urban woes. As a consequence, the impulse to protect grew strong in this sector of tourism and recreation and ultimately played a vital role in matters of conservation, particularly in matters of ecology.

Although interest in exploring Vermont's hospitable Green Mountains waned during the last quarter of the nineteenth century, yielding to New Hampshire's more challenging White Mountains where recreational hiking first took firm root in New England, important developments occurred in Vermont at about the same time. Depletion of the state's forests had peaked following the Civil War, and roughly 80 percent of the state's lands stood cleared of woodland cover, stark evidence of the warnings offered by George Perkins Marsh. In response, an effort to reclaim and protect the state's forests began to gather momentum, led initially by a wealthy eccentric from Middlebury, Joseph Battell. Struck by the area's scenic beauty and vexed by the mindlessly wasteful practices of timber companies, Battell acquired a substantial parcel of land in Ripton in 1865, including Bread Loaf Mountain. Five years later, he opened the Bread Loaf Inn near the mountain's base, managing the hotel himself and ultimately using it to demonstrate that conserving the state's forests could generate substantial revenue from tourism.[17]

Other important land owners, including Frederick Billings, William Seward Webb in Shelburne, Marshall Hapgood in Peru, and Silas Griffith in Danby, followed Battell's example and began introducing forest conservation practices on their extensive holdings. Battell, who represented Middlebury in the state legislature, successfully introduced a bill calling for appointment of a special forestry commission, established in

1882 and staffed by Billings, Redfield Proctor, and Edward Phelps, who together emphasized education in matters of forestry practice. Formation, in 1904, of the nonprofit Forestry Association of Vermont helped to push forest conservation in the direction of policy making, beginning with legislation that same year authorizing appointment of a state commissioner of forestry and establishing a fire protection program. A bill authorizing a state nursery passed in 1906, followed two years later by legislation giving administrative structure to forestry in state government, officially the Board of Agriculture and Forestry, and establishing the position of state forester. Professionally trained forester Austin Hawes received the inaugural appointment to that job in 1909. By 1923, the Vermont Forest Service had become a separate government agency.[18]

Battell, however, continued to acquire large tracts of woodland, becoming, in the process, the state's largest land holder. He and other like-minded timber owners including Hapgood and Charles Downer, all of whom believed that public ownership of woodlands offered the surest means to protect against corporate abuse, began making gifts of land to the state. In 1910, Hapgood donated the summit of Bromley Mountain and its adjoining range, and a year later Battell did the same for Camel's Hump Mountain, stipulating that it be used as a public park in perpetuity. Battell's estate also left enormous land holdings along the central ridge of the Green Mountains to Middlebury College, and these lands later became key segments of the Green Mountain National Forest, established in 1928. Gradually, a network of state forests grew from these and other land acquisitions. Similarly, a state park system acquired legislative sanction in 1929, five years after a generous gift of land at Mount Philo.[19]

At the local level, towns began establishing municipal forests after 1915, and the state's forestry service soon provided assistance. Although Vermont's municipal forestry program progressed only slowly for several decades, lagging behind more energetic programs in Massachusetts and New Hampshire, the legislature amended the state's enabling law in 1945 and began reimbursing towns for half the price of lands acquired for forests. A separate law enacted in 1951 required communities to include propositions for municipal forests in warnings for annual town meetings. Vermont's forest service subsequently created positions for two full-time municipal foresters, the only New England state to do so. With both financial and technical assistance mandated by law, Vermont set a progressive standard in municipal forestry, and its program soon flourished, surpassing Massachusetts and New Hampshire after 1960 and lingering into the mid-1970s, more than a decade after local conservation commissions had replaced town forest committees in many parts of New England.[20]

GREEN MOUNTAIN CLUB

Thus, a context for protecting Vermont's forests had begun to emerge by 1910, the year that James Taylor, headmaster of the Vermont Academy for Boys at Saxtons River, organized a meeting of nearly two dozen men in Burlington, who together formed the Green Mountain Club. By that time, a number of footpaths had been blazed to mountain summits in Vermont, including Camel's Hump and Ascutney, where, in the latter case, a trail existed as early as the 1820s and where a stone hut had been built during the 1850s. In general, however, the trail networks and organized recreational hiking popular in other parts of the Northeast were mostly absent in Vermont, circumstances lamented by Taylor. The club he conceived emerged as the region's first organization devoted to trail building and, equally important, to the country's first long-distance through trail, what Taylor envisioned as a footpath stretching along the spine of the Green Mountains from the Massachusetts border to Canada.[21]

Taylor proved more adept at promotion than at the physical work of trail building, so others took up the cause, notably Judge Clarence Cowles, a Burlington lawyer who, with Craig Burt from Stowe, blazed and cleared the first section of the Long Trail from Mansfield to Nebraska Notch in 1911. Taylor also enlisted the support of Austin Hawes, who assigned men from his forestry staff to build portions of the trail with the goal of implementing fire control.[22]

However, the forestry program's trails typically followed lowland routes rather than traversing summits, to the chagrin of many club members in search of vistas. Will Monroe, a strong-willed New Jersey college professor who began spending summers in Vermont in 1914, launched a campaign to relocate the trail across high ridges and peaks, selecting exceptionally scenic routes with meticulous attention to detail. His legacy, the Monroe Skyline, was complete by 1926 and

This view over a small valley in Pomfret north of Woodstock, near the former location of the Prosper Ski Area, an early 1930s rope-tow recreation facility, typifies Vermont's much-prized, working landscape of forest, field, and farm. Just a few hundred yards to the south, the Appalachian National Scenic Trail crosses the valley, connecting Killington Peak to the west with Hanover, NH and the Connecticut River Valley to the east.
(Robert L. McCullough)

stretches from Mount Ira Allen south of Camel's Hump to Burnt Hill near Middlebury Gap. Today, it serves as the heart of the Long Trail, and historians Laura and Guy Waterman have called it one of the most scenic hiking trails in the Northeast. Monroe also organized the New York Section of the Green Mountain Club, encouraging hikers in the New York City area to venture north into Vermont. Meanwhile, work continued on the trail in southern parts of the state, and, finally, on the short segment in the north between Jay Peak and the Canadian border during the late 1920s. By the summer of 1930, Taylor's vision for a Long Trail had been realized.[23]

The Green Mountain Club has attracted a small but dedicated group of hikers to Vermont ever since, and club members have often played an important role in issues affecting conservation in the state.

Equally important, the ability to hike the length of the Green Mountains along a single route, or to turn eastward and continue along the Appalachian Trail, captures the imagination of many who seek refuge from the turmoil of everyday life. Access to such otherworldly places thus lends itself well to marketing strategies, and Vermont's forested mountains have joined its pastoral landscapes as devices to entice visitors from afar.

WILDLIFE MANAGEMENT

Pastoral landscapes and resurgent forests invite other very specific forms of recreational use that have influenced the course of conservation in Vermont, hunting and fishing for example. By the middle of the nineteenth century, the damaging effects of treating fish and game as common resources had become clear to many people. Deer had disappeared altogether in large sections of the state, and salmon and trout populations had dwindled as well, circumstances noted by George Perkins Marsh in a report he prepared as the state's fish commissioner in 1857. In 1866, only two years after publication of *Man and Nature*, the state established a three-person Fish Commission, leading to the adoption of a fishing code, seasonal restrictions, and regulation of equipment. By 1876, a system of local, county, and state wardens had been organized as a means to instill better wildlife management practices, and by 1892 a Fish and Game Commission had superseded the Fish Commission.[24]

Just as interest in forestry practices generated a local response in the form of town forests after 1920, so, too, did awareness about the relationships between wildlife management and landscape conservation. For example, Wilbur Bradder, a member of the state forestry staff, emphasized reclamation of abandoned farms and worn-out pastures for game and wildlife. In a plan for Proctor's town forest, and also in a remarkable 1934 drawing for a "Game, Food, and Cover Improvement Plan of a Typical Abandoned Farm and Woodland Area at Tinmouth, Vermont," Bradder called for planting poor soil with hickory and butternut, releasing cordwood borders, hedgerows, and apple orchards, preserving marsh edges, rocky spots, and marginal areas as sites for winter food—cowpeas, barberry, and wild rice, and leaving hollow oak trees standing as shelter. This site eventually became the Tinmouth Channel Wildlife Management Area.[25]

Gradually, institutionalizing the management of resources, whether forests, fish, game, or landscapes, gave structure to the conservation dialogue. It aligned users of these resources into complex, sometimes unpredictable, and often adversarial groups comprised of local citizenry, recent arrivals to Vermont, summer residents, out-of-state organizations, and agencies within state government. By the 1930s, the federal government had become an important voice as well.

Even within state government, with separate agencies struggling to represent these diverse constituencies, dialogue often is conflicted. Such struggles seem to surface frequently in the context of the state's ski industry, heavily reliant on out-of-state business, but also popular (and expensive) for those who live in Vermont. In 1934, the country's original rope tow, powered by a Ford Model-T engine, was installed at a small ski hill in Woodstock, representing the beginning of a new industry for Vermont. State forester Perry Merrill took advantage of CCC labor during the 1930s to cut ski trails in the Mount Mansfield and Smuggler's Notch areas, and in 1939 Governor George Aiken leased state-owned land on Mount Mansfield to private developers for ski-area use. The state's first chairlift was installed a year later.[26]

Debate has followed ever since, focused on finding a balance between private interest and public good, economic growth and environmental conservation. Today, that dialogue surfaces in the context of depleted stream levels caused by artificial snowmaking, preservation of wildlife habitat, and new development in the form of seasonal homes, hotels and motels, and all the trappings of an industry that appeals to fashion and demands expensive services to match.

GREEN MOUNTAIN PARKWAY

Evidence of an evolving conservation ethic among Vermonters became more clear as the result of a number of events that placed public opinion into sharp focus, at times divisively so. During the Depression years, a proposal to build a parkway among the ridges, valleys, and foothills of the Green Mountains proved to be just such a catalyst. The scheme, a device to capture more than ten million New Deal dollars for state coffers and to ease unemployment, fit neatly into the federal government's plans for developing recreation and tourism as a means to generate economic growth.

Modeled after the Blue Ridge Parkway in Virginia and North Carolina, the Green Mountain Parkway was designed as both a roadway for automobiles and a linear park, "a great scenic reservation." William Wilgus, a civil engineer for the New York Central Railroad who had retired to Ascutney, proposed the project in 1933, gaining support from the Vermont Chamber of Commerce and its executive secretary, James Taylor. Wilgus used his influence in Franklin Roosevelt's administration to generate interest in the project, and in 1934 a team of landscape architects from the National Park Service, including well-known urban planner John Nolen, retained as a consultant, joined engineers from the Bureau of Public Roads to prepare a survey of the route. Their report called for a right-of-way at least 1,000 feet wide, reserving 500 feet of parkland available for hiking and bridle trails on each side of the road. Adjoining lakes and stream valleys invited opportunity to expand the park at numerous locations.[27]

The corridor stretched the length of the state from the Canadian border to a point several miles southerly of the Massachusetts state line, where it merged with a similar proposal for the Berkshire Hills Parkway near the latter's juncture with the Mohawk Trail. The route passed through more than thirty Vermont towns and linked state parks, the newly established Green Mountain National Forest, and a proposed wilderness park at its northerly terminus near Jay Peak. In approving the proposal, Secretary of the Interior Harold Ickes and National Park Service director Arno Cammerer anticipated that the parkway would add to a national program for parks and would join the metropolitan areas of Boston and New York to recreational regions throughout the Northeast.[28]

However, after nearly three years of contentious debate, and despite what appeared to many to be very appealing aspects of the plan, Vermont voters decisively rejected the proposal in a bond referendum on Town Meeting Day in 1936. The reasons for the vote were complex and reflected, at least in part, broader political disagreements swirling around New Deal programs, public works projects funded by the National Industrial Recovery Act, and federal control of land. Suspicions over the cost of building and maintaining such a road, distrust of wealth from afar (and the influences that accompany it), and concerns about the potential impact of tourism on the state's rural qualities may have influenced many who voted as well. Yet a more subtle argument also emerged, centering on the appropriate methods for conserving Vermont's Green Mountains, or how to keep the state "unspoiled," as some Vermonters put it.[29]

Although the Green Mountain Club officially opposed the parkway, not all among its ranks agreed. Some influential figures, including founding members James Taylor and Clarence Cowles, viewed the project as an opportunity to connect the Green Mountains with the lives of ordinary Vermonters—a step in the direction of progress—or regarded the federal government as a friendly partner in forestalling inevitable development. Others, especially club leaders, thought the scheme to be poorly conceived, a damaging and unnecessary scar across the mountains that would slash forests, create fire hazards, injure wildlife, and generally impair the wilderness experience of those who hiked the Long Trail. Their concerns reflected a similar debate that had developed around the Blue Ridge Parkway's Skyline Drive in Virginia, and to some extent the influence of national figures such as Aldo Leopold, Robert Sterling Yard, and Robert Marshall may have helped to shape opinion in the Green Mountain Club.[30]

In retrospect, however, the controversy marked the beginning of a long and continuing era during which the overwhelming influence of the automobile has forced the shaping of opinion about the ways in which Vermonters should practice conservation. That dialogue endures today in numerous contexts and, in turn, helps to define the course of environmentalism in this state. Although many of these debates are no less quarrelsome than that involving the Green Mountain Parkway, growing ecological awareness has at least strengthened the underpinnings of that dialogue. The strength of this foundation will continue to be tested in unpredictable ways, most recently in the debate about wind power, where the goals of protecting both environmental health and mountainous scenery are in tangled alignment.[31]

WATER RESOURCES

During the late 1930s, dialogue in Vermont became sharply focused on another topic affecting conservation, that of water resources. By the close of the nineteenth century, an important alliance had begun to develop between forestry and watershed protection, essential for adequate supplies of drinking water. Awareness of forest-water dynamics—the important relationships between forest cover, soil condition, and collection of surface water—had steadily increased following publication of *Man and Nature*. The results of this awareness could be observed in a broad range of contexts in Vermont. Reforestation of lands designated as state forests and parks is one example, and the state nursery established in Essex in 1922 supplied many of the seedlings used in a planting campaign that remained vigorous until World War II. Moreover, many of Vermont's most successful municipal forests were created principally to protect reservoirs dependent on surface collection within watersheds. Thus, in the context of the fundamental need for clean drinking water, ecological concerns were carefully considered.[32]

Yet protection of drinking water supplies at the local level did not address many of the broader problems facing the state's streams and water bodies, specifically pollution from industry and agriculture, as well as flooding. For many decades, however, Vermonters often voiced concern about these effects in economic, rather than ecological, terms. In times of drought, for example, diminished stream levels proved to be devastating for agriculture or polluted waters limited recreational opportunities, causing a similarly direct economic effect.

Flooding was no less damaging economically, and during the late 1930s, contention began to swirl around proposals to address that problem. Shortly before Vermont's infamous 1927 flood, Congress had passed the Rivers and Harbors Act with an eye toward planning for flood control on the country's principal rivers. A year after the flood, Congress transferred authority for flood control to the Army Corps of Engineers, which recommended construction of a large number of dams throughout Vermont, with costs to be recovered through the sale of hydroelectric power by private, out-of-state utility companies. However, George D. Aiken, a freshman member of the Conservation and Development Committee in Vermont's House of Representatives, led a successful fight to defeat that proposal. Aiken, a nurseryman, farmer, and wildflower enthusiast from Putney, distrusted large utility companies that had been unwilling to invest in rural areas, and he supported local electric cooperatives instead.[33]

Nevertheless, work on several important retention dams began during the 1930s, including an especially large earth-filled dam and reservoir on the Little River north of Waterbury. Laborers from the Civilian Conservation Corps completed that project in 1938, helping to establish a record of collaboration between state and federal government. Calls for similar projects continued in response to flooding in the Connecticut River Valley during 1936 and 1938, the latter the result of a hurricane. States had been authorized to enter into flood-control compacts by federal legislation enacted in 1936, and Vermont, New Hampshire, and Massachusetts formed the Connecticut River Valley Flood Control Commission in response.[34]

By 1937, the year Aiken became governor, the commission had developed a Connecticut River Valley Flood Control Compact. The compact proposed a number of federal dams and storage reservoirs to be built by the Army Corps of Engineers and jointly funded by federal and state governments. Aiken, though, expressed concern about the loss of productive agricultural lands in river valleys, inundation of communities, loss of tax base, and lack of control over the dams and the hydroelectricity to be produced. Although the commission assented to Aiken's concerns and agreed to state control of the dams, President Roosevelt refused to relinquish management of the dams and hydroelectric power to states, principally to prevent private utility companies from gaining access to the projects, and the compact failed to gain the congressional ratification it required.[35]

Congress supported Roosevelt, reinforcing his position by passing the Flood Control Act of 1938, which allowed the federal government to acquire land necessary for flood-control dams and reservoirs, and to do so without consent from states. The battle over water resources then developed into a judicial contest pitting the federal government against state and local interests, the latter led into court by tiny Vermont almost single-handedly, albeit at the expense of critical regional concerns. Although the Supreme Court upheld the constitutionality of the 1938 law, President Roosevelt eventually relented, agreeing not to build dams in Vermont without the state's consent and giving Vermont's share of the funding to other states.

Aiken continued his uncompromising stance when he served as a U.S. senator, blocking similar proposals for dams during the early 1940s. Ultimately, Congress modified the 1938 law, agreeing to consult with states in developing flood-control projects and resolving the matter of local compensation for lost tax revenue. Dams on the Ompompanoosuc, West, Black, and Ottauquechee rivers were built between 1946 and 1961, but by forestalling construction of these projects, Aiken may have unwittingly mitigated environmental impacts. Reservoirs, which limit storage capacity in times of flooding and are unnecessary absent the generation of hydroelectricity, were eliminated in favor of a series of dry dams and catch basins. These designs added storage capacity in times of flooding and allow upstream lands to remain in agricultural use except when flooding is imminent.[36]

As with the battle involving the Green Mountain Parkway, the contest over flood control reaffirmed the perception of Vermont as independent, at least among its own citizens. The struggle also helped to foster skepticism regarding New Deal projects, an attitude that spilled over into other initiatives. Beyond Vermont's borders, fortunately, the battle also established a firm foundation for regional alliances on matters affecting water resources.

The need to control and manage Vermont's water resources comprehensively gained formal structure with creation of the Water Conservation Board in 1947. The board's mandate required collaboration with the health commission and the fish and game commission, and it was also given authority over dams not owned by public utilities. However, control of power dams remained with the public service commission. Two years later, legislative amendments required the board to classify the state's waters into four categories according to designated uses, and specified that no increase in pollution could occur without authorization from the board. The law placed authority to enforce those controls with the state's court system.[37]

In more recent decades, Vermont efforts to address water-quality issues have followed patterns set with the passage of federal legislation in the 1970s. Under the mandates of the 1972 Clean Water Act, the federal Environmental Protection Agency established a regulatory and permitting framework to control point sources of water pollution. In this framework, the states hold responsibility for implementing water quality standards. This led to a round of watershed basin planning in Vermont during the 1970s. The Department of Environmental Conservation generated plans for seventeen river basins to address point sources of pollution in the state.

More recent concerns and initiatives, for example algal blooms in Lake Champlain caused by phosphorus runoff, and the federal Clean Water Action Plan of 1998, have raised the profile of nonpoint sources of pollution in the state. As a result, another round of watershed basin planning is underway, including retyping and reclassification for all of Vermont's surface waters. The standards governing this work also require public participation in basin planning.[38] Thus, public involvement through the use of watershed councils has become a central part of watershed planning. These councils include a range of stakeholders and emphasize participation of citizens within each basin. In the case of Lake Champlain, concerns about algal blooms provide a strong link among ecology, aesthetic, and recreation interests. However, patterns of conflict between economy and ecology are also central as people debate the roles of suburban development and agricultural practices in contributing to pollution through stormwater runoff and phosphorus loading in streams and, ultimately, the lake.

AGRICULTURAL RESOURCES

When Vermont's U.S. Congressman (later Senator) Justin Smith Morrill successfully championed the Land Grant College Act of 1862, scientific farming received a much needed boost. However, Morrill's home state was slow to take advantage of the law's incentives, and the State Agricultural College did not acquire a charter until 1864, and did not join the University of Vermont until a year later. Agricultural marketing became increasingly complex and competitive following the Civil War, and specialized farming and dairying developed into the most lucrative types of farming in Vermont during the last quarter of the nineteenth century. In 1869, the Vermont Dairyman's Association formed, and a year later the state legislature established the Vermont Board of Agriculture, Manufacturing, and Mining, thus placing state government in the midst of agricultural practice.[39]

In 1880, that body became the Vermont Board of Agriculture and during the ensuing decade the board's responsibilities included regulating butter quality, investigating herds for tuberculosis, testing and licensing fertilizer, and preventing the sale of unwholesome meat. Government's increasing role in agriculture was aided by creation of an agricultural experiment station in 1886, but the Vermont State Extension Service did not emerge until 1915, two years before the government's programs, including forestry, were consolidated under a commissioner of agriculture. By this time, concerted efforts at disease eradication were well established, and preventing the contamination of milk, controlling plants harmful to productivity, and eradicating pests became key objectives for the department.[40]

Recognition of the direct relationship between agriculture and ecological abuse has steadily advanced during the twentieth century, but battles between the two continue today in various contexts, including bovine growth hormones, genetic seeding, and chemical fertilizers. Organic farming has gained increasing attention in recent years as well.

Soil conservation also developed into a principal concern, and in 1941 the state legislature finally enacted meaningful legislation to retard soil losses, establishing a soil conservation council and authorizing the designation of soil conservation districts that, today, encompass the entire state. Practices have evolved in this field as well, and efforts by the state to straighten streams during the second half of the twentieth century are now being reversed, with the goal of returning water courses to their natural, sinuous shape and creating buffer zones along these river corridors where cultivation is avoided.[41]

At the local level, Burlington's Tommy Thompson led a grassroots movement to establish community gardens. He called the effort Gardens for All, and this organization eventually became the National Gardening Association. By 1973, his project had expanded into a citywide gardening initiative, with more than a thousand garden plots scattered throughout the city. In 1988, the garden he established in the city's Intervale district was given his name. That sector of the city remains one of the few areas in a New England city where commercial agriculture continues. In addition to land under cultivation and large plots devoted to community gardens, an organic composting business has been established. Unfortunately, the costs of permitting have forced that business to close recently. Nonetheless, a network of trails continue to link the Intervale Foundation's headquarters with outlying fields and the Winooski River corridor.[42]

INTERSTATE HIGHWAYS

By no means had concerns about keeping Vermont unspoiled abated when plans for a national interstate highway system were being developed after World War II. However, by the time President Dwight Eisenhower signed the Federal Aid Highway Act of 1956, formally authorizing construction of a "National System of Interstate and

Defense Highways," the approximate location of Vermont's future interstate corridors had already been identified, but with little of the public consternation that had surrounded proposals for the Green Mountain Parkway two decades earlier.

Instead, securing Vermont's share of the economic benefits that would travel these corridors became an overwhelming force, silencing doubts about influence from afar. In fact, in the summer of 1959, about six months after the first section of highway had opened in Windham County, the highway department launched a study to determine the project's economic yield. Such prospects, and the federal mission to provide for national defense in the nuclear age by connecting major metropolitan areas, presented insurmountable obstacles to any lingering sentiments favoring isolation. The need to modernize highways to accommodate high-speed automobile traffic had become obvious to most people as well.[43]

To be fair to Vermont's citizens, another factor also contributed to the path of interstate construction. The Federal Aid Highway Act of 1944 marked the true beginning of a national interstate system of roads linking principal metropolitan areas and industrial centers. However, the word interstate was first used during that period to designate existing roads as part of a connective network. That same year, the American Association of State Highway Officials (AASHO) developed specific engineering standards for these highways, emphasizing the need for modern design. By 1945, Vermont had selected its choices for designated interstate routes, and in 1947 the Federal Works Administrator officially approved a national system, which included most of the roads Vermont had selected.[44]

By 1952, federal legislation had allocated money for construction of this interstate system, and Vermont participated in that program as well. When President Eisenhower approved the accelerated strategy in the Highway Act of 1954, plans for limited-access superhighways built on new locations became possible and routes were identified quickly. Engineers from the state highway department and the federal Bureau of Public Roads jointly selected locations, but with national objectives principally in mind. By 1956 federal and state highway engineers had enjoyed more than a decade of collaborative endeavor and were far better prepared to present these conceptions than were the few progressives who had hastily promoted the Green Mountain Parkway years before. For those Vermonters who still leaned toward skepticism, the presence of a defense-minded federal government was never very far away.[45]

Further solidifying this collaborative strategy, AASHO had already prepared design standards for interstate and defense highways and adopted them less than two weeks after President Eisenhower signed the 1956 bill. Thus, state engineers were able to begin a design phase once the details of location had been determined. The strategic skill with which this juggernaut of federal, state, and AASHO engineers advanced such an enormous public works undertaking is one of the remarkable chapters in the interstate story. Equally important, these strategies would be inherited as lessons for subsequent generations of engineers. Both the methods used today to plan, design, and build public highways, and the recently emerging public responses to some of these heavy-handed methods, are tied to the developments of the post World War II era. In Vermont, recent efforts to force highway engineering to consider the broad environmental context, cultural as well as natural, have produced new standards that allow more flexibility in the sometimes rigid requirements for highway design developed in the name of motorist safety. All are closely linked to initiatives ultimately affecting conservation as well.[46]

By the early 1970s, most of Vermont's interstate system had been completed, opening nearly every sector of the state to economic, social, and political change. Corresponding changes to the environment, both natural and cultural, reached far beyond the physical boundaries of rights-of-way. Between 1959 and 1969, for example, ten new state parks were created, with attendance rising to 125,699 in 1961, up from only 2,638 in 1944. Equally important, legislative and policy responses to these changes have shaped (and continue to shape) the course of conservation in Vermont in myriad ways. One can quickly measure the impact of the interstate system simply by scanning a list of the legislative initiatives that followed: local planning and development enabling legislation (1968), control of billboards (1968), Act 250 (1970), land-speculation tax (1973), growth management (1988), and downtown development (1997), to name just a few. Vermont's Historic Preservation Act, passed in 1975, evinced a response to similar pressures of change facing the state's built environment.[47]

BILLBOARDS: PROTECTING SCENERY, PROMOTING TOURISM

The seeds of concern about roadside clutter, sown during the 1920s when automobile travel rapidly increased, sprouted dramatically as waves of motorists hastened toward Vermont. The state's 1968 law banning billboards is a landmark policy in a decades-long debate about outdoor advertising along roadways. The debate encompassed safety, aesthetics, commercial advertising, free speech, and property rights. Nationally, it pitted the advertising industry against advocates for scenic preservation and public interests in highway safety. In Vermont, policy makers forged a compromise among these interests by recognizing a link between tourism and aesthetics. What began in the 1930s as state policy to promote safety along highways, became in the 1960s a nationally recognized policy to protect a rural and natural aesthetic central to economic stability.

In the 1930s, Vermont law regulated outdoor advertising under the police power. It provided for a system of permits for the display of advertisements in public view. The system established fees, sizes, and locations for advertisements, and restricted their display along highways, as well as prohibiting them near parks, playgrounds, and cemeteries. This approach to billboards was not unique to Vermont. The use of permits and regulations to control signage existed in local and state policies elsewhere in the country by the mid-1930s. It also was part of national policy established in the 1950s and 1960s to control billboards along the federal highway system, through the Bonus Act of 1958 and the Highway Beautification Act of 1965.[48]

Vermont became a leader in this arena by passing the 1968 state law that prohibits off-premise billboards. In addition to this prohibition, the law established an alternative system for advertising in the public view that attends to both commercial interests and aesthetic concerns. Vermont's response to billboards reflected a conservation aesthetic in the state, and continues to serve as a landmark on the national stage.[49]

What is known colloquially as a ban on billboards is a statute entitled Tourist Information Services. This title is the most succinct of several cues in the policy record about the bridging of conflicting interests that occurred through the law. In a regulatory sense, it controls outdoor advertising along roadways and visible to the traveling public. One finds prohibitions on commercial billboards in these provisions. Beyond this, the law takes a constructive approach to providing information to motorists traveling in Vermont. It creates a system of official business directional signs, overseen by a government structure, the Travel Information Council. These signs provide opportunities for commercial interests to advertise their goods and services through a regularized system of sign posts along the highways. This approach is supported with an infrastructure of tourist information centers. Policy documents issued in the years prior to the passage of the law provide greater depth of explanation of its approach and logic.[50]

In 1966, the state Central Planning Office issued a report on the importance of the aesthetics of Vermont to tourism and the economic well-being of the state. This report outlined a justification for protecting aesthetics along Vermont roads under the police power. It began by quoting George Perkins Marsh, expressing the need to respond to the advances and destructive tendencies of civilization through preservation and restoration. Thus it provided justification for what was to come next with ties to the tradition of conservationist thought in Vermont. While the report described problematic change underway during the 1960s, it did not stand against it but rather argued for controlling it. Governor Philip Hoff and the Central Planning Office staff made a plea to "preserve and enhance scenic values visually related to Vermont's highways by regulating the use of land adjacent to such highways." The report justified such regulation by linking aesthetics and economics. It stressed that Vermont's scenery is critical to tourism, that tourism is critical to the state's economy, and so preserving the scenery is critical to economic well-being. The report made a case for using the police power to protect aesthetics.[51]

The case made by the governor and Central Planning Office was paralleled in a 1967 legislative report generated by a committee convened to study outdoor advertising. The committee described the importance of the scenic appeal of Vermont to economic interests such as tourism, and emphasized the need to controls billboards. They addressed the legal viability of such controls under the police power. To this they added the importance of creating a new system to communicate information to tourists. Thus, the committee took a critical step in addressing the need to provide information to motorists.[52] The

As historian Robert McCullough has observed (Crossings: A History of Vermont Bridges, 2005), *for many writers, artists, and photographers, covered bridges symbolize the rural ideal. This wood engraving of a Vermont covered bridge, done by New England artist Asa Cheffetz in 1945, confirms that observation. As the artist himself wrote in the autobiographical text accompanying his 1940 print,* Down Montgomery Way (Vermont), *"I love this fertile land, the simple way of life, and its rugged people. I love the very temperament of the land in all its moods. It is this quality of the inner mood of the scene which (has) beguiled me . . ."*
(Boston Public Library)

legislative findings for the Tourist Information Services Law bring together concerns of tourism, economics, and aesthetics, and continue a longstanding theme of ensuring public safety. The law effectively responded to the range of interests engaged in debate about outdoor advertising. Since its passage, Vermont has served as a beacon of success in preserving scenic resources.[53]

While Vermonters controlled outdoor advertising in the state, Vermont senators carried the message to the federal level. Senator Robert Stafford, during his tenure in Washington between 1972 and 1989, and as chair of the Senate Public Works Committee between 1981 and 1987, advocated for legislation to reform the Highway Beautification Act of 1965. As described by Charles Floyd, the 1965 law was intended to address the billboard issue, but instead became "a sign-industry dominated program that is actually enriching and subsidizing the industry it was meant to regulate, serving as a protective umbrella to shield the industry from state and local governments that desire to effectively control billboard blight." As an advocate for reform of the law, Senator Stafford supported the regulation of billboards.[54] Senator Jim Jeffords, who took the retiring Senator Stafford's seat in 1990, had served on the Vermont state legislative committee that produced the 1967 report described above. When he took his Senate seat, he became a member of the Public Works Committee and supported the 1990 Visual Pollution Control Act, a reform effort that returned power to states and municipalities to control billboards.

Vermont's success in controlling billboards illustrates how people in the state have periodically blazed a trail to protect natural resources while recognizing links among conservation, aesthetics, and economics.

RED SPRUCE DECLINE ON CAMEL'S HUMP: EFFECTS OF ACID DEPOSITION

While regulating billboards was within Vermont's jurisdiction, factors out of the state's control were negatively affecting its ecology and aesthetics, for example, acid deposition. In the 1980s, acid deposition came into the national spotlight as a controversial issue. It remains a pressing problem in the United States, and other parts of the world. Pioneering research by scientists in the Northeast, specifically Vermont, New Hampshire, and New York, have contributed substantially to our understanding of the effects of acid deposition. In Vermont, a series of studies of red spruce on Camel's Hump Mountain was one of the first to implicate acid deposition in the decline of forest health. This information, along with the results of other studies, informed the policy debate over acid deposition and contributed to the passage of stricter national air quality standards in 1990.[55]

By the 1970s, the impacts of acid rain on aquatic systems were established. The discovery that acid rain also affected terrestrial systems, and more specifically forests, occurred later and was an unintentional product of ecological research carried out in Vermont. In 1965, Dr. Hubert Vogelmann of the University of Vermont and his graduate student Thomas Siccama began a long-term ecological investigation of Camel's Hump Mountain in Duxbury, Vermont. They inventoried the western slope of Camel's Hump because it had never been logged, and it represented a typical forested mountain. These characteristics made it valuable to people seeking a wilderness experience and to scientists who used the forest as a control in ecological studies. It was designated as a Natural Area in 1965, a Natural Landmark in 1968, and State Park in 1969.[56]

Despite the local protection these designations afforded, Camel's Hump was being negatively affected by broader scale pollution. Vogelmann and Siccama conducted a second inventory in 1979 that revealed disturbing information: the boreal population of red spruce, a major component of northeastern forests, had declined by about 50 percent. Another inventory in 1983 corroborated this trend. The spruce decline spanned all age classes, and could not be explained by conventional causes such as forest succession, disease, insects, or drought. Air pollution, including acid deposition, was suggested as a possible reason for the decline.[57]

Acid deposition is caused by the burning of fossil fuels, which releases sulfur and nitrogen oxides into the atmosphere. These gases combine with water to form sulfuric and nitric acids. Westerly winds send sulfur and nitrogen emissions produced by coal burning plants in the midwestern states to the Northeast, where they return to the earth as both dry (particulate matter) and wet (rain, fog, snow) acid deposition. Besides causing ecological damage, these oxide-bearing winds have an aesthetic impact. They often cause acid haze, which obscures scenic views, even on cloudless days. By the mid 1970s, acid deposition would be classified as a regional phenomenon. Levels of acidity had increased dramatically since 1955, and were thirty to forty times higher than would be expected in unpolluted precipitation.[58]

Because acid deposition fails to respect state boundaries, a national approach was needed. In 1979, fueled with evidence from studies in the Northeast, Senator Daniel Patrick Moynihan of New York introduced the Acid Precipitation Act to "determine the causes and effects of acid precipitation throughout the United States and to develop and implement solutions to this problem." The act passed Congress in 1980 and established the National Acidic Precipitation Assessment Program, a ten-year research effort.[59]

In a 1982 article in *Natural History*, Dr. Vogelmann described the damage to red spruce on Camel's Hump and its correlation with acid rain. He and Dr. Richard Klein set up the Acidic Deposition Research Project in the Botany Department at the University of Vermont. After several years of study, they came to understand that acids combined with nutrients, such as calcium and magnesium, leaching them from the soil at accelerated rates, and making them less accessible to trees. At the same time, studies conducted by scientists such as Dr. Arthur Johnson of the University of Pennsylvania suggested that winter injury, possibly caused by a weakening of the trees due to acid rain, might be a major contributor to red spruce decline. While not definitive, these results helped to inform the nation of the problem.[60]

Educating the public was important. Midwestern coal burning plants were a powerful presence in Washington D.C. and emission regulations would result in financial losses for that industry, as power plants would have to be modified to conform to stricter standards. The Reagan administration, strongly pro-industry, refused to acknowledge that acid rain was caused by air pollution until 1986, when international reports and reports from the National Academy of Sciences established a definite link between midwestern industrial emissions and environmental damage in the Northeast. The Vermont ecological studies contributed important scientific information to efforts to reform national policy in the Clean Air Act amendments of 1990.[61]

Another decade would pass before the mystery of red spruce decline was solved. In 1999, Dr. Donald DeHayes of the University of Vermont and his colleagues established a direct link between winter freezing injury and red spruce decline. They found that acid mist depletes spruce needles of their membrane calcium, resulting in cell destabilization. This leads to increased vulnerability to winter freezing damage.[62] Similar processes may occur in other species. What began as an unintentional discovery on Camel's Hump in 1979 led to a greater understanding of acid deposition, its effects, and how it can be prevented. In 2001, Vermont's Senator Jeffords proposed the Clean Power Act to substantially reduce power plant emissions. However, although reintroduced in subsequent legislative sessions, the bill has not been passed. Efforts to promote the ecological and aesthetic interests of Vermont and the Northeast have been blocked by economic interests in other parts of the country.

ACT 250: A FORUM FOR DEBATE ABOUT DEVELOPMENT AND CONSERVATION

By the 1960s, dialogue about how to keep Vermont unspoiled had escalated along a number of fronts as development pressures advanced. Vermont's Land Use and Development Law (Act 250), one of the most visible pieces of legislation to emerge from this dialogue, was adopted in 1970 and reflects the tensions entailed in managing growth while attending to its environmental and social impacts. Act 250 also reflects Vermont's culture of decentralized politics, which emphasizes direct involvement of citizens and town government in setting a course for the future.[63]

Act 250 established a two-part framework for addressing development, statewide plans, and a permit system for residential, commercial, and industrial development. The plans were intended to provide direction and guidance to manage growth. Had these plans been established, they might have provided a proactive approach to managing growth and its impacts. While people spent considerable time and energy developing them, the goal of adopting a statewide land-use plan was never realized. This failure is attributed to disagreements about the role of the state in exerting regulatory or "top-down" authority through such a plan.

The permit system was put in place and has a regulatory effect in the reactive sense of responding to, and placing conditions on, individual proposals for development. It is carried out through a structure of nine regional district commissions. These commissions review development applications and decide whether to grant permission, with or without conditions, based on ten criteria from the statute. The criteria attend to the impacts of proposed development on water, air, soil, highway congestion, municipal services, aesthetics, historic sites, and natural areas. They also require that development conforms to regional and local plans.[64]

Concern about the conflict between economic growth and conservation dominated discussions leading to Act 250's passage. In 1968, the Vermont Planning Council, made up of citizens, state legislators, and state agency commissioners and chaired by Governor Philip H. Hoff, issued a statement in *Vermont's Future: Vision and Choice, The Vermont*

State Framework Plan. It conveyed a sense of urgency about the need to respond to growth:

> The pace of growth and change in Vermont demands consideration of basic issues and illumination of alternative paths to the future. We cannot delay such consideration . . . We cannot continue to abuse and destroy our environment . . . By permitting the deterioration of Vermont's natural beauty and resources, we are committing an inexcusable offense against our children and their children . . . We must create new opportunities for productive endeavor in livable communities.[65]

The council's statement brought to the fore questions about aesthetics and environmental conservation, and also highlighted social disruptions such as shortages of social services, loss of sense of community, antagonism between newcomers and rural Vermonters, and an undermining of the traditional town meeting.[66] The roots of the plans in Act 250's framework are evident in the council's recommendations to develop a statewide approach to managing growth to ensure protection of the landscape and cultural traditions in Vermont.[67]

Governor Deane C. Davis took office in 1968 and provided leadership for the drafting and passage of Act 250. He appointed a Commission on Environmental Control headed by State Representative Arthur Gibb, with a membership that included legislators, state agency personnel, developers, and conservationists. The commission's 1970 report called for regulations to control the impacts of large subdivisions and development, and recommended the adoption of a comprehensive plan to which such developments would conform. It highlighted areas above 2,500 feet in elevation for special protection, reflecting concerns about proposed ski area developments in southern Vermont. Act 250 strictly limits development above 2,500 feet and protects water supplies including headwaters and areas above 1,500 feet, thus recognizing the fragile ecology of high elevation areas and their importance to clean water.[68]

Act 250 relies on citizens for its implementation. The district commissions are volunteer citizen boards. The law also established an

Environmental Board, which oversaw the permit system and served as the first point of appeals to district commission decisions. Similar to the district commissions, the Environmental Board consisted of citizens appointed by the governor who, except for the chair, volunteered. Although the district commissions and the Environmental Board retain professional staff, decision-making power lies in the hands of citizens. According to David Heeter, the approach of the district commissions was "essentially one of negotiation rather than hard and fast application of criteria." The Environmental Board's current *Guide to Act 250* emphasizes that, during the 1990s, the board and district commissions "began encouraging parties to mediate their issues rather than go through extended adversarial hearings." The guide suggests that "because citizens have a voice in the Act 250 process, applicants often work with neighbors and other interested citizens and groups to address the concerns of people who will be affected by a proposed development."[69]

As part of reforms to environmental permitting in Vermont, in 2004 the legislature revised the approach of using quasi-judicial boards in the executive branch for first appeals of agency and district commission decisions. The reforms removed the appeals function from the jurisdictions of the Environmental Board and the Vermont Water Resources Board, and placed it with an expanded Environmental Court. The two boards merged into a single Natural Resources Board, which oversees the administration of Act 250 and Vermont's water laws. The Environmental Court had been established in 1989 as an environmental division within the judicial system to address a backlog of environmental cases.[70]

Over the three and a half decades since Act 250's inception, policy makers have made many adjustments to it, and debated issues about its implementation that include to whom it applies, how its criteria should be interpreted, who should be involved in decision processes and at what stages, how formal the process should be, and what responses to infractions are appropriate. These questions affect the framework in which dialogue occurs about balancing the goals of development with those of conservation.

What set Act 250 apart from other laws passed in the 1960s and 1970s was its intention to take a statewide view of guiding growth and protecting ecological and aesthetic values. However, as Richard Brooks observes, with the failure to adopt a statewide plan in the early 1970s, "Vermont forfeited a golden opportunity to guide growth on a statewide level with a heightened awareness of the ecology of the state."[71] While policy makers did not succeed in adopting a statewide plan, they did establish, through the permit system, a public forum for the governors, legislators, agency planners, and citizens of Vermont to engage in debate about development and conservation that continues today. Notably, Act 250 emphasizes the role of citizens as decision makers and parties to the debate. Circumstances have changed substantially since 1970 and Vermont would do well to consider what additional reforms are needed for Act 250 to promote conservation goals in the twenty-first century.

CONSERVATION AND AFFORDABLE HOUSING ALLIED

During the post-Interstate era, a sense of urgency has shaped the dialogue surrounding conservation and the economic, social and political stresses confronting those who seek to keep Vermont unspoiled. That sense of urgency has produced a range of initiatives. Many have been practical minded and effective in some measured way; tax laws and administrative organization of government agencies fall into this category. Others bear the mark of enlightened policy that should continue to shape public outlook far into the future. The Vermont Housing and Conservation Trust Fund Act of 1987 is one such landmark initiative.[72]

Vermont's alliance between affordable housing and conservation grew from pressures common to both. The same forces that threatened consumption of open land, disrupted ecological balance, or weakened traditional patterns of settlement, also increased the price of real estate, making the prospect of homeownership for some Vermonters increasingly difficult. An alliance took shape, led by the Vermont Land Trust, the Affordable Housing Coalition, the Vermont Natural Resources Council, and the Low Income Advocacy Council. These organizations formed the Housing and Conservation Coalition in 1986, which pressed for the passage of the law during Governor Madeline Kunin's administration, and generated substantial funding for coalition participants, even during times of economic constraint.

The Vermont Housing and Conservation Board, legislatively chartered as a body politic and corporate, and thus partially insulated against political whim, exercises administrative oversight of the trust fund, which receives allocations from the state legislature. The board is authorized to support projects that meet the collaborative goals of affordable housing, retention of agricultural lands and active farms, protection of wildlife habitat, and preservation of historic properties or recreational lands. The fund has had farreaching effects, evident in a diversity of projects, and also in an increasing number of nonprofit community land or housing trusts and other coalition organizations, which expand opportunities for innovation. Coalition efforts assume an interdisciplinary quality, demonstrating the ability of conservationists to confront social conditions in tangible ways, and potentially expanding the reach and depth of conservation in the process. What began as an experiment during the Kunin administration became institutionalized during the 1990s when Howard Dean became governor. Dean played a principal role in some of Vermont's largest land-conservation projects, helping to solidify and expand the coalition in the process.[73]

FRAGMENTATION OF FARM, FOREST AND COMMUNITY

In addition to the landmark initiatives that have distinguished conservation efforts in Vermont during the post-interstate era, the state has also employed numerous devices or strategies already in place in other parts of the country. Many of these were intended, either directly or indirectly, to confront the fragmentation of working farms and forests, or to aid communities trying to thwart the undesirable economic, ecological, and aesthetic consequences of sprawl. Property tax incentives have been part of the strategy, in one form or another, throughout most of the twentieth century. As early as 1912, for example, Vermont had allowed owners of fully stocked forests, with growth not yet fifteen years old, to be taxed at the value of the land alone. Owners of land in agricultural or forest use had also been given authority to negotiate tax stabilization contracts with towns, with tax determined by the value of the use rather than by market value. That practice acquired more formal structure with passage of the Use Value Appraisal (or Current Use) law in 1978. The law became effective in 1980 and has been refined by amendment several times since.[74]

Increasing property values also contributed to speculation by absentee owners, who pushed the already spiraling price of land still higher and aggravated the widening social divisions that had accompanied arrival of people from afar. Moreover, Act 250's jurisdictional limits prevented the law from reaching a large percentage of the housing projects occurring during this inflationary period. In an aggressive response, Vermont's legislature adopted a land-gains tax, which took effect in 1973 and, bolstered by subsequent amendments, rendered the outlook for short-term speculative land investment substantially less attractive.[75]

Yet tax measures offered scant help for a rapidly changing timber economy and industry, circumstances that threatened the sustainability of working forests in northern Vermont. Spurred by the potential for sale and liquidation cuts of large holdings owned by international timber companies, the governors of New York, Vermont, New Hampshire, and Maine appointed a region-wide Northern Forest Lands Council to guide the future of twenty-six million acres of forest stretching across the northern tiers of all four states. The council, aided by congressional funding obtained through the efforts of Senators Patrick Leahy and Warren Rudman from Vermont and New Hampshire, respectively, conducted the Northern Forest Lands Study between 1990 and 1994. In a report titled *Finding Common Ground: Conserving the Northern Forest*, the council addressed the need for stewardship of private forest lands, tax policies, protection of exceptional resources through public acquisition, water quality and biodiversity, rural communities, and the important role of tourism and recreation to the economy. Concluding with a request that the four states implement its recommendations, the council then dissolved. A nonprofit coalition based in Vermont, the Northern Forest Alliance, soon formed and today tries to advance the council's vision. Among that alliance's most recent plans is an effort to revive the state's municipal forests, an undertaking titled simply, Town Forest Project. Separately, the Vermont Urban and Community Forestry Council has launched a number of inventive initiatives since its inception during the early 1990s.[76]

Urban and community forestry is one program among many administered by the state's Agency of Natural Resources. That agency's current composition dates from the post-interstate era, emphasizing natural resource management, pollution abatement, and wildlife protection, as well as regulations implementing federal policy standards through state-level plans. By 1970, Vermont's conservation agencies had been consolidated into the Agency of Environmental Conservation, later renamed the Agency of Natural Resources, with responsibilities that include managing forests and parks, fish and wildlife, air and water quality, and waste disposal and recycling. As a statutory party, the agency plays a vital role in Act 250 permit applications.[77]

In both agency and nonprofit sectors, attention is being given to sustaining Vermont's rural working landscape, and forests are central in that effort. In the mid-1990s, Vermont reconvened its Forest Resources Advisory Council to address issues of clear cutting of timber and application of herbicides in forest management through aerial spraying. Public hearings, roundtable discussions, and field visits ensued, leading to two legislative initiatives: (1) a framework for forest plans on private lands that limits heavy cutting in the absence of management plans, and (2) a moratorium on aerial herbicide spraying.[78] In 1996, the Agency of Natural Resources began drafting a plan for public land acquisition, viewed as an essential aspect of assuring forest sustainability, notwithstanding opposition from the private sector. The agency's Land Conservation Plan, released in 1999, emphasizes goals of biodiversity, conservation, and recreation in future public lands acquisition, exchange, or disposition.[79]

Wilderness designations, in the minds of many, also help to keep Vermont unspoiled. In 1984, the Vermont Wilderness Act designated four wilderness areas in the Green Mountain National Forest: Breadloaf, Big Branch, Peru Peak, and George D. Aiken. Those four supplemented two areas previously designated by the Eastern Wilderness Act of 1975: Bristol Cliffs and Lye Brook. Among these six locales, encompassing about 59,600 acres, Breadloaf Wilderness is the largest, at more than 21,000 acres, and stretches from Middlebury Gap northward to Lincoln Gap—most of it above 3,000 feet in elevation. Attempts by the U.S. Forest Service to produce a management plan for Breadloaf culminated in a draft proposed in 1996, but efforts to refine that plan continue. The consortium Vermont Wilderness Association, loosely affiliated with a number of environmental groups, including the Wilderness Society and the Sierra Club, has played a vocal role in supporting designation of these areas and continues to work toward improved management plans. Two more wilderness areas were designated in a congressional bill adopted in December 2006: The Joseph Battell

Wilderness, 12,300 acres in Goshen, Hancock, Ripton, and Rochester; and the Glastenbury Wilderness, 22,400 acres. Including additions to existing areas, the bill designated more than 40,000 acres of wilderness.[80]

Recognition that sustaining Vermont's working landscape also depends on protecting traditional patterns of settlement led to passage in 1998 of the Historic Downtown Development Law, which recognized that economically strong downtowns represent long-term investments in public and private infrastructure. The law encourages development and planning for a cohesive core of commercial and mixed-use buildings and to avoid strip development. The bill created a Downtown Development Board and established methods for towns to designate downtown development districts, village centers, and new town-center development districts. Various incentives are offered to towns that comply, including priority for state and federal financial aid, state tax credits to rehabilitate historic buildings, planning grants, financial aid for transportation projects, and technical and financial assistance for contaminated sites.

Related bills were enacted in 2006. One focused on creating designated growth centers, and a second provided tax and economic incentives for development in growth centers. The first bill defines "growth centers" and supports local planning with through three components: a municipal plan, local bylaws that implement the plan, and a budget to fund the plan. Planning is to be coordinated among local governments, regional planning commissions, the state's Department of Housing and Community Affairs, and the Natural Resources Board (formerly the Environmental Board). Key, too, is a process by which towns can designate growth centers that will accommodate a substantial part of the towns projected growth over a twenty-year period.

Its companion bill authorizes municipalities to enact Tax Increment Finance Districts within designated downtowns, village centers, and growth centers. In these districts, a portion of the property tax revenues attributable to new development are reallocated from the state education fund to pay for infrastructure improvements in eligible towns. The bill also encourages affordable housing by revising jurisdictional thresholds for Act 250 in designated growth centers, and seeks to facilitate compliance with various Act 250 criteria for projects in growth centers. Together, the initiatives hopefully will strengthen local and regional planning, use state resources efficiently by directing investment to areas designed to accommodate growth, maintain Vermont's rural landscape and economically vital downtown centers, attain energy efficiency, promote housing opportunities, and make the regulatory process more predictable.[81]

In 1998, Smart Growth Vermont (formerly the Vermont Forum on Sprawl) was established to encourage economic vitality in community centers and preserve Vermont's working landscapes. This nonprofit organization was funded by the Orton Family Foundation and headed initially by Jack Ewing, former chair of the Environmental Board. Public education is one of the organization's primary goals, and toward that end they provide technical assistance, advocacy, and partnerships with businesses, other nonprofit organizations, state agencies, and local government groups. The organization also played a principal role in passage of the state's growth center legislation and works with communities to implement the law.

Finally, the Vermont Law School and its Environmental Law Center in South Royalton have contributed to the expanding dialogue on conservation, not just in Vermont, but also in national and regional contexts. The school offers comment through several forums, including the *Vermont Law Review* and the *Vermont Journal of Environmental Law*.

KITCHEN-TABLE DIPLOMACY

Focus on Vermont's landmark conservation initiatives of the past provides a useful historical context for informed decision making in the present. Yet today, the pace of environmental change (cultural as well as natural) is escalating and we can ill-afford to linger in the past. Conservation historians, in addition to building upon the visible foundations of environmental policy making, must also explore less tangible evidence of emerging trends (or at least turn earth with an eye to possibility) that may hold promise in years to come.

Thus, with a look toward the future, at least three aspects of a shifting conservation ethic in Vermont (and elsewhere, too) deserve consideration. One is the recognition that solutions to environmental problems cannot be confronted separately from other human concerns: poverty, education, housing, health care, employment, economy, and transportation, to name some. Vermont's alliance among proponents of affordable housing and conservation is a farsighted initiative that acknowledges this truth, suggesting good reason for those representing differing poles to align in similar fashion.

Second, following logically from the first, the complexity of human concerns in relation to the environment demands an interdisciplinary approach. The gaps between disciplines remain wide and barriers formidable. The champions of one cause are often not conversant with those whose missions point in other directions, albeit with compatible destinations. Yet these same divides also can be viewed as uncharted territories ripe for exploration by those who are willing to learn the language of other disciplines. Some efforts, yet to bear fruit, are occurring at the University of Vermont where a program in Conservation Leadership is being proposed, drawing from curricula available in programs as diverse as forestry, agriculture, and historic preservation.

Finally, voids in the web of responses to environmental stresses left unfilled by governmental programs are being stitched together by a growing array of informal associations, local committees, nonprofit cor-

This view of the Vermont State Capitol, situated at the base of Hubbard Park before a backdrop of a red pine plantation and a deciduous forest sprouting from abandoned pastures, illustrates how long forest and community have been interdependent in Vermont. In fact, masons used material from the pasture's stone walls to construct the observation tower that now provides visitors a view of such distant summits as Camels Hump and Mount Hunger. (Robert L. McCullough)

porations, educational institutions, and other community or regional groups. Such affiliations are often comprised of members whose interests are directly imperiled or who have incentive to persevere in the quest for solutions. These groups offer a forum for thoughtful, interdisciplinary dialogue, a foundation for any meaningful success. The nonprofit Moosalamoo Association in Goshen is a perfect example. That group successfully fashioned alliances among often conflicted users of the Green Mountain National Forest, and in 2004 won a World Legacy Award for destination stewardship. The same bill that designated the Battell and Glastenbury wilderness areas in 2006, also established the Moosalamoo National Recreation Area. Tony Clark, one of the founders of the group, attributes their success to what he describes as "kitchen-table diplomacy." For Clark and others, finding the solutions to problems seems much easier when the people involved are all sitting around a kitchen table. Any recipe for keeping Vermont unspoiled will surely benefit from that advice.

NOTES

1. Dorothy Canfield Fisher, "Vermonters," in Federal Writer's Project, Works Progress Administration, *Vermont: A Guide to the Green Mountain State* (Boston: Houghton Mifflin Company, 1937): 3.

2. Bureau of Publicity, Department of State, *Unspoiled Vermont. Come to Vermont—A Vacation Paradise Forever Unspoiled*, 1933. See also, Publicity Service, Vermont Department of Conservation and Development, *Only a Step to Unspoiled Vermont*, 1939.

3. George Perkins Marsh, *Man and Nature*. (New York: C. Scribner, 1864.) The 1874 edition was retitled: *The Earth as Modified by Human Action*. See also David Lowenthal, *George Perkins Marsh: Versatile Vermonter*. (New York: Columbia University Press, 1958); and David Lowenthal, *George Perkins Marsh, Prophet of Conservation*. Foreword by William Cronon. (Seattle: University of Washington Press, 2000.) Subsequent citations from Lowenthal will be from the latter.

4. Lowenthal, *George Perkins Marsh*, 269–270.

5. Lowenthal, *George Perkins Marsh*, 273–275.

6. Franklin Hough, "On the Duty of Governments in the Preservation of Forests," in American Association for the Advancement of Science, Forestry Committee, *Cultivation of Timber and the Preservation of Forests*. Reports of the Committee of the House of Representatives, 43rd Congress, 1st Session, 1873–74, Serial 1623, Report No. 259. Washington, D.C.: GPO, 1874. See also Lowenthal, *George Perkins Marsh*, 303.

Hough's successors, Nathaniel Egleston and Bernard Fernow, both relied heavily on Marsh's study. Fernow, in particular, continued to advance the dialogue centering on forestry science and the critical role that forest cover plays in sustaining soil condition and assuring the mechanical retardation of runoff, a dialogue begun by Marsh. See Bernard Fernow, *Forest Influences*. U.S. Department of Agriculture, Forestry Division, Bulletin No. 7. (Washington, D.C.: G.P.O., 1893.)

7. See "An act to encourage the growth of timber on western prairies," Chapter 277, Section 4, Volume 17, *United States Statutes at Large* (March 1871 to March 1873), enacted March 3, 1873. See also "An act to repeal timber-culture laws and for other purposes," Chapter 561, Section 24, Volume 26, *United States Statutes at Large* (December 1889 to March 1891), enacted March 3, 1891. Supplemented by Chapter 2, Vol. 2, Supplement to the Revised Statutes of the United States (1892 to 1901), enacted June 4, 1897.

8. Winks, *Billings*, 303–305. See also Laurel Ginter, "Building Billings," in *Vermont* (Winter 1984): 2–7; and Douglas Kent Lehman, "An Oration in Stone. The Billings Library at the University of Vermont," M.A. Thesis, University of Vermont, 2004.

9. Rutland Railroad Company, *Heart of the Green Mountains. Souvenir Edition*, H. A. Hodge, ed., (Rutland, Vermont: Rutland Railroad Company, 1897); and Central Vermont Railroad, *Summer Homes Among the Green Hills of Vermont and along the Shores of Lake Champlain*, (St. Albans, Vermont: St. Albans Messenger Job Print), 1892. See also William C. Lipke and Philip N. Grime, eds., *Vermont Landscape Images. 1776–1976*. (Burlington, Vermont: Robert Hull Fleming Museum, 1976.)

10. Lipke, *Vermont Landscape Images*.

11. Vermont Department of Agriculture. *Resources and Attractions of Vermont: with a List of Desirable Homes for Sale*, (Montpelier, Vermont: Watchman Publishing Co., 1891); Vermont Department of Agriculture, *A List of Desirable Vermont Farms at Low Prices*, (Montpelier, Vermont: Watchman Publishing Co., 1893); and Vermont Department of Agriculture, *Vermont. A Glimpse of its Scenery and Industries*, Victor I. Spear, ed., (Montpelier, Vermont: Argus and Patriot Print, 1893.) See also *Dona Brown, Inventing New England. Regional Tourism in the Nineteenth Century*, (Washington, D.C.: Smithsonian Institution Press, 1995): 142–150.

12. Mortimer Proctor and Roderic M. Olzendam, *Vermont. The Unspoiled Land*, (Rutland, Vermont: The Tuttle Company, n.d., c. 1915.) Reprinted by the Vermont Publicity Service in 1916, under the title *The Green Mountain Tour Through Vermont, the Unspoiled Land.*

13. Jan Albers, *Hands on the Land. A History of the Vermont Landscape*, (Cambridge: M.I.T Press for the Orton Family Foundation, 2000): 256–57.

14. Dorothy Canfield Fisher, *Vermont Summer Homes: An Open Letter*. (Montpelier: Vermont Publicity Service, 1932.) The booklet was reprinted through 1941. See also Sherman, et al., *Freedom and Unity*, 520–522.

15. Lipke, *Vermont Landscape*, 44–45. See also Mary Hart Bort, *Art and Soul. The History of the Southern Vermont Arts Center*, Margot Page, ed. (Manchester, Vermont: Southern Vermont Arts Center, 2000.)

16. 10 V.S.A., Chapter 151, Section 6001, et seq.; 24 V.S.A., Chapter 117, Section 4301, et seq.

17. Michael Sherman, Gene Sessions, and P. Jeffrey Potash, *Freedom and Unity: A History of Vermont* (Barre, Vermont: Vermont Historical Society, 2004).

18. Andrew and Edith Nuquist, *Vermont State Government and Administration: An Historical and Descriptive Study of the Living Past,* (Burlington, Vermont: Government Research Center, University of Vermont, 1966): 354–357. See also Sherman, et al., *Freedom and Unity,* 377–380.

19. Sherman, et al., *Freedom and Unity,* 379–380.

20. For Vermont's municipal forest legislation, see Public Acts of 1945, No. 86, Public Acts of 1951, No. 74, and Public Acts of 1977 (Adjourned Session), No. 253. See in general 10 Vermont Statutes Annotated, Sections 2651–2655. For appointment of municipal foresters, see Vermont Department of Forests and Parks, "Municipal Forests," in *Biennial Report* (1957–1958): 26–28. See also Robert McCullough, "A Forest in Every Town," *Vermont History* 65 (Winter 1996): 5–35.

21. Laura and Guy Waterman, *Forest and Crag. A History of Hiking, Trail Building, and Adventure in the Northeast Mountains,* (Boston: Appalachian Mountain Club, 1989): 353–373.

22. Waterman, *Forest and Crag,* 353–373

23. Waterman, *Forest and Crag,* 353–373, see esp. 362.

24. Christopher Klyza and Stephen C. Trombulak, *The Story of Vermont. A Natural and Cultural History* (Hanover, New Hampshire: University Press of New England/Middlebury College Press, 1999): 94–95.

25. See drawing for a "Game, Food, and Cover Improvement Plan of a Typical Abandoned Farm and Woodland Area at Tinmouth, Vermont," dated February 1934 by Wilbur Bradder. The plan is on microfilm and as available at the Vermont Department of Public Records, Middlesex. See also Bradder's "Forest Type Map. Proctor Town Forest," 1933, available at the Vermont Department of Parks and Recreation.

26. Sherman et al., *Freedom and Unity,* 465–467.

27. National Park Service, *Green Mountain Parkway. Reconnaissance Survey, 1934.* (Washington, D.C.: National Park Service, eight-page printed pamphlet, 1934.)

28. Vermont State Chamber of Commerce. *Col. William J. Wilgus Explains Proposed Green Mtn. Parkway.* Burlington, Vermont: Vermont State Chamber of Commerce, Informational Bulletins on State Problems, Bulletin No. 3, August 28, 1933. See also Hal Goldman, "James Taylor's Progressive Vision: The Green Mountain Parkway," *Vermont History* 63 (Summer 1995): 133–157.

29. Hanna Silverstein, "No Parking: Vermont Rejects the Green Mountain Parkway," *Vermont History* 63 (Summer 1995): 133–157, at 152. See also Goldman, *Taylor's Vision.*

30. Klyza and Trombulak, *Story of Vermont,* 100–101. In 1937, a year after the referendum, Vermont's legislature offered a measure of compromise by authorizing

the State Highway Board to establish a scenic corridor through the state's central region, north to south, relying mostly on existing roads. Portions of the corridor had already been designated as Route 100, and today that highway stretches between the Canadian and Massachusetts borders. However, it is confined to valleys.

31. In 1965, the Vermont Department of Highways joined numerous other state agencies and the U.S. Forest Service in a short-lived attempt to resurrect the Green Mountain Parkway, using nearly an identical route but calling it the Vermont Scenic Parkway. Although enthusiasm for the project soon waned, passage in 1977 of the Scenic Road Law finally recognized the important relationship between transportation corridors for automobile travel and protection of the state's scenic rural landscapes. Legislators gave local citizens and officials the responsibility for designating scenic roads and authorized the state's Scenery Preservation Council, created in 1966, to develop selection criteria and to design maintenance standards.

32. Vermont Commissioner of Forestry, *Biennial Report of the Commissioner of Forestry. 1922–1924*, (Rutland, Vermont: Tuttle Company, 1924): 16. Vermont's first state nurseries were established in 1906 in Burlington and at what became the Charles Downer State Forest in Sharon. Public Laws of Vermont (1915), No. 24, Section 1–5, enabled towns to establish municipal forests.

33. Sherman, et al., *Freedom and Unity*, 446.

34. Sherman, et al., *Freedom and Unity*, 447–455.

35. Sherman, et al., *Freedom and Unity*, 447–455. See also Charles H.W. Foster, *Experiments in Bioregionalism. The New England River Basins Story*, (Hanover: University Press of New England, 1984): 143–145.

36. Nuquist, *Vermont State Government*, 370–374.

37. Nuquist, *Vermont State Government*, 370–374.

38. Department of Environmental Conservation, Vermont Water Quality Standards § 1–02 D1 and 4. Montpelier, Vermont: Agency of Natural Resources.

39. Nuquist, *Vermont State Government*, 337–351.

40. Nuquist, *Vermont State Government*, 337–351.

41. Nuquist, *Vermont State Government*, 337–351.

42. Beret Halverson and Jim Flint. *Patchwork. Stories of Gardens and Community in Burlington, Vermont*. (South Burlington, Vermont: Community Works Press, 2005, for the Friends of Burlington Gardens.) See also Sam Bass Warner, *To Dwell Is to Garden. A History of Boston's Community Gardens*. (Boston: Northeastern University Press, 1987.)

43. Vermont Department of Highways, *Immediate Economic Benefits Resulting from the Construction of the Interstate System in Vermont*. Montpelier, Vermont: Vermont Department of Highways in cooperation with the Bureau of Public Roads, 1961. Typewritten report. See also Vermont Department of Highways, *Vermont's State Highway Needs and 12-Year Construction Program*. (Montpelier, Vermont: Vermont Department of Highways, 1960.) Printed pamphlet.

44. Vermont State Highway Board, *Sixteenth Biennial Report* (1951 to 1952).

(Montpelier, Vermont: Capital City Press, 1952.) 37 (mention of interstate highways). See also Vermont State Highway Board, *Eighteenth Biennial Report* (1954 to 1956). (Montpelier, Vermont: Capital City Press, 1956.) 14–17.

45. Vermont State Highway Board, *Nineteenth Biennial Report* (1956 to 1958). (Montpelier, Vermont: Capital City Press, 1958.) 26–31.

46. American Association of State Highway Officials, *A Policy on Design Standards. Interstate System.* (Washington, D.C.: American Association of State Highway Officials, July 12, 1956.) Revised April 12, 1963. See also State of Vermont, Agency of Transportation, *Vermont State Standards for the Design of Transportation, Construction, Reconstruction, and Rehabilitation on Freeways, Roads, and Streets.* (Montpelier, Vermont: Vermont Agency of Transportation, 1997.)

47. Laws of Vermont (1975), No. 109. See also 22 V.S.A., Chapter 14, Sections 701 et seq. See also Nuquist, *Vermont State Government*.

48. 36 V.S.A., Chapter 331, Sections 7676–7698; 36 V.S.A., Chapter 332, Sections 8338–8358 and 8395–8398. See also Bonus Act of 1958, 23 U.S.C. Chapter 1, Section 131, and Federal Highway Beautification Act of 1965, Public Law 89–285 (October 22, 1965), 79 U.S. Statutes, Section 1028. For historic district laws in Charleston, South Carolina, and New Orleans, Louisiana, see Charles B. Hosmer, Jr., *Preservation Comes of Age: From Williamsburg to the National Trust. 1926–1949*. (Charlottesville: University of Virginia Press, 1981.)

49. Vermont's off-premise billboard law is found at 10 V.S.A. Chapter 21, Section 481, et seq.

50. See 10 V.S.A. Chapter 21, Section 488 and 486.

51. Allen Fonoroff, *The Preservation of Roadside Scenery Through the Police Power*. (Montpelier, Vermont: Central Planning Office, 1966): frontispiece, not paginated; 3; 9; and 19.

52. Legislative Council of the State of Vermont, *Report of the Committee to Study Outdoor Advertising (Proposal No. 21)*. (Montpelier, Vermont: Legislative Council of the State of Vermont, 1967): 2.

53. For the requirements of Vermont's Tourist Information Services Law, see specifically 10 V.S.A. Chapter 21, Section 481 et seq. See also Richard Brooks and Peter Lavigne, "Aesthetic Theory and Landscape Protection: The Many Meanings of Beauty and their Implications for Design, Control, and Protection of Vermont's Landscape," in *U.C.L.A. Journal of Environmental Law and Policy* 4 (1985): 129–172.

See also David A. Carson, "Billboard Blight: Is the Aesthetic Quality of Vermont's Landscape in Jeopardy after Metromedia?" in *Vermont Law Review* 9 (1984:2): 341–371; Harvey K. Flad, "Country Clutter: Visual Pollution and the Rural Roadscape," in *Annals of the American Academy of Political and Social Science, No. 533* (1997): 117–129; Charles F. Floyd, "Requiem for the Highway Beautification Act," in *Journal of the American Planning Association* 48 (1982:4): 441–453; Charles F. Floyd and Peter J. Shedd, *Highway Beautification: The Environmental Movement's Greatest Failure*. (Boulder, Colorado: Westview Press, 1979); John A. Jakle and Keith A. Sculle, *Signs in America's Auto Age: Signatures of Land-*

scape and Place. (Iowa City: University of Iowa Press, 2004); and William H. Whyte, *The Last Landscape*. (Garden City, New York: Doubleday & Co., Inc., 1968.)

54. Charles F. Floyd, "Requiem," 441–453; and Floyd and Shedd, *Highway Beautification*, xii–xiv.

55. D. H. DeHayes, P. G. Schaberg, G. J. Hawley, G. R. Strimbeck, "Acid Rain Impacts on Calcium Nutrition and Forest Health," *Bioscience*, 49 (1999:10): 789–800.

56. Vermont Department of Forests, Parks and Recreation, *Camel's Hump State Park*. Last retrieved 9/11/06 from www.vtstatepraks.com/htm/camels. See also H. W. Vogelmann, *Natural Areas in Vermont: Some Ecological Sites of Public Importance, Report 1* (Burlington, VT: Vermont Resources Research Center, Vermont Agricultural Experiment Station, University of Vermont, 1964); H. W. Vogelmann, *Vermont Natural Areas, Report 2*. (Montpelier, VT: Central Planning Office and Interagency Committee on Natural Resources, 1969); H. W. Vogelmann, "Catastrophe on Camel's Hump," *Natural History*, 91 (November 1982): 8–14; R. Mello, *Last Stand of the Red Spruce* (Island Press: Washington D.C., 1987); and H. W. Vogelmann, G. Badger, M. Bliss, and R. M. Klein, "Forest Decline on Camel's Hump, Vermont," *Bulletin of the Torrey Botanical Club*, 112 (1985): 274–287.

57. T. G. Siccama, M. Bliss, and H. W. Vogelmann, 1982, "Decline of Red Spruce in the Green Mountains of Vermont," *Bulletin of the Torrey Botanical Club*, 109 (2): 162–168. See also H. W. Vogelmann et al., *Catastrophe*, 985.

58. Vogelmann, *Catastrophe*, 1982. See also F. H. Bormann, "Air Pollution Stress and Energy Policy," in Carl Reidel, ed., *New England Prospects*. (Hanover, New Hampshire: University Press of New England, 1982): 85–140. See also G. E. Likens, F. H. Bormann, "Acid Rain: A Serious Regional Environmental Problem," *Science*, 184 (1974): 1176–1179; G. E. Likens and T. J. Butler, "Recent Acidification of Precipitation in North America," *Atmospheric Environment*, 15 (1981:7): 1103–1109.

59. Library of Congress. (2006). *Acid Precipitation Act Cosponsors*. Last retrieved 9/11/06 from http://thomas.loc.gov/egibin/bdquery/D?d096:33:.temp/~b KCKC:@@@P. See also Environmental Proctection Agency. *Acid Rain in New England, a Brief History*. (Washington, D.C.:, G.P.O., 2006.) Last retrieved 9/11/06 from http:www.epa.gov/ne/eco/acidrain/history.html.

60. Vogelmann, *Catastrophe*; and Mello, *Last Stand*.

61. Mello, *Last Stand*. See also Vermont Agency of Natural Resources, Water Quality Division. 2006. *Acid Rain: The Vermont Perspective*. Last retrieved 9/11/06 from www.anr.state.vt.us/dee/waterq/bass/htm/bs_acidrain-vt.htm.

62. DeHayes et al., *Acid Rain*.

63. Vermont Planning Council, *Vermont's Future: Vision and Choice, the Vermont State Framework Plan*. (Montpelier, Vermont: Vermont Planning Council, 1968.) Act 250 is found at 10 V.S.A. Chapter 151, Section 6001 et seq.

64. 10 V.S.A. Chapters 151, Sections 6041–6047; 6081–6092; and 6086. For a discussion about the state's failure to enact a statewide land-use plan, see David

G. Heeter, "Almost Getting it Together in Vermont," in *Environmental and Land Controls Legislation*, D. R. Mandelker, ed. (New York: Bobbs-Merrill Company, Inc., 1976): 323–391. See also Richard O. Brooks, *Toward Community Sustainability: Vermont's Act 250*. (South Royalton, Vermont: Serena Press, 1997); specifically Vol. 2, *The History, Plans and Administration of Act 250*.

65. Vermont Planning Council, *Vermont's Future*, 45 and unpaginated preface.

66. Vermont Planning Council, *Vision and Choice*, 4–21, 33, and 34.

67. Vermont Planning Council, *Vision and Choice*, 37 and 39–42.

68. For Act 250's controls on subdivision and development, and the need for a development plan, see David Heeter, *Getting it Together*. For discussion regarding the protection of high-elevation areas, see Hubert Vogelman, James W. Marvin, and Maxwell McCormack, *Ecology of the Higher Elevations in the Green Mountains of Vermont: Report to the Governor's Commission on Environmental Control*. Burlington, Vermont: Vogelman, et al., *Vermont Natural Areas, Report 2*, at 6. See also 10 V.S.A., Chapter 151, Sections 6001(3); and 6086(a)(1)(A).

69. David Heeter, *Getting it Together*, 372. See also Vermont Environmental Board, *Act 250: A Guide to Vermont's Land Use Law*. Montpelier, Vermont: Vermont Environmental Board, n.d, 3 – 4.

70. 4 V.S.A. Chapter 27, Section 1001 et seq. For a discussion regarding creation of Vermont's Environmental Court, see Matthew Witten, "Administrative Enforcement Still on Hold," in *Environmental Monitor* 1 (1990:3): 1–2.

71. Brooks, *Toward Community Sustainability*, xi–6.

72. 10 V.S.A. Chapter 15, Section 301 et seq.

73. J. Libby, "The Vermont Housing and Conservation Trust Fund: A Unique Approach to Affordable Housing," *Clearinghouse Review*, 23 (1990:10), National Clearinghouse for Legal Services, 1275–1284. See also D. Bradley, and J. Libby, "Vermont Housing and Conservation Board: A Conspiracy of Goodwill among Land Trusts and Housing Trusts," in C. Geisler and G. Daneker, eds., *Property and Values: Alternatives to Public and Private Ownership*, (Washington DC: Island Press, 2000): 336; P. M. Dennis, "A State Program to Preserve Land and Provide Housing: Vermont's Housing and Conservation Trust Fund," in E. Endicott, *Land Conservation through Public/Private Partnerships*, (Washington D.C.: Island Press, 1993); and G. Hamilton, "Report of the Governor's Commission on Vermont's Future: Guidelines for Growth," Montpelier, Vermont, 1988. See also Vermont Housing and Conservation Board, "Vermont Housing and Conservation Board: Mission and History." Last retrieved 9/11/06 from www.vhcb.org/Mission.

74. Laws of Vermont (1912), No. 40, *An Act Relating to the Taxation of Young Timber*; and No. 41, *An Act to Exempt Reforested Lands from Taxation and to Promote the Growth and Maintenance of Wood and Timber Lots*. See also:32 V.S.A. Chapter 124, Section 3751 et seq; Vermont Agency of Natural Resources, *Use Value Appraisal Program Manual*, (Waterbury, VT: Vermont Department of Forests Parks and Recreation, 2006); and T. Daniels, *Vermont's Current Use Program: An Evaluation of Policy Options and Their Land Use Implications*, (State of Vermont, Joint Fiscal Office Reports, 2002.)

75. 32 V.S.A., Chapter 236, Section 10001 et seq.

76. Northern Forest Lands Council, *Finding Common Ground: Conserving the Northern Forest.* (Concord, New Hampshire: Northern Forest Lands Council, 1994.)

77. B. Whittaker, "Natural Resources" in Michael Sherman, ed., *Vermont State Government since 1965*, Burlington Center for Research on Government and the Snelling Center for Government, University of Vermont, 1999: 485–508. A sweeping reorganization of the Agency of Natural Resources is currently underway.

78. D. Haight and C. Ginger, "Trust and Understanding in Participatory Policy Analysis: The Case of the Vermont Forest Resource Advisory Council," *Policy Studies Journal* 28 (1999:4), 739–759.

79. Vermont Agency of Natural Resources, *Lands Conservation Plan: A Guide to the Acquisition, Exchange, and Disposition of Agency Lands*, (Waterbury, VT: Vermont Agency of Natural Resources, 1999.) See also J. M. Hurly, C. Ginger, and D. Capen, "Property Concepts, Ecological Thought, and Ecosystem Management: A Case of Conservation Policymaking in Vermont," *Society and Natural Resources*, 15 (2002): 295–312.

80. Larry Anderson, "The View from Breadloaf," in *Wilderness* (Spring, 1993), 11–19. The New England Wilderness Act was signed into law on December 1, 2006.

81. Senate Bill 142, *An Act relating to creation of designated growth centers and downtown tax credit program*; and Senate Bill 165 (the tax incentive and economic development bill), were signed into law by Governor James Douglas in May 2006.

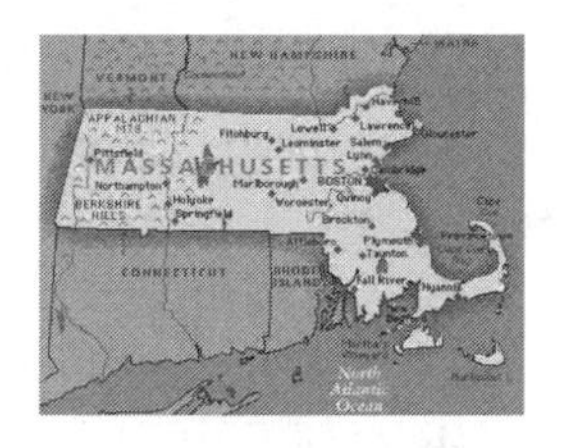

ENVIRONMENTAL CONSERVATION IN MASSACHUSETTS

A TWENTIETH-CENTURY OVERVIEW

CHARLES H.W. FOSTER

THE CONCEPT OF ENVIRONMENT

Taken from its most immediate French roots, *en* (in) and *viron* (circle), the environment is literally that which lies within a circle. It is the sum total of all that surrounds a particular entity. In such cases, it includes not only the physical setting, but also the systemic and even human factors (economic, political, social) that affect an entity's development and well-being. Strictly speaking, environment has no meaning in and of itself — only as it relates to a particular entity or a particular place (e.g., the environment of the right whale in Massachusetts Bay).

That said, the word environment has been popularized to take on more of a descriptive meaning. Despite the above, environmentalists are perceived to be those who have a concern only for the portion of the environment that is natural. What is called environmentalism has become both the theory and the practice of providing for these concerns in policies, programs, and activities.

Card-carrying environmentalists are to be found in governmental agencies and nonprofit organizations — at local, state, regional, national, and even international levels. But environmentalism is also practiced individually through personal activities (e.g., recycling) and acts of proper land and resource management (e.g., stewardship).

Environmental enthusiasts are likely to encounter a profusion of terms. Natural resources, for example, are generally deemed to be materials occurring in nature that are potential sources of use or wealth. The classic natural resources are soil, water, forests, and wildlife (including fish). They are often categorized by location (e.g., inland and coastal resources) or by type (e.g., forest resources). Important resources of a nonmaterial nature, such as air and energy, have been added to the natural resource list in recent years. It is important to remember that nature herself rarely honors such distinctions. Natural resources are invariably interrelated and often interactive. A good environmentalist, therefore, is concerned with the environment as a whole.

Ecology is another commonly misused word. Despite its popular distortion, ecology remains simply a science concerned with the relationship between organisms and their environment (their *ecos* or home). Those relationships occur within defined ecosystems. Ecosystem management is what ecologists use to safeguard the health and integrity of such systems both now and for the future. An ecologist is an environmentalist, but an environmentalist is not necessarily an ecologist.

Older environmentalists often prefer to speak of conservation, the management and wise use of natural resources. Again, thanks to popular distortion, because of their willingness to accept the use of resources, conservationists are often labeled as destroyers rather than savers of the environment, a distinction that is more perceived than real. In a state like Massachusetts, the material use of resources has been an accepted practice for centuries.

THE PRACTICE OF ENVIRONMENT

State and federal natural resources managing agencies, such as the U.S. Forest Service and the National Park Service, have been in existence for more than a century. Since the early 1970s, protecting the environment has become one of the added responsibilities of government. For example, the United States Environmental Protection Agency (EPA) was created in 1970, the United Nations Environment Programme (UNEP) in 1972, and the Massachusetts Executive Office of

Environmental Affairs (EOEA) in 1974 (renamed the Executive Office of Energy and Envirnomental Affairs in 2007). These and other agencies carry out (or oversee) programs for the development, use, and protection of the environment.

In Massachusetts, similar governmental units are authorized at regional or municipal levels. For example, the state now has sixteen conservation districts organized along county lines. Since 1957, any city or town has been permitted to establish a conservation commission charged generally with the oversight and coordination of activities relating to natural resources. All 351 cities and towns now have at least one official environmental agency.

So pervasive has the concept of environment become that special procedures are now required for any public project affecting the environment. A National Environmental Policy Act (NEPA) and a companion Massachusetts Environmental Policy Act (MEPA) set forth the requirements. Under MEPA, for example, any proposed action significantly affecting the environment is subject to a formal environmental impact report describing its environmental effects, the measures being taken to minimize environmental damage, any adverse effects that cannot be avoided, and reasonable alternatives to the proposed action. These reports must be made available to the public.

Yet, despite the profusion of environmental procedures and policies, there is no accepted legal definition of the environment itself. Agencies tend to have generalized mission statements (e.g., the EPA's is "to protect public health and safeguard and improve the natural environment — air, water, and land — upon which human life depends"). In other cases, an agency's mission is determined by its authorized programs and activities (e.g., the EOEA's twenty-nine statutory functions). In these cases, the environment tends to be defined by what needs to be done rather than what it is.

Environmental responsibilities are constantly being modified as new issues arise and new legal requirements and programs are added. Confusing though this may seem to be, it is only symptomatic of the environment's essentially dynamic nature — a process of constant change.

With these preliminary thoughts in mind, it would be timely to turn to the attributes of the Massachusetts environment and the steps taken to ensure its well-being.

THE NATURE OF MASSACHUSETTS

Massachusetts is one of the thirteen original British colonies. It was the sixth to ratify the US Constitution. Influenced by the famous Mayflower Compact, drawn up by the sea-weary Pilgrims in Provincetown Harbor on November 11, 1620, Massachusetts has preferred to call itself a commonwealth rather than a state, a matter of historic rather than legal significance. The preamble to its state constitution, believed to be the oldest written document of its kind still in use, makes this distinction by calling for a body politic formed by voluntary association and characterized by mutual covenants for the common good. Always freethinkers, Massachusetts citizens have developed a way of life built upon a blend of tradition and innovation.

Viewed environmentally, Massachusetts is like a miniature map of the United States. To the west lie the scenic and resource rich Berkshire and Taconic Mountain ranges, the equivalents of the western Sierras and Rocky Mountains. Massachusetts' Mississippi is the Connecticut River, rising on the Canadian border and traversing four states before its confluence with Long Island Sound 280 miles from its point of origin. East of the Connecticut Valley are fertile uplands that slope gradually to the state's 1,500 mile Atlantic shoreline.

Only 5.3 million surface acres in extent, Massachusetts ranks forty-sixth among the fifty states in land area. But with more than six million in population, it is the third most densely populated state in the nation. Despite an extensive state forest and park system, and similar tracts held by municipal and nonprofit agencies, three out of every four acres is still in private ownership.

From the air, Massachusetts appears to be mostly woods and water. Two-thirds of its land area is currently in forest. As for water, there are almost one million acres of salt and fresh water within Massachusetts— some 70 percent lying within the state's extended coastal jurisdiction. Twenty-eight riverine or coastal basins drain its upland, flowing typically from north to south as a result of the great continental glacier that overspread the region some 10,000 years ago. Deposits of sand and gravel left behind by the ice sheet are today sites of important groundwater sources, such as the sole-source aquifer that underlies much of Cape Cod. Water is renewed annually by some forty-four inches of precipitation received evenly throughout the year. Given the fact that

Massachusetts extends farther into the Atlantic Ocean than any other state except North Carolina, it is subject to episodes of intense storm activity (e.g., hurricanes) that further replenish its water capital virtually annually.

The Massachusetts landscape has been described as a patchwork of widely varied terrain. Its western mountain ranges extend generally northward, culminating in the 3,491 foot Mount Greylock near the Vermont–New York border, a peak Herman Melville affectionately termed "The Most Excellent Purple Majesty." These Berkshire and Taconic ranges are of great antiquity. Fossil remains found at the top of the mountains bespeak the presence of a primeval sea perhaps a half billion years ago.

An exception to the generally north-south orientation is the Holyoke Range of the mid-Connecticut Valley, which runs roughly east and west. Formed some 200 million years ago during volcanic activity, the region once hosted perhaps 150 species of dinosaurs, whose tracks can be seen even today embedded in the sandstones and shales. A mere 10,000 years ago, glacial Lake Hitchcock, extending some 160 miles north from Rocky Hill (CT) to Lyme (NH), left its own mark on the Connecticut Valley, depositing copious quantities of clay and silt and leaving dunes fifty feet high as the waters drained and the westerlies blew sand from the no longer submerged lakebed.

Still another prominent landform occurs in central Massachusetts —the series of abrupt elevations the Indians called *monadnocks*. These are isolated peaks composed of rocks resistant to erosion. Although their elevation above sea level is quite modest, their relief (distance above the surrounding eroded plain) can be quite striking. 2,000 foot Mount Wachusett in Princeton (Worcester County) is a case in point.

And certain portions of Massachusetts are marked by what are called *drumlins*—small, rounded hills, oriented in the northeast-southwest direction of the glacier, that are composed of resistant rock capped by deposits of clay and gravel—what one enterprising Harvard freshman defined as "a hill of a lot of till" (as reported by geologist Father James W. Skehan of Boston College). The most famous of these in the Boston area are Bunker and Beacon Hills, which are actually a combination of drumlins and glacial moraines. Unaware of the ecological implications, our pragmatic ancestors set to work to use these formations as fill in order to convert the surrounding marshes to buildable land.

Around all of these high elevation points occur landscapes of both variability and versatility. For example, Massachusetts stands at the vegetative transition zone of New England's broad-leaf and coniferous tree species. Some ten large ecological regions have been identified; at least twenty-seven subregions have been described. In what the Massachusetts Audubon authors characterized as its principal "rooms," the "nature of Massachusetts" occurs in four major regions (the sea and shore, the central peneplain, the Connecticut River Valley, and the western highlands) and in twenty-three major habitat types.

Encountering all this abundance, the colonists wasted little time in putting the Massachusetts environment to work. Forests were cleared, communities were established, and fish and wildlife were used to sustain settlements until the land began producing agricultural crops. The native population of perhaps 100,000 Indians in southern New England in the early 1600s was displaced rapidly by waves of colonists from England, increasingly restive about religious intolerance in their homeland and anxious to make a fresh start in the New World. They brought with them commitment, experience, ingenuity, and skill. By the time of the American Revolution, Massachusetts was an acknowledged political and economic force in the struggle for independence.

Yet, before the seventeenth century came to a close, the new republic would already be taking steps to conserve its resources. For example, the Plimoth and Massachusetts Bay colonies, as early as 1650, enacted ordinances to regulate forest harvesting and control the use of fire. Water resources-related conservation and use commanded priority attention—especially in the so-called Great Ponds (natural bodies of water ten acres or more in extent) and the coastal foreshore (the area lying between the high and low tide marks). Measures to manage and allocate the fisheries are among the oldest laws on the statute books (for example, the General Court passed an act regulating the mackerel fishery even before it established courts of law in the new commonwealth). Massachusetts decreed a closed season on deer in 1694 and required local communities, in 1739, to enforce the law through local deer wardens. And the growing concern for proper soil conservation practices was reported by such travelers as Yale University President Timothy Dwight, who visited New England during the nineteenth century, and *Farmers Monthly Visitor* journal editor, Isaac Hill.

As we turn to the particulars of conservation, it would be wise to remember the characterization of Massachusetts offered by the editors of *The Historical Atlas of Massachusetts* (Richard W. Wilkie and Jack Tager) — "a place where tradition thrives in the midst of innovation, and continuity persists in the midst of change."

EARLY EVENTS AND EFFECTS

During the eighteenth century, the fledgling commonwealth moved rapidly to expand its agricultural base. As the land was cleared and settlements expanded, the practice was increasingly one where the ownership of land became individual rather than communal. At first, each holding was devoted exclusively to subsistence. Later, commonwealth farmers were able to use and exchange their surpluses to meet the growing demands societally for fuel and wood materials, meat and dairy products, and specialty crops such as tobacco, thereby steadily increasing their individual net worth and freeing up funds for investment. However, due to the lack of transportation, these markets remained primarily local.

By 1860, nearly 70 percent of the land statewide had been cleared and was in agricultural use. But with the development of roads, canals, and railroads in the early to mid-1800s, supplemented by the nearly five hundred miles of municipal street railways in place by 1890, Massachusetts farmers were at last free to abandon their farms and either move their families into the developing factory towns or head west to the promise of more extensive lands and markets. The effect on Massachusetts's land and natural resources would be profound.

As the Harvard Forest has so ably documented in its extraordinary dioramas, the forest rebounded with a vengeance, reseeding the abandoned farms with largely white pine, a species ideally suited to reclaiming cleared land and a tree crop materially different from the mixed and multiaged forest that had been there before. Wildlife began to return, especially the large, broad-ranging species such as deer, moose, and bear and the habitat-specific beaver, fisher, and wild turkey. By the turn of the twentieth century, nearly two-thirds of the state would be back in forest, a remarkable reversal in barely half a century that would trigger a massive new period of forest harvesting and utilization during the early 1900s. Ironically, the subsequent overcutting and accompanying burning of the forest would seed another important crop — conservation.

As if human influences were not enough, Massachusetts would be ravaged by a series of natural disasters during this era. Hurricane storms would leave a legacy of floods and devastation. The 1938 hurricane, for example, blew down some three billion board feet of timber in north-central Massachusetts; the 1927, 1933, 1944, and 1955 storms breached dams and stimulated an era of concern for water resources and flood control throughout the state and region. Although man introduced, Massachusetts forests would be ravaged by insects and disease — for example, the gypsy moth, the chestnut blight, and the Dutch elm disease. Each disaster would contribute in its own way to the state's developing conservation ethos — for example, greater concern for shade trees and their community values, attention paid to the economic potential of other tree species, such as oak, pine, and red maple and, in the case of the gypsy moth, a statewide effort to control the use of pesticides.

Conservation would also be affected politically, socially, and economically by the forces of change transforming the nation as a whole. For example, international conflict would leave its mark on Massachusetts more than once — the Revolutionary War and its heritage of self-reliant independence; the War of 1812 that diverted leadership and capital away from overseas commerce to home-grown, mercantile development; and World Wars I and II with their material demands, yet their eventual transformation of the Massachusetts economy away from manufacturing to the modern industries of finance, high-tech, and education that service us today.

On the social front, the state would go from a society of largely English Europeans to a flood of immigrants who now outnumber the traditional Yankees by a margin of twelve to one. One effect of this diversity has been the changing attitudes and relationships with respect to land and natural resources — less concern with its physical productivity than its recreational, esthetic, and ecological attributes. How land is used, and under what constraints it should be owned, have become the paramount conservation issues today.

Periods of social stress, advocacy, and reform would also leave their mark on the Massachusetts conservation scene. These have been manifest in the founding of the state's respected medical, educational, governmental, and nonprofit institutions. As an example, most of the governmental conservation institutions we now rely on were not in place at the turn of the twentieth century. They were stimulated first by concerns over the condition of the resource and later by well-connected private organizations pledged to their creation. Once started, however, the proliferation of governmental agencies proceeded at a steady pace — so much so that in 1918 a constitutional convention was called to authorize the consolidation of governmental functions into no more than twenty departments. It was at this time that the state's first Department of Conservation was created. Yet, as the twentieth century dawned, the conservation initiative remained firmly in individual and private hands. Government was there to help but was far from the dominant force it is today. Even when agencies were created, the practice was to place them under the supervision of prominent individuals.

A good example was agriculture, one of the state's earliest land conservation related agencies (1852), overseen by an unpaid board of agriculture composed of the chairmen of the county agricultural societies founded in the 1790s. At that time, virtually every landowner could claim to be a farmer. Those nearest to Boston were what Tamara Plakins Thornton has called "cultivating gentlemen," members of the Brahmin and mercantile elite emulating the British tradition of landed gentry. Yet, Massachusetts had also become the third state to accept the provisions of the federal Morrill Act and to found an Agricultural College (1867). Thus, the tradition of the industrious, yeoman farmer, needful of extension services in crop growing, animal husbandry, and soil conservation was alive and well despite its peculiar congruence with those for whom the ownership of land was primarily a symbol of wealth and power.

The latter, as early as 1792, had formed the private Massachusetts Society for Promoting Agriculture to provide educational assistance and offer premiums for the successful practice of agriculture. So extensive was its influence base that the MSPA's legislative charter was signed by none other than Governor John Hancock and its first meeting, held appropriately on April 19, 1792, was chaired by Revolutionary War firebrand Samuel Adams.

Another example was the state's fisheries and wildlife agency. In 1865, the Commissioners of Inland Fisheries were constituted by the legislature with two unpaid citizens serving as board members. Two issues drove its formation. One was the need to address impediments to

Internationally-renowned physician-scientist John Charles Phillips of Wenham, MA, president of the Massachusetts Fish and Game Association and the Massachusetts Conservation Council, and an early leader in the 1920s of the New England wildlife conservation movement, is shown dressed for a day's bird shooting in northern New England's rugged second growth coverts.
(JOHN CHARLES PHILLIPS II)

anadromous fish runs (e.g., shad and salmon) due to the construction of industrial mainstream dams; the other the new concept of fish culture making its way from Europe to the United States. Functions related to game were added in 1886. Here again, prominent individuals and established private organizations kept a watchful eye on governmental practice.

For example, Civil War veteran and businessman Theodore Lyman III served as a Massachusetts fish commissioner and was the first chairman of the New England Fish Commission. Physician-sportsman John C. Phillips, in later years, was similarly instrumental for wildlife. His Massachusetts Fish and Game Association, the oldest incorporated conservation organization in the United States (1873) and a veritable Who's Who of the Massachusetts elite, provided the institutional advocacy. The turn of the century was what the association, in its institutional history, characterized as "that lush period of American life when men took what they could get of the generous resources of a still abundant land."

And outdoorsmen of another sort would organize and come to play a significant role in the impending conservation movement. Inspired by the British tradition of walking in the countryside, hiking clubs began to appear as early as 1850 (e.g., Cyrus Tracy's Exploring Circle in Lynn), characterized by Robert L. McCullough as organizations that encouraged "walking, climbing, exploring, and observing nature." By 1876, a Northeast-wide Appalachian Mountain Club would be formed for the dual purposes of "geographical exploration" and recreation. Over time, inspired by such citizen leaders as businessman-philanthropist Thomas D. Cabot and Boston attorney William A. King, and staff professionals C. Francis Belcher and Thomas S. Deans, the AMC developed a formidable conservation presence. Much of this came about from simple proximity, for office space was made available in the club's buildings on Joy Street for such citizen conservation organizations as the Massachusetts Forest and Park Association, the Conservation Council, the Conservation Law Foundation, and the Environmental League of Massachusetts.

EARLY TWENTIETH-CENTURY CONSERVATION

Despite these established entities, one of the first major land conservation initiatives of the new century would be concerned with the seemingly perilous status of the Massachusetts forest, beset as it was by the effects of overcutting and fire. As William A. King has written, two prominent Bostonians, looking out the train window on a bleak December day in 1897 as they traveled to their downtown offices, remarked how ragged the woods looked. There and then they were inspired to form the Massachusetts Forestry Association to "introduce judicious methods in dealing with forests and woodlands, arouse and educate a public intent, promote the afforestation of unproductive lands, and encourage the planting and care of shade trees . . ." For the first third of the ensuing twentieth century, the later-renamed Massachusetts Forest and Park Association would be the dominant force in the state for land conservation.

An early target for attention was western Massachusetts's Mount Greylock where commercialization and exploitation had already laid bare the east slope and was impending to the north, thereby threatening the state's highest peak and the inspiration it provided for writers, artists, and the residents of surrounding communities. By the end of its first year of existence, the fledgling association had managed to persuade the legislature to authorize the creation of a Mount Greylock state reservation, the first unit of what would ultimately become the eighth largest state forest and park system in the United States.

Under the leadership of its remarkable executive director, Harris A. Reynolds, the MFPA would go on to advocate the creation of the office of state forester (1904), the reforestation of private lands (1908), and the establishment of a State Forest Commission (1914) to begin assembling units of a new state forest system. Now fully credible and accomplished, the MFPA would also prove instrumental in reducing and reorganizing the number of state conservation agencies and achieving the creation of the commonwealth's first consolidated conservation agency, appropriately named the Department of Conservation.

In the hands of another influential group of Boston Brahmins was a companion conservation seed ready to be planted, one predicated upon the growing interdependence of human health and well-ordered human communities. As Gordon Abbott, Jr. has chronicled, twenty-four-year-old Charles Eliot, an apprentice in the office of the renowned landscape architect and planner Frederick Law Olmsted, began publicly advocating a role for parks and open space in improving the physical and psychological well-being of urban residents. What was needed, he said, was an incorporated association to acquire and hold "well-distributed parcels of land free of taxes, just as the public library holds books and the art museum pictures — for the use and enjoyment of the public."

On May 21, 1891, after prodding by such luminaries as State Board of Health Chairman Henry Wolcott, Boston attorney George Wigglesworth, and business leader Charles Francis Adams, the Trustees of Public Reservations — now simply The Trustees of Reservations (TTOR) — received its legislative charter. TTOR would become the first land trust in the United States and the model for others throughout the world.

In a third corner of patrician Boston, two redoubtable ladies would start another important conservation organization — the first of the national family of Audubon societies. Harriet Lawrence Hemenway and Minna B. Hall, as historian Stephen Fox has observed, came from a heritage of reform. The high fashion of millinery made from bird parts and feathers conflicted with their concern for wild creatures and the natural environment, leading them to spearhead a crusade for bird conservation. The first meeting of what would become the Massachusetts Audubon Society took place in February 1896 in the living room of the Hemenway's Back Bay home. By 1916, the donation of a 225-acre tract of meadow and forest in Sharon adjacent to Moose Hill would be the first unit of what would later become a statewide system encompassing more than 30,000 acres of Audubon sanctuaries and reservations.

But in the early part of the twentieth century, despite the influence of the Theodore Roosevelt administration and the national attention to conservation, this promising movement would fall upon difficult times. World War I interrupted progress toward an integrated, effective program, diverting manpower, resources, and public attention away from conservation toward the exigencies of a global conflict. In 1927, the country underwent a major economic depression. The latter, curiously, did much for land conservation, for the Franklin D. Roosevelt administration, as part of its New Deal for America, elected to create a national Civilian Conservation Corps (CCC) to provide work and vocational training in conservation for unemployed single men.

By July 1933, as William H. Rivers has reported, thirty-one camps had been established in Massachusetts housing more than 6,000 young men and World War I veterans. By the close of the CCC program in 1942, a total of fifty-three camps had been constructed statewide.

Coincident with the forest and recreational development work carried out by the CCC crews, the legislature saw fit to authorize the acquisition of an additional 40,000 acres of state forest land.

Although the CCC era has come and gone, its effects are still evident, manifest in not only the heritage of public forests, parks, and reservations it left behind, but also the conservation ethos that so drives the environmental movement today.

THE WORLD WAR II ERA

Promising though these accomplishments seemed to be, the peculiar rise and fall of fortunes that so often characterize the conservation movement would occur again. In the late 1930s, the conflict in Europe and later the Pacific sector would herald another virtual decade of global unrest and conflict. As America went off to war again in 1941, there would be fewer people at home to advance the conservation agenda, and even fewer resources available to support the programs that seemed so promising. However, those left behind continued to carry on.

In December 1935, the Massachusetts Forest and Park Association presented an initiative petition to the legislature signed by more than 23,000 registered voters proposing the purchase of a half million more acres of state forest lands over the next ten years. In October 1940, the MFPA hosted a national conference in Springfield (MA) to celebrate the Massachusetts-initiated town forest movement, now 1,500 in number, located in thirty-four states, and three million acres in extent.

On another front, the Massachusetts Fish and Game Association persuaded the legislature to authorize a complete revision and codification of the fish and game laws (1929), sponsored the first New England Game Conference (1930), the model for what would later become the North American Wildlife and Natural Resources Conference and, in 1945, helped win legislative approval for the establishment of a five-man administrative Fish and Wildlife Board and the creation of an Inland Fisheries and Wildlife Fund that would dedicate sportsman license fees and associated federal revenues to conservation purposes.

Meanwhile, the Trustees of Reservations' acquisitions were proceeding steadily, aided by its 1929 Report of the Committee on the

Needs and Uses of Open Spaces and the subsequent 1933 Massachusetts Landscape Survey. The latter was a study carried out in cooperation with the state Department of Conservation that identified some seventy specific sites of scenic and historic interest needing to be protected. Although not to be officially authorized until 1980, the so-called Bay Circuit, a circumferential greenway around Boston first proposed in the Trustees' 1891 metropolitan parks report, was included in both the 1929 and 1933 guideline planning documents.

In 1936, as public interest in natural resources continued to increase and be manifest in more and more citizen citizen organizations, conservation leaders took the unusual step of forming a delegate-based, statewide, Massachusetts Conservation Council to coordinate their relationship with each other and to settle on collective representations to government and the legislature. Designed simply as a clearinghouse, and deliberately lacking a competing office and staff of its own, the council was chaired on a rotating basis by its member organizations.

The council offered one important service to all — access to the lobbying capabilities of the Massachusetts Forest and Park Association, functions that most nonprofit organizations were legally barred from carrying out. The council was also recognized by the increasingly powerful National Wildlife Federation as the federation's principal state affiliate, a connection that gave Massachusetts its first voice in the growing debate over federal natural resources policy.

WATER RESOURCES

By 1926, the Commonwealth would begin undertaking what would become the most significant single-water event of the twentieth century, the construction of Quabbin Reservoir in central Massachusetts designed to serve the future water supply needs of Boston and more than forty other cities and towns — what historian Fernon L. Nesson has called the story of "great waters" and biologist Thomas Conuel the creation of an "accidental wilderness."

Identified as early as 1895 as the ultimate Boston water supply source, the diversion of the Ware and Swift river's watersheds in 1939 would eventually require the abandonment of four established towns of great antiquity — Enfield, Dana, Greenwich, and Prescott — and the relocation of six other town boundaries. By legislative act, the Swift

A panorama of changes in the landscape from the 1930s to the 1970s is seen from a common vantage point facing what used to be the town of Enfield on the Prescott Peninsular in central Massachusetts, all brought about by the construction of Quabbin Reservoir, the main water supply for metropolitan Boston. The 75,000 acres of managed forest and water, termed by writer Thomas Conuel an "accidental wilderness", are highly prized today for hunting, fishing, and nature appreciation. Virtually all of New England's landscape today is the product of human-induced change.
(Photos by Les Campbell of Belchertown, MA)

River reservoir would be given an Anglicized version of the Nipmuck Indian name, *Qaben*, meaning the place of "the meeting of waters." The result of the construction would be a thirty-nine square mile reservoir holding more than 400 billion gallons of water, requiring seven years to fill, at a cost of $50 million.

The acquisition of 120,000 acres of reservoir land and contributing watershed would create a haven for wildlife, including rare and endangered species; the largest professionally managed forest in the commonwealth; and a virtual paradise for hikers, naturalists, and fishermen. It would also trigger a twenty-first century debate over the role of forested landscapes in preserving critical ecosystem processes, and the Harvard Forest's subsequent vision of a statewide system of similar wildlands, surrounded by privately owned but development-restricted woodlands, all guided by locally constituted woodland councils.

But Quabbin, the most major diversion in the history of the commonwealth, would also trigger a number of other important water-resource initiatives — among them the 1967 pumped storage project at Northfield Mountain, a combined energy-generating and water-supply project poised to skim flood flows from the Connecticut River and divert them into Quabbin when necessary. Ironically, it would also lead to passage of the 1983 Interbasin Transfer Act and the 1985 Water Management Act, measures that would help guard against special interest projects and subject all future diversions of this magnitude to the prior approval of the state's Water Resources Commission and Department of Environmental Protection.

POSTWAR CONSERVATION

As Massachusetts conservation leaders returned from WW II military service, the favorable conditions of a booming postwar economy and the pent up desire of Americans for outdoor and leisure opportunities brought about an astonishing era of conservation progress. Accustomed to governmental leadership during the wartime years, the public was willing to endorse and accept a wide range of natural resource initiatives under federal auspices.

For example, a national Outdoor Recreation Resources Review Commission, established in 1958 with businessman/philanthropist Laurance S. Rockefeller as its chairman, led the way by proposing a major effort to renovate federal, state, and local park programs and to earmark special revenues to support the expansion of outdoor recreation lands and facilities. Cooperative state and federal action also came into vogue in the early 1950s, with special federal-state agencies established for economic development, planning, and river basin coordination. Those charged with conservation responsibilities found themselves received warmly by the general public.

At the state level, the infusion of federal funds proved irresistible, leading to the establishment of important new federal parks and refuges and the availability of federal grant funds to expand and improve state and local facilities. Such new national units as the Cape Cod National Seashore and the Minuteman National Historic Park and, later, the Sudbury and Silvio Conte National Wildlife Refuges, would come about largely from the combined prodding of state and local interests. Massachusetts conservation leaders were in the forefront of many of these national reforms, but equally eager to advance measures of their own.

For example, as the University of Massachusetts's Andrew J. W. Scheffey has reported, an obscure Massachusetts bill in 1957 authorizing cities and towns to establish conservation agencies of their own, and giving them access to state and federal funds, would lead to a municipal conservation commission movement that would ultimately spread throughout the entire Northeast, from Maine to Pennsylvania. The state's once modest Department of Conservation, reorganized in 1953 as a Department of Natural Resources, would now have one thousand volunteers to help cope with the new responsibilities.

But as the returning veterans began to reacquaint themselves with their natural resources, a new era began — an awakening of concern for the actual condition of the environment. Attention swung away from simply the use of resources to a correction of the abuses that threatened their continued well-being, such as air and water pollution, solid waste disposal, and toxic and hazardous substances. Lobbied by increasingly vocal and effective national conservation organizations, Congress was persuaded to establish major statutes and programs to curb these abuses. Most employed a federal-state approach wherein the federal government would set the standards and provide a measure of funds, leaving it to the states to remediate the actual damage. An independent entity, the Environmental Protection Agency, was created in 1970 to administer these programs. A similar approach governing the use of land was hotly debated but ultimately left to the discretion of the states themselves.

Concern for the environment began to arise in other arenas. For example, the future of the once 1.8 billion-acre public domain in the West was addressed by the 1964 Congressionally authorized Public Land Law Review Commission. The bulk of these lands was ultimately withdrawn from disposal to be managed under the provisions of the Federal Land Policy and Management Act of 1976. The need to ensure a component of permanent wilderness was resolved by the Wilderness Act of 1964. And in 1973, special federal legislation was enacted to safeguard rare, threatened, and endangered plant and animal species in an unusual move to ensure biodiversity protection on private as well as public lands. Easterners, such as those in Massachusetts, would be in the vanguard of these national representations.

Even long-accepted uses of natural resources, such as in agriculture and forestry, began to come under increasing scrutiny for their possible environmental effects. In 1969, by passage of the National Environmental Policy Act, Congress took the unprecedented step of requiring all federal agencies to subject their policies, programs, and projects to comprehensive environmental impact analysis and to make the results public.

MASSACHUSETTS INITIATIVES

In 1959, after one of the first comprehensive assessments in the country, conducted by the private engineering firm of Edwards & Kelcey, a twenty year, $100 million state parks expansion program was authorized by the legislature with the support of the citizen conservation community. The leader in that fight was commissioner of natural resources and future governor, Francis W. Sargent.

Of particular note during this period was the arrival on the scene of a remarkable former insurance executive, Allen H. Morgan, who in 1957 would be asked to take over the operation of the Massachusetts Audubon Society. Having started the Sudbury Valley Trustees two years earlier with the help of six of his neighbors, Morgan was convinced that Audubon itself should move boldly away from its historic preoccupation with birds and play an active role in environment and land conservation. Under his direction, Audubon began aggressively courting gifts of land, adding to its membership support base, expanding its educational efforts in the schools, and reaching out to partner with public agencies.

The first notable collaboration would be aimed at preserving the

Thoreau's House, Concord, MA – Prepared from field sketches made at the Walden Pond State Reservation and commissioned for the 1938 Heritage Press edition of Walden is Thomas Willoughby Nason's wood engraving of the cabin Henry David Thoreau so famously built and occupied in the Concord woods for more than two years during the 1840s. A prominent New England artist, and a friend of Robert Frost, Nason was later characterized by biographer Walter Muir Whitehill as one of two New England poets who shared and expressed the same vision of New England, one through words and the other through wood.
(Boston Public Library)

Thoreau-famed environment of the Sudbury and Concord River valleys. In 1959, Audubon's James Baird would be enlisted to help document the need for preservation of the marshes. Morgan and Massachusetts Commissioner of Natural Resources Charles H.W. Foster would ultimately persuade the U.S. Fish and Wildlife Service and the state legislature to approve an expansion of the small Great Meadows National Wildlife Refuge in Concord northward into Sudbury and Wayland, the first new Massachusetts refuge acquisition in some two decades.

In 1962, the public-private partnership went to work on a larger scale. As evidence mounted of a vital relationship between estuarine marshes and the well-being of the state's recreational and commercial fisheries, Foster took the bold step of seeking special legislation to impose without compensation restrictions on the use of privately owned coastal marshes, an authority that would later be extended to the state's inland wetlands. Foster mobilized the fisheries community to press the legislature and the governor for action. Morgan organized and chaired a special Wetlands Action Committee of the Massachusetts Conservation Council. They would be joined by *Gloucester Times* editor Gordon Abbott, Jr., later to become director of the Trustees of Reservations, to see that a steady drumbeat of supportive press comment and educational materials reached the public.

Prompted by the formation of the Conservation Law Foundation in 1966, a series of legislative enactments important to conservation were proposed and won by a coalition of environmental lawyers working primarily outside of government. Modeled after Michigan's pioneering "Sax Law," an environmental citizen suit statute was enacted in 1971 granting individuals the right to sue project proponents who threatened significant damage to the environment. This right, coupled with the later passage of the Massachusetts Environmental Policy Act (1972), would ensure that government and its citizenry would henceforth work collaboratively to prevent and ameliorate environmental damage.

In 1972, members of the same environmental coalition would win passage of important amendments to the state constitution. The so-called environmental bill of rights would first clarify the intent of the 1918 Constitutional Convention by upholding the right of the commonwealth to control the use of private land in the public interest by acquisition or regulation. It would go on to prohibit the adverse use of lands and easements acquired for conservation purposes without a two-thirds affirmative vote of each branch of the General Court.

Also in 1972, Massachusetts enacted its own Massachusetts Environmental Policy Act, going beyond the federal NEPA provisions to actually require project agencies to avoid or remediate environmental damage.

ENERGY

Although not directly environmental in nature, the deregulation and restructuring of the electric power industry has to be considered a major benchmark in Massachusetts conservation. During the 1960s, for example, frequent power cutbacks and even blackouts caused the regionally interconnected power companies to seek expanded and more reliable generating and transmission capacities. The prospects of augmentation through the import of hydroelectric power from Canada, and additional pumped storage capacity at home, seemed promising. However, the turbulent 1970s produced a number of other shocks such as fossil fuel price increases, high inflation, stricter pollution controls, and the controversy over nuclear power. There would be growing tensions with environmentalists at virtually every turn.

By the 1980s, the trend toward deregulation that had started in California had made its way to the Northeast. New England Electric circulated a plan called Choice New England, negotiated a new approach in Rhode Island, and ultimately stimulated the Massachusetts Department of Public Utilities to devise an electric industry restructuring plan for Massachusetts. Legislation to that effect was adopted by the General Court in 1997. Under the new law, utility monopolies were broken up, consumers were allowed to choose their own electric suppliers, use of new and cleaner energy sources was encouraged, and consumers were given an initial rate cut. A special one-half mill charge on all electric consumers was authorized to be set aside in a Renewable Energy Trust Fund to support the development and promotion of natural gas, solar energy, and wind power.

Challenged by referendum in 1998, not because of the renewable energy provisions but continued use of potentially hazardous and polluting nuclear and coal-fired plants, the law withstood repeal through the efforts of an unusual coalition of energy experts, civic, and industry leaders abetted by a segment of the environmental community. Mainstream conservation organizations such as the Conservation Law Foundation found themselves pitted against such populist groups as the Sierra Club, Ralph Nader, the Massachusetts Clean Water Coalition, and the Massachusetts Public Interest Research Group (MassPIRG). The divisions among environmentalists would late resurface during the contentious debate over the siting of wind generating stations. Energy was certain to become a major, if not the paramount issue for conservation in the twenty-first century.

LAND CONSERVATION

In the meantime, in the early 1960s, a creative young lawyer, Robert Lemire, had begun working on another important conservation front — the conservation of land through private rather than public initiative. The case in point was the hundred-acre Wheeler Farm in Lincoln adjacent to the Revolutionary War road over which the dead British soldiers were carried by cart to the town's graveyard. Without a new approach, the land would simply be subdivided into expensive houselots. With the help of fellow lawyer Kenneth Bergen and farmer/selectman Warren Flint, Sr., Lemire and others created the Rural Land Foundation, Inc. and, in 1965, with guarantees of $10,000 each from prominent members of the community and a loan from Boston's State Street Bank, acquired the entire property. By subsequently carving out and selling just eleven well-sited houselots, the foundation was able to more than recoup its investment and reserve the remainder of the acreage for permanent conservation — "without one cent of public expenditure."

In 1978, an innovative young commissioner of food and agriculture, Frederic Winthrop, set about to accomplish farmland conservation on a statewide basis. He persuaded a reluctant legislature to authorize an Agricultural Preservation Restrictions (APR) program wherein farmers could convey to the state their development rights without having to sell out the entire farming operation. By permitting owners to cash in on a portion of their appreciated land value, farmers were able to access funds for capital improvements and make the agricultural operation affordable for a new generation of young farmers. By

1993, this popular program had managed to secure more than 50,000 acres of the state's most productive farmland.

Intrigued by the Lincoln and APR experiences and other creative applications of the law, a conservation-minded Boston tax attorney, Kingsbury Browne, Jr., in 1980, sought a sabbatical from his law firm to spend a fellowship year at the Cambridge-based Lincoln Institute of Land Policy. At Browne's urging, the Institute subsequently agreed to sponsor an experimental Land Trust Exchange to encourage and facilitate land conservation by local land trusts. From that early beginning, a Washington-based Land Trust Alliance has emerged that now services more than 1,500 land trusts nationwide, including 140 in Massachusetts alone. Local land trusts nationally have reportedly helped save more than nine million acres of land already.

REORGANIZATION

But as of 1969, the state was found to have in place a total of forty-four distinct agencies relating to natural resources and the environment, many reporting directly to the governor. Duplication of functions, overlapping of responsibilities, and operating inefficiences were found everywhere. In bipartisan action, the legislature moved to authorize the creation of a cabinet system of government, including a new Executive Office of Environmental Affairs (EOEA), whose first secretary would be former Commissioner of Natural Resources Charles H.W. Foster, giving each cabinet officer two years to submit a comprehensive reorganization plan for legislative approval.

In historic fashion, Foster and the EOEA turned to its citizen constituency for ideas, creating eight major task forces to help devise the underlying policies. In 1974, the reorganization was approved, the only one of the ten cabinet agencies to be fully reorganized. The resultant agency, would remain in existence, structurally intact, for the balance of the century.

The reorganization act would not only spell out twenty-nine specific program functions, carefully balance management and protection responsibilities but, in some instances, would add to the agency's arsenal of conservation tools and techniques. One such example was the authority to identify and designate Areas of Critical Environmental Concern (ACEC), aspects of the environment that would remain in nongovernmental ownership but be subject to oversight by the EOEA. By 1993, some twenty-five ACEC orders, covering 170,000 acres statewide, had been promulgated. The 1974 reorganization would herald the final phase of Massachusetts's twentieth century environmental conservation history.

THE CABINET ERA

Economist Evelyn Murphy took office in 1975 as the first EOEA secretary following reorganization. She hardly had time to become acquainted with her agency before issues began to surface. The establishment of Boston Edison's Pilgrim nuclear plant on the South Shore, fraught with uncertainties and possible hazards, was one such occasion. Another was the federal government's proposal to drill for oil on Georges Bank. If that was not enough, the *Argo Merchant* managed to spill much of its oil cargo into Buzzards Bay on her watch. Nevertheless, Secretary Murphy was able to undertake two benchmark actions of her own.

One was the establishment of the Office of Coastal Zone Management, an agency that would henceforth guide conservation and development in the state's coastal communities and near-shore waters and preempt independent action by the federal government. Her second accomplishment was to encourage the Metropolitan District Commission (MDC) to fix leaks and halt waste in its water supply distribution system before seeking additional sources of drinking water, an effort that would succeed only when the MDC was replaced by the Massachusetts Water Resources Authority in 1985. In this she had the strong support of a statutory Water Supply Citizens Advisory Committee drawn from individuals opposed to further diversions from the Connecticut River. As it turned out, this redevelopment of existing sources would prove sufficient well into the twenty-first century.

Succeeding Murphy in 1979 would be John Bewick, the third successive secretary with Ph.D.-level credentials. Bewick would set his sights on improved business/environment relations. Among his priorities were the proper siting of hazardous waste processing facilities, an accelerated pace of sewage treatment plant construction, and the environmental amelioration of proposed runway improvements at Logan Airport. Despite what many viewed as the Governor Edward King administration's pro-business bias, Bewick also moved greenly to safeguard barrier beaches, encourage a statewide eagle restoration program, and have the state acquire the unspoiled South Beach on Martha's Vineyard. Even more memorable was his courageous and skillful leadership in winning passage of returnable bottle legislation in the face of relentless opposition from the business community.

Bewick's successor, attorney James S. Hoyte, would be the one to finally set in motion the cleanup of Boston Harbor. After negotiating a pollution settlement with the Environmental Protection Agency and the federal court in 1983, Hoyte would preside over a new Massachusetts Water Resources Authority charged with water supply and waste disposal functions for the entire Boston metropolitan area. As part of that agreement, Hoyte would see to it that the two million dollar federal fine imposed by the settlement remained in Massachusetts under the control of an independent Massachusetts Environmental Trust so that the restoration of Boston Harbor's environment could begin promptly. In addition to advancing his special interest in urban problems, such as affirmative action and environmental justice, Hoyte would also launch a determined effort by the EOEA to accelerate open-space protection, particularly in western Massachusetts.

Taking over from Hoyte as secretary in 1987, and believing that government should lead, not just follow, activist John P. DeVillars would plunge into environmental affairs both literally and figuratively, later diving into the Charles River to illustrate its potential for recreation after cleanup. Among significant firsts would be his advocacy of and involvement in the state's first development cabinet, one that would meet weekly, and his sponsorship of a statewide Youth Environment Corps. DeVillars also became directly involved in the controversial siting of the Third Harbor Tunnel and the Charles River crossing of the Central Artery. His particular concern for the redevelopment of formerly polluted sites in low-income areas would lead him to later found and run a private consulting firm, the Brownfields Recovery Corporation. And his restless, creative leadership would subsequently win him a term as New England regional administrator of the federal Environmental Protection Agency.

Susan F. Tierney, following DeVillars in 1991 as the first of three successive secretaries to serve Republican administrations, would successfully deploy her reputation, contacts, and skills as a management and business consultant. Her first challenge would be to implement initiatives set in motion by her predecessors, such as placing MEPA on a firm footing, spurring the new Massachusetts Water Resources Authority to begin actually remediating pollution, and carrying out the far-reaching environmental agreements for the Big Dig, Boston's bold move to replace its unsightly overhead highway structures with buried roadways and a new Charles River bridge crossing. Tierney would also have to cope with the state and regional provisions of the newly enacted federal Clean Air Act, especially the requirement for a new generation of clean cars.

In the meantime, despite the impediments of the state's unremitting economic downturn, capital funds would have to be sought for the parks, wildlife, and open space initiatives so favored by Governor William Weld and the statewide wetlands protection that needed to be sustained. In the latter case, Tierney would break new ground in invoking the historic waterways authorities under Chapter 91 and in becoming the first secretary to utilize the unique authorities of her office to designate and oversee the use of so-called Areas of Critical Environmental Concern (ACEC). In later years, Tierney would return to commonwealth service as chairperson of the Massachusetts Ocean Management task force, a group that would recommend the nation's first state-based system of ocean zoning.

Tierney's successor, Trudy Coxe, came to the secretaryship in 1995 after leading a successful fight in Rhode Island for the conservation of Narragansett Bay. Transferring those regional interests to Massachusetts, she repositioned her agency to make watershed management its central organizational focus. She would also establish the first statewide wetlands banking program in the nation, an effort to put private development to work creating rather than destroying wetlands. Among her most difficult assignments would be the resolution of a dispute with the now federal DeVillars over an expensive, EPA-mandated, drinking water treatment plant below Wachusett Reservoir that environmentalists felt was less desirable than simply buying up more of the natural watershed.

Addressing what was fast becoming a central policy issue in Massachusetts and the nation as a whole—how to achieve environmentally responsible growth—Coxe found ways to encourage open space-based community planning and, in a particularly creative move, organized a Guns and Roses coalition of sportsmen and environmentalists to win support for a $400 million environmental bond issue.

Just before the turn of the twenty-first century, former state senator and staunch outdoorsman, Robert A. Durand, would become the eighth EOEA secretary. A registered Democrat appointed by Republican Governor Paul Celluci, Durand would seize upon his assignment with unabated enthusiasm, setting in motion a serious commitment to biomapping and biodiversity conservation and the new EIC educational curriculum (Environment as an Integrating Context), sponsoring a series of community based "Bioblitzes," and setting a goal of 200,000 additional acres of land for state acquisition. Before he left office, Durand saw to it that the EOEA received the largest, legislatively approved environmental bond issue ($743 million) in history.

NONGOVERNMENTAL EFFORTS

Lest this account leave the impression that citizen-based conservation had fallen by the wayside during the cabinet era, there is abundant evidence to the contrary. Here are a few examples.

The Connecticut River Watershed Council (CRWC), a broad-based, citizen-led, river advocacy organization headquartered in Greenfield (MA) and founded in 1952, has helped stage a vigorous watershed movement throughout the entire New England region. Like so many nonprofits, the impetus for CRWC came in response to a perceived threat — in this instance, the various proposals for federal authorities to develop public power and spur economic development within the valley. In 1950, members of the New England congressional delegation had pending three separate authority proposals for the Connecticut River, and five legislative measures to study the resources of New England's rivers and watersheds.

The region's investor-owned utilities, concerned about competition from public power, were quick to promise support for the CRWC. A virtual Who's Who of valley business, civic, academic, and environmental leaders was recruited. An eight-point agenda was agreed upon, ranging from improving agriculture to providing better recreation facilities. An award-winning film, *Your Valley, Your Future* was produced and shown to receptive audiences throughout the valley. Underlying the programmatic aspects was the theme expressed by CRWC's New

Hampshire Vice President Lawrence Rathbun: "Democracy had its birth in the Connecticut River Valley. The people of this valley are very much interested in keeping it alive and healthy."

Of particular interest was the assistance given the fledgling CRWC by the established Brandywine Valley Association in Pennsylvania. Clayton M. Hoff, the executive director of the Brandywine, served as an initial professional advisor until the CRWC's first director, Elmer R. Foster, was engaged in December 1952. Hoff's background as the former manager of a large industrial plant in the Brandywine Valley in 1937, and a leading player subsequently in reducing pollution and advancing environmentally responsible development, resonated well with the CRWC founders. New England's proclivity for reaching out to others for guidance served it well in this as well as other instances.

In recent years, the CRWC's prominence as a watershed association has tended to be overshadowed by others — the Farmington and Housatonic in Connecticut, the Nashua and Charles River in Massachusetts, and the Penobscot in Maine, to name just a few. Nevertheless, the CRWC still maintains a formidable presence in the valley. High on the list of achievements was its successful campaign in 1998 to have the Connecticut designated presidentially as one of fourteen American Heritage rivers and its continuing efforts to work cooperatively with the Silvio O. Conte National Fish and Wildlife Refuge in promoting community level land conservation. This valley clearly still has a promising future ahead.

Spurred initially by the controversy over the proposed development of the east slope of Mount Greylock, Massachusetts's highest peak — the "Most Excellent Majesty" termed by Herman Melville — lawyer-activists in 1966 moved to create the Conservation Law Foundation of New England (CLF) to provide an independent environmental advocacy organization for the six New England states, one of the first of the national public-interest law firms. As former Executive Director Kelly McClintock recalled, the early emphasis was on getting the new environmental statutes and regulations enacted and operational. Not until later would litigation become a major part of CLF's program and activities.

Moving from Beacon Hill in 1970 to briefly co-office with two other regional organizations, the New England Economic Research Foundation and the New England Natural Resources Center, CLF took off on its own in the 1970s, waging successful legal battles over such issues as the proposed four-lane, divided highway through Franconia Notch (NH), the use of destructive logging practices in northern New England's federal roadless and wilderness areas, the proposed Big A dam on the Penobscot River in Maine, and the pollution of Boston Harbor. A major issue from the outset has been the prevention of overfishing off the coast of New England.

With the advent of Douglas O. Foy in 1977 as chief executive officer, CLF branched off into larger, long-term issues such as mitigating the effects of Boston's Big Dig and other major New England highway and transportation projects, and the development of alternative energy sources to replace New England's aging and polluting coal-fired power plants (the so-called Filthy Five). Over time, these kinds of interventions have earned CLF a reputation not only as a vigorous defender of the environment, but a party willing to engage in direct dialogue with prospective polluters. This has led it to explore how the forces of social and economic development, and the resources of government, can be used to actually improve the environment.

Recent examples include the creation of the Massachusetts Environmental Trust in 1988 to redirect pollution penalty money to the improvement of the environment of Boston Harbor, and its current controversial stance in favor of the proposed energy wind farm in Nantucket Sound. In 2002, Foy took the ultimate step by accepting Massachusetts Governor Mitt Romney's invitation to serve as the commonwealth's first supersecretary coordinating the combined program activities of transportation, housing, economic development, and environment.

With respect to land conservation, the Berkshire Natural Resources Council, formed in 1961 by George S. Wislocki, for thirty-four years its president, and newspaper publisher Donald B. Miller of the *Berkshire Eagle*, has advanced to become today the largest single private landowner in western Massachusetts (14,000 acres). Other significant regional actions relating to land conservation include those of the Mt. Grace Land Conservation Trust in the north Quabbin area, the Valley Land Fund and the Kestrel Trust in the Connecticut River Valley, the Essex County Greenbelt Association and the Sudbury Valley Trustees in the Boston metropolitan area, and the Martha's Vineyard Land Bank and Nantucket Conservation Foundation in the Cape and islands region.

A promising latecomer, the Massachusetts Land Trust Coalition, barely ten years old in 2000, has grown to nearly one hundred local land trusts with a combined membership approaching 100,000. These organizations have helped augment the well-established pioneers—the Appalachian Mountain Club, the Trustees of Reservations, and the Massachusetts Audubon Society—who continued to grow and be effective. The thirty-year-old Massachusetts Association of Conservation Commissions, representing every city and town in the commonwealth and specializing in wetland protection and land conservation, now typically holds annual meetings with up to a thousand or more commissioners in attendance.

Perhaps the most prominent illustration of continued citizen-based action has been that of the Environmental League of Massachusetts (ELM), the legal successor to the early Massachusetts Forest & Park Association and the Massachusetts Conservation Council. Reorganized in 1988 and 1993, by 2000 ELM had become the primary lobbying organization for the environment in Massachusetts. Headed by a respected former assistant secretary of environmental affairs, James Gomes, ELM routinely advocated in the areas of land use, toxics use reduction, recycling, water resources protection, and funding for environmental programs. As both a supporter and occasional critic of the EOEA, it made certain that the tradition of citizen involvement so central to the history of environmental conservation in Massachusetts remained alive and well.

THE CHANGING ENVIRONMENT OF ENVIRONMENT

What then is the import of this admittedly abbreviated view of Massachusetts conservation history during the twentieth century? Let us start with a review of what has been said thus far.

As the turn of the century drew near, Massachusetts leaders came home from a Civil War that had taken a terrible toll of human life and natural resources. Their return was to stimulate incomparable economic vitality and activity, great social change, and a virtual intellectual renaissance during the last third of the nineteenth century. It would also be a time of marked material wealth and leisure as the extravagancies of the so-called Gilded Age would later illustrate. Conservation was one of the new ideas that appealed and took hold for several reasons.

First, in the face of cutover and burned over forests, fished out streams, declining game coverts, dammed and polluted rivers, the needs were incontrovertible. The early twentieth century Massachusetts leaders, either still owners or not far removed from the land, found these needs to be compelling.

Second, nature appreciation, a movement that had started in Europe, was all the vogue in Massachusetts. It was in 1906, for example, that the journals of Henry David Thoreau would be first discovered, edited, and published.

Third, since new institutions were being formed to deal with the societal problems of health, education, culture, and commerce, why not conservation too? There was interest, leadership, and wealth available — often from the same sources that were behind the commonwealth's social and economic development. This led to individuals willing to serve in multiple capacities.

Interestingly, as our brief history has demonstrated, the approach for conservation would follow a path similar to those of other institutions — private leadership and then the creation of an interlocking directorate of influentials to be certain that the new institution would be carefully controlled. Later, some functions would be allocated to government but, as we have seen, even these institutions would be placed under substantial citizen control. During the first third of the new century, relations between the governmental and private entities would remain cordial — often mutually supportive — but there was little dispute over who was actually in charge.

During the first and second quarters of the century, governmental conservation would start to come into its own. Depression conditions would make that change welcome, but it also became obvious that certain national conservation issues, such as the overall state of the forest and wildlands, the decline of the continental waterfowl population, and the loss of vital topsoil throughout America's bread basket, were beyond the scope of just private remediation. The result was the founding of the great federal conservation agencies — for example, the Forest Service, National Park Service, Fish & Wildlife Service, and Soil Conservation Service — and, to an appreciable extent, their mirror images at the state level.

In post-World War II America, as we have seen, the federal presence proliferated to the point where it seemed to predominate. State functions similarly expanded but in a largely subordinate position to the authorities and resources of the federal government. For the most part, the private conservation organizations welcomed this development, aware that their own resources were insufficient to address the magnitude of the problems they perceived. On national conservation issues, a curious "iron triangle" of influence developed consisting of private conservation leaders, federal agency heads, and members of Congress. The result was an unprecedented flow of statutory and programmatic authorizations in the early 1970s. But by the last quarter of the twentieth century, important changes would start to take place in environmental conservation both in Massachusetts and the nation as a whole.

First, the principal leadership would shift significantly away from the nongovernmental to the governmental sector. As the advent of the cabinet system in Massachusetts would illustrate, government had clearly come into its own without the historic constraints of private influence. Environmental leadership, too, had begun to move away from the classic Yankee to embrace those of different gender, ethnic, and economic circumstances and persuasions—in short, to mirror society as a whole.

Second, public support for the environment had begun to wane. In many circles, rather than an imperative, environment had become an impediment. Environmental regulations, procedures, and requirements were increasingly perceived to be obstacles to economic progress and threats to the fundamental American concepts of personal entrepreneurship and independence, especially in the ownership and use of land. In principle, public support for the environment remained strong but, in practice, it had become increasingly suspect.

Third, those who had been in the forefront of the crusade in years past—the organized nonprofit conservation institutions—were undergoing a metamorphosis of their own. Once led by charismatic individuals, whose foot soldiers were largely volunteers, the organizations were becoming increasingly specialized and professionalized, run by administrators rather than those merely there to save the world.

Fourth, as the subject's complexity was uncovered, the environmental movement had become internally fragmented, often more concerned with the individual parts than the whole. It was not only perceived to be a special interest in a political sense—often one intensely partisan—but it harbored internal interests that did not always resonate well with each other. Coalition building around common objectives was becoming the exception rather than the rule.

Finally, the environmental agendas, due in part to the nature of the organizations, were preoccupied with financial needs. On the governmental side, appropriations were declining while statutory obligations remained undiminished or even expanded. On the private side, programs were chosen often with a careful eye to the prospect for individual and charitable gifts—namely, issues that underscored the most vivid threats to the environment. This had led the movement's message to become increasingly shrill and extreme, a far cry from the precedents of reason and accommodation that had marked the early history of conservation.

In states like Massachusetts, emboldened by their new cabinet status, state agencies were starting to bristle at being told what to do by federal agencies and being held hostage by insufficient program support. Increasingly, they were electing to seek funds and authorities of their own. No longer beholden to private organizations, they would occasionally consult and confer but at the end, more than likely, follow an independent path dictated by their dependence upon the governor and legislature. But like their federal counterparts, state agencies were becoming increasingly subject to a public disaffected by what many perceived to be the unwarranted intrusion of government, via taxation, law, and regulation, into provinces that were regarded as largely private business. On the environmental side, their unpopularity rested as much on their lack of regulation as overregulation.

CIVIC ENVIRONMENTALISM

As the 2002 Massachusetts gubernatorial election drew near, State Senator Richard T. Moore, the president of the state chapter of the American Society for Public Administration, called for a series of thoughtful papers from Massachusetts's nongovernmental leaders on programs the new administration should undertake. A group of seven individuals, drawn from as many different institutions, responded with recommendations for a commitment to what they called *civic environmentalism.*

As the authors' stated (*Memos to the Governor*, 2003), civic environmentalism is designed to address the new and less traditional business of the environment, such as protection of whole ecosystems on an interconnected local to global scale. Rather than just imposing restrictions, it would make extensive use of nonregulatory tools such as education, technical assistance, and subsidies. It would seek out alternatives to confrontation through the use of mediation, consensus building, and conflict resolution. And it would cast government in a new role as participant rather than simply proscriber of environmental programs and policies.

Civic environmentalism would be used as a bridging concept to engage localists of all kinds in environmental affairs, thereby leading to an increased capacity for people to govern themselves. It would replace coercion from afar with the encouragement and exercise of personal responsibility at home. It would make people, not government, the primary drivers and eventual beneficiaries of environmental decision making. Most importantly, by conceiving of an environment that was ecologically healthy, but also economically and socially vibrant, civic environmentalism would make the environment relevant not just to those in rural and suburban settings, but also to the more diversifed inhabitants of urban areas.

ENVIRONMENTAL CITIZENSHIP ACADEMY

In 2003, intrigued by these recommendations, a private foundation enabled the University of Massachusetts (Boston) to constitute an experimental Environmental Citizenship Academy to explore the training of individuals for this new kind of civic engagement. As Alice E. Ingerson has reported, nearly forty individuals, with little or no prior knowledge of the environment, were recruited from existing Boston neighborhood organizations and voluntarily committed themselves to a program of six to ten case-based community classroom sessions. Over a six month period, a profile emerged of what, hopefully, will be tomorrow's environmental citizen.

To begin with, a citizen, by birth or naturalization, was defined as one who receives the legal protection of the state and nation and, in return, pledges loyalty and a willingness to assume a share of their obligations. Thanks to the many different aspects mentioned at the outset of this paper, there are now ample opportunities for citizenship to be practiced on behalf of the environment. In addition to providing the formal institutions with political support and tax-derived funds, an environmental citizen may choose to become active in one or more nongovernmental capacities.

In the past, it has been traditional to view environmentalism as an adversary process. People for the environment are often pitted against those who would disturb its natural or productive qualities for economic reasons. Where land and other resources are plentiful, such as in rural or suburban areas, there is often room to accommodate both sets of objectives. But in the state's urban areas, the choices become more difficult.

Thus, a new concept of the environment needs to emerge — a form of citizenship that construes the environment more broadly, recognizing the legitimacy of other human needs and aspirations, such as education, housing, jobs, dependable transportation, safe neighborhoods, and social equity. Rather than opponents, the new environmentalists should become participants in the process of balancing the use of resources. Beyond a proper understanding of the environment, there are fundamental human qualities needed for this form of environmental citizenship. The academy graduates summarized these as follows.

First, to be effective, environmental citizens need to be self-assured and self-reliant — open-minded in recognizing what they do not know and prepared to put acquired knowledge to work in actual practice. This will require a willingness to seek out and consider a full range of possible options.

Second, the environmental citizen must be attuned to and appreciative of the role of history, for all environmental problems have a past as well as a present. Getting to know the true context is usually the first step toward finding an acceptable solution.

Third, environmental citizens must be aware of their role in the natural world and the value of having a personal connection with one or more particular places in order to focus their commitment and actions.

Lastly, an effective environmental citizen is one who must be adaptable in social and political domains, able to accept the representations of others, willing to accommodate responsible solutions to problems, and then see them through to completion.

With such citizens, the prospect for Massachusetts conservation in the twenty-first century can only be a promising one. It echoes the joint declaration made in 1999 by more than three hundred college and university presidents that the challenge of the next millennium will be the renewal of our own democratic life and the reassertion of our commitment to active, effective, social stewardship.

REFERENCES

Abbott, Gordon, Jr. 1993. *Saving special places: a centennial history of The Trustees of Reservations, pioneer of the land trust movement.* The Ipswich Press. Ipswich, MA.

Applegate, M. Richard. 1974. *Massachusetts Forest and Park Association: a history, 1898–1973.* Massachusetts Forest and Park Association. Boston, MA.

Barske, Carolyn. January 21, 2002. "Oral history of the Massachusetts Executive Office of Environmental Affairs." Massachusetts Historical Society. Boston, MA.

Benson, A.E. 1942. *History of the Society for Promoting Agriculture, 1892–1942.* Meador Publishing Co. Boston, MA.

Bosso, Christopher J. 2005. *Environment, Inc: from grassroots to beltway.* University Press of Kansas. Lawrence, KA.

Burns, Deborah E. and Lauren E. Stevens. 1988. *Most excellent majesty: a history of Mount Greylock.* Berkshire Natural Resources Council. Pittsfield, MA.

Caldwell, Lynton Keith. 1998. *The National Environmental Policy Act: an agenda for the future.* Indiana University Press. Bloomington, IN.

Connor, Sheila. 1994. *New England natives: a celebration of people and trees.* Harvard University Press. Cambridge, MA.

Conuel, Thomas. 1981. *Quabbin: the accidental wilderness.* University of Massachusetts Press. Amherst, MA.

Cook, Harold O. (with Lewis A. Carter). 1961. *Fifty years a forester.* Massachusetts Forest and Park Association. Boston, MA.

Danielson, Diane K. 1992. "Environmental regulation in Michigan and Massachusetts: two states with two different solutions to the same problem." *Boston College Environmental Affairs Law Review* 20(1): pp. 99–133.

Dawson, Alexandra and Sally Zielinski. 2006. *Environmental handbook for conservation commissions* (10th edition). Massachusetts Association of Conservation Commissions. Belmont, MA.

Donahue, Brian. 1999. *Reclaiming the commons: community farms and forests in a New England town.* Yale University Press. New Haven, CT.

Dube, Denise. October 12, 2000. "The great land grab." *Boston Globe* (*Globe* West Section). Boston, MA.

Ells, Stephen F. April 1969. *Massachusetts open space law: government's influence over land-use decisions*. Open Space and Recreation Program for Metropolitan Boston (Volume 5). Metropolitan Area Planning Council. Boston, MA.

Federal Writers' Project, Works Project Administration. 1937. *Massachusetts: a guide to its places and people*. Houghton Mifflin Company. Boston, MA

Foster, Charles H.W. (ed.). 1998. *Stepping back to look forward: a history of the Massachusetts forest*. Harvard University Press. Cambridge, MA.
  Bond, Robert S. "Professional forestry, forestry education, research";
  Foster, Charles H.W. "An historical overview";
  Fox, Stephen "Massachusetts contribution to national forest conservation";
  King, William A. "The private forestry movement in Massachusetts";
  McCullough, Robert L. "Town forests: the Massachusetts plan";
  O'Keefe, John F. and David R. Foster "Ecological history of Massachusetts forests";
  Rivers, William H. "Massachusetts state forestry programs."

Foster, Charles H.W. 1991. *Yankee salmon: the Atlantic salmon of the Connecticut River*. University Press of New England. Hanover, NH.

French, Jonathan (ed.). 1986. "Boston's water resource development: past, present, and future." Proceedings of the Boston Society of Civil Engineers Section session, October 29, 1986. American Society of Civil Engineers. New York, NY.

Hall, Peter Dobkin. 1974. "Family structure and class consolidation among the Boston Brahmins." Ph.D. dissertation, State University of New York. Stonybrook, NY.

Hamilton, James T. and Henry Lee. May 1987. Rapporteur's report: executive session on Northeast electric power policy. Presentation by Peter Bradford, chairman, Maine Public Utilities Commission. Discussion Paper Series E-87-06, Kennedy School, Harvard University. Cambridge, MA.

Ingerson, Alice E. June 30, 2004. Environmental Citizenship Academy: final project report. University of Massachusetts. Boston, MA.

Jorgensen, Neil. 1977. *A guide to New England's landscape*. The Globe Pequot Press. Chester, CT.

Kaynor, Edward R. 1976. Connecticut River water resource decision making. Publication No. 83, Water Resources Research Center, University of Massachusetts. Amherst, MA.

Lambert, John H. November 16, 1992. Interview with former Massachusetts chief forester John H. Lambert. Conducted by William H. Rivers, Douglas Poland, and Alden Cousins. Unpublished.

Leahy, Christopher et al. 1996. *The nature of Massachusetts*. Massachusetts Audubon Society. Addison-Wesley Publishing Co. Reading, MA.

Lewis, George. 2004. Sudbury Valley Trustees: 50 years of conservation. Sudbury, MA.

Lyman, Charles P. 1992. *The Massachusetts Society for Promoting Agriculture: the years 1942–1992*. The Ipswich Press. Ipswich, MA.

MA Department of Agricultural Resources. 2006. "The history of the Department of Food and Agriculture, 1852–2002." Available at www.mass.gov/agr/150/dfa-history.htm.

MA Department of Food and Agricultural Resources. 1976 *A policy for food and agriculture in Massachusetts*. Boston, MA.

MA Division of Fisheries and Game. 1964–65. Centennial Series, *Massachusetts Wildlife*. Westboro, MA.

Bridges, Colton H. Mar.-Apr. 1965. "The fisheries program";

Chaplin, Bryant. Nov.-Dec. 1964. "Your right to know";

Pushee, George. Sep.-Oct. 1964. "The game program";

Chaplin, Bryant. July-Aug. 1964. "Fish and Game Division: a unique stateAgency"

MA Executive Office of Environmental Affairs. November 1972. Citizen task forces on environmental reorganization: synopses of the task force reports. Institute for Man and his Environment, University of Massachusetts. Amherst, MA.

MA Executive Office of Environmental Affairs. January 6, 1975. Special report of the secretary relative to the further reorganization of said executive office. House No. 5366. Boston, MA.

MA Executive Office of Environmental Affairs. April 2000. "The state of our environment." Boston, MA.

MA Executive Office of Environmental Affairs. 2004. "Massachusetts water policy." Boston, MA.

Massachusetts Fish and Game Association. 1937. "A brief history." Boston, MA.

Massachusetts Forest and Park Association. January 1948. "The first fifty: 1898–1948." Boston, MA.

Massachusetts Historical Society. 2006. New England Environmental Archive. Boston, MA.

Journals of Theodore Lyman, III

Personal papers of Charles H.W. Foster

Oral histories of the secretaries of environmental affairs (1971-2001).

MA Office of the Secretary of State. 1998. Background information for voters. Referendum No. 4, Electric Utility Industry Restructuring. Boston, MA.

Massachusetts Society for Promoting Agriculture. 1942. *Centennial year: 1892–1942*. The Meador Press. Boston, MA.

May, Ernest R. 1964. "The Progressive era." *Life History of the United States*, Volume 9 (1901–1917). Time Incorporated. New York, NY.

McCullough, Robert. 1995. *The landscape of community: a history of communal forestry in New England*. University Press of New England. Hanover, NH.

McDonald, Angus. October 1959. "Early American soil conservationists." USDA Miscellaneous Pubications No. 449. Washington, DC.

Mead, Edwin D. January 1930. "The meaning of Massachusetts." New England Quarterly 3 (1): pp. 25–54.

Metropolitan District Commission. July 11, 1995. Quabbin watershed: MDC land management plan, 1995–2004. Boston, MA.

Moore, Richard T. (ed.). 2003. Memos to the governor: management advice from the commonwealth's experts in public administration and policy. 1st Books Library. Bloomington, IN.

Morgan, Allen H. Undated. "A history of the Massachusetts Audubon Society." Unpublished draft. Personal papers of Allen H. Morgan. Massachusetts Historical Society. Boston, MA.

Museum of Comparative Zoology and Harvard Archives, Harvard University. Various dates. Personal papers of John C. Phillips. Cambridge, MA.

Nesson, Fern L. 1983. *Great waters: a history of Boston's water supply*. University Press of New England. Hanover, NH.

Peirce, Neal R. 1976. *The New England states: people, politics, and power in the six New England states*. "Massachusetts." W.W. Norton & Co. New York, NY.

Quarles, John. 1976. *Cleaning up America: an insider's view of the Environmental Protection Agency*. Houghton Mifflin Co. Boston, MA.

Rural Land Foundation. 1977. *The first twelve years*. Lincoln, MA.

Scheffey, Andrew J.W. 1969. Conservation commissions in Massachusetts. Conservation Foundation. Washington, DC.

Skehan, James W. 2001. *Roadside geology of Massachusetts*. Mountain Press Publishing Co. Missoula, MT.

Spear, Robert J. 2005. *The great gypsy moth war: the history of the first campaign in Massachusetts to eradicate the gypsy moth, 1890–1901*. University of Massachusetts Press. Amherst, MA.

Thornton, Tamara Plakins. 1989. *Cultivating gentlemen*. Yale University Press. New Haven, CT.

Vaughan, Gerald F. April 2002. *Pioneer in wildlife conservation: Reuben Edwin Trippensee, 1894–1997*. Publication No. 02-01, The Environmental Institute, University of Massachusetts. Amherst, MA.

Weiskel, Peter K. et al. 2005. "Water resources and the urban environment, lower Charles River watershed, Massachusetts, 1630–2005." Circular 1280, US Geological Survey. Denver, CO.

Wilkie, Richard W. and Jack Tager (eds.). 1991. *Historical atlas of Massachusetts*. University of Massachusetts Press. Amherst, MA.

Wyner, Michael. December 16, 1999. "Trustees remember Rural Land Foundation beginnings." *Lincoln Journal*. Lincoln, MA.

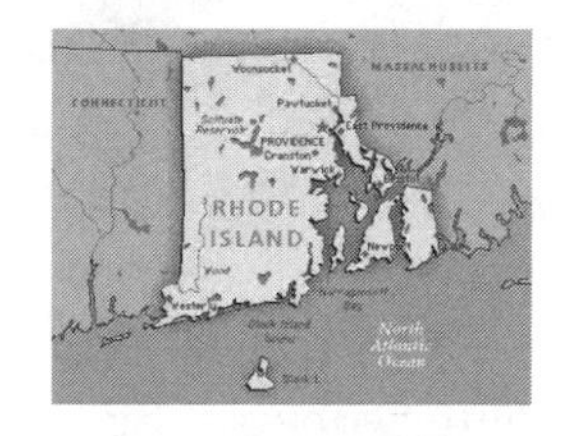

THE RHODE ISLAND CONSERVATION STORY

BACK FROM CRISIS

Peter B. Lord

At the onset of the twentieth century, Rhode Island cities were booming industrial powerhouses that had attracted thousands of immigrants and rural farm families to a wide range of skilled and unskilled jobs. Walk around downtown Providence one hundred years later and you get some idea of the great wealth that was created during the so-called Gilded Age. Some of the city's grandest buildings—City Hall, the State House, the Providence Train Station—were built during the booming years of the late nineteenth century. But that economic success came with an enormous price. It is difficult to explain to someone today just how stressed and battered the Rhode Island landscape was at the turn of the century. Most of the trees were gone. Most Rhode Islanders lived in large, dirty cities and worked in vast mills or machine shops. Abandonment of small farms was so common the state began marketing the farms to get them back into operation.[1] Rivers and streams were filled with sewage and industrial waste. There were few parks or other public places. Communicable diseases were rampant.

Today, Rhode Island's relatively clean waterways, expansive beaches, and broad tracts of preserved forests and farm fields make it an attractive destination for tourists from all over the world. Tourism pumps nearly $5 billion into the state's economy annually and supports nearly 10 percent of its workforce.[2] Slightly more than one million people live in a state that covers about 1,000 square miles — unless you count its bodies of water, which add a few hundred more square miles. As the nation's smallest state, Rhode Island is constantly used as a unit of measure — it is the size of the 2007 wildfires in California or an ice mass that has floated free in Antarctica.[3] Rhode Island still has its busy cities, surrounded by sprawling suburbs. But it has preserved thousands of acres of woodlands on the Connecticut border along with much of its natural treasure, Block Island. The state, with more than 400 miles of shoreline and a major estuary that cuts deeply into the state's core, is known as the Ocean State. Nowhere in Rhode Island can one be more than twenty-five miles from the bay's shoreline.

Rhode Island was founded in 1636 by Roger Williams, a Puritan who was banished from Massachusetts for questioning local authority figures. When he founded Rhode Island on the shores of Narragansett Bay in what is now East Providence, his goal was to create a free com-

The Slater Mill, America's first successful water-powered cotton spinning mill, and a forerunner of what would be the advent of New England's famed Industrial Revolution, is located on the banks of the Blackstone River in Pawtucket, Rhode Island. This complex of former mills, early industrial communities, natural resources, and landscapes, stretching upstream into Massachusetts, is now contained within a Congressionally authorized, federal-state, National Heritage Corridor.
(National Park Service)

munity of seekers of the truth and a haven for those persecuted elsewhere for their conscientious beliefs. He was willing to accept the will of the majority when it came to civil issues, as long as his neighbors agreed that "no man should be molested for his conscience."[4] In the following decades, Rhode Island became a haven for other religious dissenters, including Calvinists, Quakers, and Jews.[5] With safe harbors and freedom to do business, the economy in communities that became Rhode Island was fueled by trade. By the time of the Revolutionary War, Newport was one of the top seaports in the colonies and it was deeply involved in the slave trade.[6] "Rhode Islanders realized that, if they intended to grow up with the country, they had to agree that the business of the colony was basically commerce; all else was subservient to the goal of keeping cargoes flowing in and out." (McLoughlin)[7] After the war, Rhode Islanders continued to prosper by sending their large fleet to do business in ports all over the world. Such long-distance trading was lucrative, but also risky. The War of 1812 hastened Rhode Island's diversion of capital from shipping into manufacturing, and Rhode Island entered an era of heavy industry, using water power to provide an inexpensive away to run the early mills.[8] Using technology learned in England, the first successful cotton-spinning machine in America was built and put to use in Slater Mill on the Blackstone River in Pawtucket.[9] The mills put many unskilled women and children to work while the legislature gave the mills preferential treatment over farmers and fishermen, much to the detriment of the environment. McLoughlin noted:

> Farmers might sue for compensation when mill dams flooded their land, but they could not stop the dams from being built or continuing in operation. Fishermen could require that dams be opened at designated times to allow fish to pass through (up or downstream), but they could not obtain laws preventing manufacturers from polluting the streams. Within a short time, most rivers and many fishing areas and shell beds became polluted with industrial waste from cleaning, bleaching and dyeing the cotton. Nor did the state give any thought in the nineteenth century to regulating working conditions or safety hazards in the mills. Nor was it

> concerned about the living conditions in the factory towns. In the interest of progress, it chose to overlook, not oversee.[10]

While the textile mills expanded, other industries developed in Providence — metal casting companies, steam engine manufacturers, dozens of jewelry makers, and machine tool shops. The population of Providence soared and it became the economic and political hub of the state. Rhode Island benefited greatly from the Civil War — the population and assessed value of Providence doubled in the 1860s due to the profits of the textile and manufacturing industries.[11] Afterward, Rhode Island and the rest of the nation enjoyed the Gilded Age as prosperity came to the wealthy and the middle class. Writes McLoughlin: "The half century following the Civil War was the halcyon era of Rhode Island. In those years, the system worked. Not for everybody, but for enough. Never before had Rhode Island as a whole been so prosperous, so attractive a place to live and to invest in."[12] All of which helps explain what an enormous challenge Rhode Island faced as it entered the twentieth century.

In the next several decades, the state would watch its industrial backbone spiral into decline. The textile jobs went south. The machine shops went abroad. Steam engines were replaced by more modern technology. At the same time, there would be growing awareness of the need to restore the environment. The changing awareness came in fits and starts. It took decades and there were remarkable setbacks. Rhode Islanders would engage in that restoration with the independence, the inventiveness, and the boldness that they used to create their earlier industrial successes. They would create new institutions — a heritage corridor and a coastal council. They would borrow ideas such as land trusts and conservation commissions from others, and quickly embrace them across the state. They cared so much about their villages and neighborhoods — Rhode Islanders will say they are from the villages of Matunuck or Riverside rather than the towns of South Kingstown or East Providence — that they have overwhelmingly passed every bond issue proposed for open-space preservation. And they would rely on leaders such as U.S. Senator John H. Chafee, who inspired through personal charisma and good will; and Alfred L. Hawkes of the Audubon Society, who was tireless in delivering the conservation message; and

Capt. John R. "Rob" Lewis, another great communicator who refused to accept his neighbors' feelings of hopelessness towards saving Block Island from developers. The restoration of Rhode Island continues to be a work in progress, but it already provides useful lessons for others. Because Rhode Island really put itself in a bad way environmentally 100 years ago, and it has come a very long way since then.

A PERIOD OF COMPLACENCY

Ironically, the lowest point in the state's natural resources depletion occurred during the Victorian Era, when many of the state's middle-class and leading citizens devoted countless hours and expenses to collecting birds and mammals and insects and mollusks. This transition period spawned some amazingly foresighted conservation programs and groups. It showed just how successfully people can pull together when a crisis is obvious.

Based on pollen data and historical accounts, scholars believe that Rhode Island was about 95 percent forested when Europeans first arrived. The state overflowed with a wide range of natural resources — wildlife was abundant and local waterways were filled with fish. European settlers cleared most of the trees for farming, reserving only small portions of the farmsteads for woodlots. The first statewide forest survey in 1767 found only 31 percent of the state's land forested.[13] By the early 1800s, many prominent citizens were raising concerns over the loss of that remaining portion of woodlands. Mill owner Zachariah Allen planted oaks and chestnuts on forty barren acres in Smithfield to show how land could be improved by planting trees.[14] He kept records of his profits and losses for fifty-seven years. But with rising industrial demands, the cutting continued, so by the end of the nineteenth century, Rhode Island's forests were at their lowest ebb. Even the white pine trees that had grown back in abundance on farms that were abandoned during the height of the Industrial Revolution in the 1800s became a marketable crop with the spread of portable, steam-powered saw mills in the 1870s and rising demand for boxes and paper bags.[15] Throughout the late 1800s and early 1900s, Rhode Islanders continued to abandon their farms to work in the factories of the Industrial Revolution. Census Bureau data helps show what happened — in 1850 (the first year such data was collected) 554,000 acres in Rhode Island, more than half the state, were classified as farmland.[16] The number dropped steadily in all but one of the subsequent decades. By 1900, farmland dropped to 456,000 acres. In 1925, it was down to just 309,000 acres and finally, in 1969, it was down to 69,000 acres.[17] The number of farms during the same period dropped from 5,000 to 6,000 in the late 1800s, to about 1,000 in the 1960s.[18] Continued migration to urban areas and mechanized agriculture caused farm populations to drop from 28,000 to 4,000 during the same period.[19]

By the early 1900s, farms were such an unattractive choice for living and working that the Rhode Island State Conservation Commission began publishing catalogues of available farms to try to draw people into returning to the land. One early catalogue listed 251 abandoned farms and posited, like an ad pitchman, that life on any farm was preferable to life in the city. Charles W. Brown, superintendent of the Rhode Island Natural Resources Survey, wrote that two-thirds of the state's land area was devoted to agriculture in 1910, but only one quarter of that land was actually being cultivated. The people who used to till Rhode Island's farms had migrated west, or to the state's mill villages, Brown wrote. Consequently, "Rhode Island offers to any one seeking a rural home splendid opportunities at the present time."[20]

According to a history by the Rhode Island Department of Environmental Management, in 1887 Bernard Fernow, chief of the U.S. Department of Agriculture Forestry Bureau, reported: "Forests in the strict sense of the word can hardly be said to exist in this state. Although 24 percent is reported covered with wood, it is mostly coppice (thickets of small trees), which here and there may be said to rise to the dignity of forests, especially on the western borders." The peak of lumber production in Rhode Island was at about the turn of the century, with 33 sawmills operating.[21] The industry began to decline at this point because of a scarcity of "suitable trees."

In 1873 there was a serious, but brief economic depression called the Panic of 1873. It prompted civic leaders in Providence to begin taking steps to ensure the continuity of the heavily industrialized economy.[22] Business leaders moved to diversify the economy by participating in international expositions. They launched civic improvements such as water supply systems and sewers, new roads, and harbor improvements. They also moved to consolidate industries and lower wages. That helped lead to creation of the labor unions, which later helped transform the economy and life for working people. Between 1886 and 1893, the business leaders expanded Roger Williams Park from 100 acres to 462 acres.[23] In the final years of the century, Providence built its downtown library, the police station, the train station and its first high-rise office buildings as well as thirty new school buildings.[24] The state completed its white marble State House in 1904. Several major churches and temples were erected as well.

An example of the kinds of challenges facing urban leaders in the late 1800s comes from the 1895 annual report of the State Board of Health. Woonsocket's water superintendent wrote about efforts to set aside land for a new reservoir for the city. But people were dumping raw sewage along streams running into the new reservoir, and it wasn't against the law:

> Twice during the past year a party owning land adjoining the brook between the pumping station and the new reservoir, has allowed it to be used as a dumping ground for soil carts from Manville. Your superintendent notified the State Board of Health, and through its efficient secretary, Dr. Swarts, and the health officer of Lincoln, the nuisance was immediately abated. I would recommend that the city government has an act passed at the next session of the legislature protecting the waters of Crook Falls brook from pollution; this is an important matter and should have immediate attention.[25]

Providence was drawing its drinking water from the Pawtuxet River while trying to persuade homes and factories to limit their discharges of "excreta and household wastes" into the river. "Much of the manufacturers' wastes, consisting of chemicals, dye stuffs and wool scourings, are cared for to a certain extent, but not in such a way as to allow the minimum amount of extracts to reach the river,"[26] said a state health official. It should have been no surprise that epidemics of diphtheria and typhus were commonplace. The leading causes of death in 1895 were infant cholera (496 cases), pneumonia (665), consumption (705), dysentery (124) and diphtheria (133).

City officials knew they faced terrible pollution problems, but for many years the solutions only managed to relocate the pollution, not get rid of it. For instance, in 1874, Providence built forty-three miles of sewers and connected 3,000 buildings. But all that raw sewage was drained untreated into the Woonasquatucket, the Providence, and the Moshassuck rivers.[27] Along with household sewage, the rivers were dumps for vast amounts of waste from the mills and other heavy industries in Providence. Providence opened its Fields Point treatment plant in 1897, but raw sewage continued to be dumped into rivers and fields outside the city. The three rivers "were allowed to degenerate into areas of blight, while the tidewater shores of the Providence and Seekonk rivers were equally neglected," wrote John Hutchins Cady,[28] an architectural historian in Providence.

By 1900, Providence was one of the world's top ten industrial centers, according to historians Patrick T. Conley and Paul R. Campbell.[29] It was a world leader in the manufacture of tools, files, engines, screws, and silverware. And it was number one in the nation in the production of jewelry and woolen and worsted goods.[30] Ironically, during this same period of intense industrial activity, many Rhode Islanders shared an avid, possibly even perverse, passion for collecting remnants of the state's natural resources. Providence built its Roger Williams Museum of Natural History in the 1890s after one of its former citizens, John Steere, donated a collection of his specimens.[31] Horace F. Carpenter donated a collection of mineral specimens and catalogued every mollusk growing in Rhode Island.[32] James M. Southwick, the museum's founder and first director, also operated a natural history store in Providence and published a monthly journal of natural history news in Rhode Island.[33] "These people loved nature," said museum curator Marilyn Massaro. "They couldn't go home and turn the Discovery Channel on. So they collected the things that interested them. So much of the habitat was denuded by farming. But a typical Victorian parlor was chock full of natural history ornaments — in the carpets, in cabinets, on shelves."[34] It became an era for building great natural history museums, observed Massaro.

Massaro writes of Providence's Museum of Natural History, "In June, 1896, when the museum opened, the public beheld a building intended not merely to house a collection of mounted mammals and birds. It was a monument to science and the embodiment of a national, indeed an international movement dedicated to natural history collecting, both amateur and professional. The movement led to the construction of museums in major cities around the nation. These typically were grand structures reflecting Victorian exuberance. Many were built in urban parks created as part of the City Beautiful Movement then reaching its peak in America."[35]

Historian Mark V. Barrow, Jr. did a paper on the specimen dealers of that era and offers this explanation for their context:

> Underlying this impressive growth in interest in natural history was a series of profound transformations—economic, technological, social, and cultural. Between 1860 and 1900 the population of the United States more than doubled, the gross national product quadrupled, and the total number of individuals living in urban areas increased more than five times. A vast, newly constructed rail system helped create the first truly national markets, while revolutions in the publishing industry combined with reforms in the postal service to decrease the cost and increase the availability of printed materials of all sorts. The wide diffusion of romanticism, once the province of a highly educated cultural elite, provided an intellectual framework both for critiquing the new urban-industrial order as well as for appreciating the aesthetic and spiritual values associated with wild nature and the outdoors. At the same time, a growing middle class with increased leisure time began to experience limited relief from the stern grip of the work ethic.[36]

As Barrow and other scholars have observed, at the end of the century more rapid changes occurred that influenced relationships with the natural world. The Victorian period gave way to the Arts and Crafts movement that emphasized cleaner design lines and looked askance at the clutter with which Victorian homeowners enjoyed filling their parlors. People increasingly began to suggest humans "ought to be kinder and gentler in their dealings with the natural world."[37] Tensions grew between those who saw nature as nothing more than a resource or commodity that existed only for human needs, and a growing romanticism that found "an aesthetic and spiritual dimension that defied simple economic valuation."[38] Those tensions remained unarticulated, Barrow says, until the rise of the Audubon movement and the growth of birdwatching around 1900.

GROWING CONCERNS FOR THE LAND

At the turn of the century, the public was looking for change. Teddy Roosevelt was elected president in 1901, and the prodigious big-game hunter also emerged as the country's greatest conservationist president. There also was a growing public awareness of the need to better manage forests in Rhode Island and elsewhere. On Christmas Day 1900, ornithologist Frank Chapman, an early officer of the Audubon Society, organized the first Christmas Bird Count. His goal was to get people to count and appreciate wild birds, rather than to take part in the traditional Christmas "side hunts," the practice of groups choosing up sides and going out and shooting as many as birds and mammals as they could find.[39]

In Rhode Island, the Aububon era began on Oct. 20, 1897, with a meeting of about thirty people at the home of Mrs. Henry R. Chase at 133 Brown St., in Providence.[40] Massachusetts had formed its own Audubon Society a year earlier. All were fed up with the slaughtering of birds for their feathers, and the gathering of their skins and eggs for collections. At the Providence meeting, Dr. Hermon C. Bumpus, a biologist at Brown University and assistant director of the Marine Biological Laboratory at Woods Hole, was unanimously elected president. Board members were all women. Also sixteen honorary vice presidents were chosen, and they were men and women. The group's initial goals, as stated in its bylaws, were: "The promotion of an interest in bird life, the encouragement of the study of ornithology, and the protection of wild birds and their eggs from destruction."[41] The group's founders pledged to give out information on the economic value of birds, to protect against the use of wild birds and their plumage for adornment, and to, above all, "awaken the interest in children in bird life and thus educate them to humane and gentle sentiments."[42]

In its early years, the society focused on the last goal, educating children about birds. Grammar schools held essay contests on birds and participated in Arbor Day activities. Children could become Audubon

Society members with a payment of only ten cents. Dr. Bumpus quickly affiliated the Rhode Island group with pro-bird organizations throughout the region. In 1905, at a meeting in Cambridge that eventually led to creation of The National Audubon Society, Dr. Bumpus was chosen to be chairman of the National Committee for Bird Protection.[43] He continued to lobby for federal laws protecting birds.

As part of the local society's efforts to educate children, in 1914 it hired Miss Elizabeth Dickens to teach bird studies to children on Block Island, a then rural place about twelve miles off Rhode Island's shores. She became a legend. She not only taught generations of island children, Miss Dickens also recorded personal bird observations nearly daily until her death in 1963. Her observations are considered "one of the best, and longest, diaries of observances of birds in any one location in the world."[44]

Pressure from so many local groups and the public prompted Congress to pass the Lacey Act in 1900. It forbade the interstate trade in any bird protected by a state. In 1913, Congress passed the Migratory Bird Act, putting those birds under the protection of the federal government. The success of those two laws left many bird groups wondering what to do. Roland Clement, a longtime executive director of Rhode Island Audubon, observed:

> All of these state groups sort of relapsed into being social organizations interested in a little bit of birding and a little bit of talk about the need for conserving birds, but the crusading spirit was passed. Therefore, they all sort of retrenched a great deal. In fact, during the Depression, most of the conservation organizations, and the state Audubon Societies, were knocked out completely. They just disappeared. Rhode Island was one of the few survivors, thanks to Alice Hall Walter.[45]

Clement said Walter kept the Rhode Island society going for about twenty-five years. Married to a noted Brown University zoologist, Walter was also a national figure in her own right. She wrote a book on birds in a Chicago park and maintained an active correspondence with Frank Chapman, one of the Audubon pioneers.[46]

In 1906 the Rhode Island General Assembly created the Rhode

Island Forest Commission. The following year the commission appointed Jesse B. Mowry the state's first forest commissioner. Mowry lobbied for stronger forest protection laws, fire suppression, and forest conservation and management. His first report to the legislature has been widely quoted:

> It is a well-known fact to most of you that the timber which once covered our hillsides, ameliorating our climate, beautifying the landscape, protecting the watersheds and constituting one of the most valuable natural resources of the state now has nearly all disappeared before the woodsman's axe. It follows, therefore, that the protection and rapid growth of the succession of young sprouts and seedlings is a problem of interest and importance to the people.[47]

Also, shortly after the turn of the century, the General Assembly created a commission to develop a metropolitan park system to benefit the inhabitants of the teeming cities.[48] The membership seems to have reflected the perceived importance of the mission: the commission included the presidents of Brown University and the Rhode Island School of Design, the mayors of all the metropolitan area cities and the head of the chamber of commerce. In the commission's first report, it noted that the state's metropolitan areas—Pawtucket, Central Falls, East Providence, Cranston, Warwick, Johnston, North Providence, Lincoln, Barrington and Cumberland—doubled in population every twenty-one years throughout the nineteenth century.[49] By 1900, there were 318,095 people in the metropolitan area, 57,779 in the cities of Newport, Woonsocket, and Westerly, and just 52,682 scattered throughout the rest of the state. No other metropolitan area in the country contained such a disproportionate majority of a state's population, the report noted. The commission predicted (wrongly) that the population in the metropolitan area would grow to one million by 1935 and one and a half million by 1950. (Apparently, and not surprisingly, the commissioners couldn't imagine the advent of suburbs.) The commission was very aware of the rapid changes in the metropolitan area: "There are active men yet living who remember when Providence was a village and cows roamed over stubby meadows from Stewart Street to Pawtuxet, interrupted only by a few stone walls and rail fences, and when New York was a smaller town than Providence is today."[50] But the commission also concluded that the time was right to turn "waste spaces," in the cities—the high points, the river valleys, the swamps—into parks. And interestingly, as noted by colleagues in Massachusetts and New York, the commission argued the parks were necessary for economic success:

> Without fresh air and happy surroundings, the two greatest factors for efficacy of labor and contented people, the industries of Providence now pre-eminent in those directions that call for skill and art of the workers, would inevitably decline. It sure means everything to the health and moral welfare.[51]

In just their first year of work, the commissioners designed a system of wide boulevards radiating from the city center and then connected with a ring of parks that would be developed on the outskirts. Some of their plans succeeded, such as the development of Roger Williams Park in Providence and Lincoln Woods State Park in Lincoln. Some are still being pursued today, such as creating greenways along the Woonasquatucket River. Development beat out park creation in many areas in Johnston and North Providence. And some ideas failed instantly, such as the concept of turning a grove of stately pines near Omega Pond in East Providence into a park. The trees were cut before the report was completed, causing the commissioners to lament:

> The price of the lumber obtained by the owners was doubtless, attractive to them, but its value in the form of boards and timber compared with its greater value to 300,000 people when standing in lofty trunks of the beautiful pines seems like a miserable pittance.[52]

The commissioners requested a $500,000 bond to buy parkland, and they pleaded with the legislature to start securing public land along the shores of Narragansett Bay, "the state's most valuable asset," where no publicly own land existed and summer houses were being built, one after another.[53] They quoted an excerpt from a speech by President Theodore Roosevelt in calling for legislative action:

> The overcrowding of cities and the draining of country districts are unhealthy and even dangerous symptoms in our modern life . . . If when children grow up, they are unhealthy in body and stunted or vicious in mind, then that race is decadent; and no heaping up of wealth, no splendor of momentous material prosperity, can available in any degree as offsets. Public playgrounds are necessarily means for the development of wholesome citizenship in modern cities. It is desirable that vacant places be reserved in sections of the city which are destined to be built up solidly.[54]

In its third and fourth annual reports, the Parks Commission used a cartoon showing a little girl in a "Parkless City." The commissioners wrote:

> The mass of humanity in the cities makes it apparent why reasonably thinking men have been forced to consider the question of air holes or open spaces in the shape of parks, promenades and boulevards. Already, London, Edinburgh, Glasgow, Paris, Frankfurt and New York have taken steps to furnish breathing spaces in their densely populated portions—tearing down solid blocks of buildings at enormous cost—but nothing costs a city like disease! Next to the public school system, nothing is as necessary to the wellbeing of the people as an adequate system of parks and public reservations.[55]

The Parks Commission continued its work until 1935, when the so-called Bloodless Revolution occurred. Democrats, representing immigrants and the working classes, took control of state government and replaced its commissions with state departments. But before it was done, the commission had extended its efforts beyond the metropolitan area and accepted gifts of land in rural towns that would one day become additional state parks. In its final report, the commission discussed plans for what became Burlingame State Park in Charlestown, a rural town on the state's south shore, as well as Goddard Park in Warwick, Haines

Park in Barrington, and parkland in Point Judith. Burlingame, in fact, now a 2,672 acre park and campground, was named after Edwin A. Burlingame, chairman of the commission in the 1930s.[56] Also in the 1930s, the commission was using workers on relief, as many as seven hundred, to work on twenty ballfields and parks across the state.[57]

Shortly after World War I, John L. Curran, a lawyer and sportsman, organized the Rhode Island Fish and Game Protective Association. It grew to be one of the largest sportsmen's clubs in the state and it lobbied the legislature for stiff conservation laws. It helped end the market-hunting boom in Rhode Island and pave the way for limiting daily catches of freshwater fish, enactment of shellfish laws, and rules to halt pollution. Honored for his work at a dinner in 1961, Curran recalled that in the early years "conservation was like a bowlegged girl who is

At the turn of the twentieth century, downtown Providence was the focus of attention in Rhode Island, with urban improvement rather than land conservation given the highest priority. It was not until later that business leaders were able to persuade legislators to establish citizen-led commissions to create public park and recreation opportunities for working people and their families in order to provide "fresh air and happy surrounding" and to ensure "the efficacy of labor and contented people."
(Providence Public Library)

never invited to a party."[58] He said in those years the idea of preserving natural resources was far-fetched, especially to market-hunters who would shoot up to one hundred grouse a day, and sell them to New York restaurants for as much as three dollars apiece.

Also in the 1920s, decades of drinking water shortages and contamination problems in Providence finally prompted the city to take bold action. Instead of drawing water from the polluted Pawtuxet River, the city went upstream to the town of Scituate, and built a dam 3,200 feet long and 100 feet high.[59] As the reservoir slowly filled, it flooded several villages. The water supply board annexed thousands of acres to protect the surrounding watershed. The foresight of city officials from that era is still appreciated. They protected thousands of acres of now pristine forestland and the reservoir now is the main source of water not just for Providence, but for most of Rhode Island.

In 1932, the state created its first state forest, the George Washington Memorial Forest in Glocester.[60] In response to the Depression and the loss of jobs in the mid-1930s, the federal government also created the Civilian Conservation Corps, which planted trees, built fire towers, and cut trails and roads through the forests in Rhode Island, as well as many other states. Nationally, preservation of forests began with the passage of the Forest Reserve Act of 1891.[61] Six years later Congress passed the 1897 Organic Act, which gave the U.S. Department of the Interior General Land Office and the U.S. Geological Survey three management goals for forest reserves: to improve and protect public forests, to secure favorable water flows, and to provide a continuous supply of timber.[62] In 1905, the Forest Service was created within the Department of Agriculture and it assumed responsibility for what were soon named the national forests. In 1911, the Weeks Law enabled the federal government to buy forest lands in the East.

The losses of more forest habitat and coastal features in the hurricane of 1938 are difficult to calculate. Forestry officials said many abandoned farms had become overgrown with white pines—thousands of which were toppled in the storm. Rhode Island lost its lobster hatchery in Wickford, which was considered the oldest and most successful in the world.[63] The oyster growers lost $300,000 of their oysters. They simply washed away or were covered with silt. The state's trap fishermen lost all of their nets and gear. The state's marine laboratory at Fort Kearney on Jamestown was destroyed. But in the midst of all that destruction, a new attitude toward Rhode Island lands seemed to be emerging.

USE WITHOUT ABUSE

Harold L. Madison, field executive of the Rhode Island Audubon Society, described three periods of changing attitudes about natural resources in Rhode Island. From 1600 to 1900, he said the attitude was "complacency, accompanied by need, greed, ignorance, larceny, and waste."[64] From 1900 to 1940, he said concern was growing, "taking the form of protection of wildlife, preservation of forests, restoration of water areas, and flood prevention. From 1941, he added that the attitude was "conservation, use without abuse by way of research, education, legislation, and appropriation."

After World War II, many changes were underway. One was a greater appreciation for conservation. On May 7, 1950, Rhode Island Conservation Day was celebrated with a mile-long parade in Johnston, R.I.[65] It was sponsored by the Rhode Island Wildlife Federation and the Johnston Wildlife Restoration Club. The parade concluded at the Johnston Memorial Park, where six trees were planted.

One of the leaders in that period was Ruth M. Gilmore, a Providence woman who worked with the poor before books such as *The Grapes of Wrath* and *Tobacco Road* turned her into a conservation advocate.[66] She cofounded the Rhode Island Conservation Workshop, a cooperative environmental college education project of the Rhode Island Wildlife Federation, the Rhode Island College of Education, the State Department of Agriculture and Conservation, and Rhode Island State College. She also took part in national and international educational programs on conservation. Gilmore once wrote that her campaigning for sound conservation practices got her "blackballed" by the American Forest Products Industries, Inc. (AFPI), an industry group. They said she was "pink," she recounted, and that they also had the clout with editors of magazines and textbooks to check them for "accuracy."[67] In a letter to a friend she joked about a chilling act of censorship by the forest industry.[68] The letter showed her energy and humor. She wrote about traveling to Cincinnati for the Izaak Walton League Convention and meeting someone at the Cosmo Club to discuss environmental issues over cocktails. Because of a forestry booklet she had written, she says that the head of the industry association referred to her in public meetings as "that damn Ruth Gilmore." She wrote that it was clear to her that the industry placed some of its members into leadership positions of all "so-called conservation organizations" so they could influence policy. She wrote: the Pennsylvania Forestry Association "was not aware of this and got sort of a rude awakening." The association was going to print a conservation article by Gilmore when, at the last minute, industry people insisted on seeing it. The industry warned that if the article were published, its two hundred members would drop out of the forestry association and hundreds of dollars in advertising would be dropped. So, Gilmore wrote, "The poor editor got caught between the devil and me. The devil won of course." She said the industry group had gone even further—"and am I laughing. It seems they have actually had me blackballed with *Colliers* and *Saturday Evening Post*! Me—who can't even write! I love it. AFPI is determined to discredit the conservation movement." She passed on one other hot tip. She said she learned that "Secretary Marshall is hell bent to let oil men get into the new 40,000 acre withdrawal from the Condor Refuge. The last day I was in Washington, via the grapevine, I heard that at a cabinet meeting April 10 he ridiculed wildlife preservation and said 'what are a few birds when there might be oil there?'"[69] Gilmore's letter to John H. Hinman, president of International Paper Co., was just as pointed. She praised a publication by the company that she said provided facts without trying to make people believe that protection from fire will cure all forest ills. She said that it was sharply divergent from the publications of the industry group, the American Forest Products Industries, Inc., which she said "prates of forest conservation" while trying to make "people believe that if they just forget about forests and leave everything to the forest industries, the United States will have abundant forests forever."[70]

Gilmore wrote about Marshall to another relentless champion for the Rhode Island environment in the 1950s and 1960s—Prof. Donald J. Zinn, a marine ecologist and professor of zoology at the University of Rhode Island. Zinn volunteered for years with the National Wildlife Federation and was a longtime board member. He wrote two textbooks, hosted a weekly television show on the environment in the 1960s, and was an advisor to numerous organizations working on conservation.[71] Records at the U.R.I. Library also show Zinn tirelessly working at every level of society to improve the ecology of Rhode Island. He started a junior conservation club at a local high school, judged science fairs, corresponded with the state's congressional delegation to express his views on environmental legislation, and even voiced concerns to the head of the National Wildlife Federation about the new washing machines reaching the market:[72]

> I have been looking at the fantastic advertisements and listening to the even more fantastic claims over the radio by the makers of washing machines and similar gadgets, and wondering when, if ever, they are going to recognize the growing crucial water situation in this country.

Zinn proposed writing to manufacturers or creating a prize to encourage engineering innovations. He wrote letters defending the environmental record of President Eisenhower, who he described as being "on the plus side." The preceding Democrat administrations, he wrote, were "halting, fearful, and do-nothing until forced at best."[73] In 1957, he wrote to a federal agency, the U.S. Soil Conservation Service, to complain about "dirt merchants" in South Kingstown who buy cheap land, sell off the topsoil, mine the gravel, and then go away, leaving "land that once supported agricultural crops or wood nothing but a pitted and ragged terrain." In a lecture in 1954, Zinn took Rhode Island's cities to task for doing a poor job treating their sewage. Woonsocket, Warwick, East Greenwich, Middletown, Wakefield, and Newport are all bad, he said in his talk.[74] East Greenwich dumps into East Greenwich Cove, he said. Meanwhile, "enforcement of the shellfish laws has become a big farce." He said a brook in Middletown that ends at the beach is little more than a sewer fed by "trailer camps and other pollution sources." Newport, he said, pours four million gallons of raw sewage into its harbor every day while the state finally convinced the Navy to stop adding another two million gallons of its raw sewage. Zinn said, "Our antipollution law is now seven years old, and all of the above communities are breaking that law." Providence, he said, was "in a class by itself." He said it was pouring fifty million gallons of sewage into the bay from a worn out plant. But at least they had a thirty-five million dollar plan to upgrade the plant. He praised the work of the Blackstone Valley Authority, East Providence, Warren, Bristol, Jamestown, Westerly, Quonset, U.R.I., the Howard Complex, and Wallum Lake. He said Cranston was outstanding. He concluded by saying:[75]

> It seems to me that our job is to "unmudify" general thinking in this matter. We must work constructively with those interests that recognize the need for wise use of our water resources, and sell by any and all means, including scare promotion if necessary, to the ignorant and lazy in mind, that it is in their best interests to consider water uses other than their own — or even water uses in general. In some instances, it may be necessary to induce persuasion in the form of rigidly enforced regulations. It can be done. Intelligent water conservation is everybody's business.

Mathias P. Harpin wrote a series of stories for the *Providence Journal* in the summer of 1945 about the condition of the state's natural resources, with grim conclusions. He said timber experts worried that the state's forests were disappearing. More trees were cut than planted, and fires, insects, and disease caused additional harm. He said the state had 90,000 acres of forest, but it was poorly maintained. Too many people lived too close to natural lands, creating more threats to nature, he wrote. Power dredging wiped out the state's quahog beds. Some 50,000 bushels of scallops were taken in 1940 from Little Narragansett Bay and Salt Pond, wiping out the scallop market.[76]

In 1950, Democratic senators from Rhode Island, New York, and Connecticut introduced legislation to study and draft a plan for the integrated "use, conservation, and development of natural resources" in the seven Northeast states.[77] Their bill was inspired by President Truman, who made several speeches to Congress and at White House press conferences pointing to "real and serious problems" of soil and forest conservation in the Northeast, as well as conservation and management of water to prevent floods, to provide for domestic and industrial water supplies and to furnish low-cost hydroelectric power. The bill called for creation of a seven-person commission. The bill was not passed, but later that year, President Truman ordered federal agencies to conduct a natural resources survey of the Northeast.[78]

> The study would include power, navigation, flood control, forest management, fish and wildlife conservation, mineral development, industrial and municipal water supply, pollution control, recreation, and soil conservation.[79]

A commission representing several federal agencies and the states was formed. It held a series of public hearings and information sessions around the region. Truman asked Congress to appropriate $350,000 for the survey.[80] One of the more unusual proposals to the commission came from two Rhode Island state legislators and Henry Ise, chief of the Rhode Island Division of Rivers and Harbors. Ise, and Cranston Sen. Leon W. Brower and Rep. Nelson F. Duphiney of Pawtucket suggested spending $40 million to create an inland waterway from Westerly to Wickford, making extensive use of the South County salt ponds.[81] The waterway would be far cheaper than the $250 million spent on inland waterways ranging from Texas to Virginia, the legislators said. The waterway would provide a safer route for shipping than the coastal route, exposed to open and often stormy seas. Some five hundred people attended the commission's meeting in Springfield, MA, as it listened to ideas in major cities throughout the region. Much of the testimony focused on hydro power in the region, with interests pushing for a public power authority such as the Tennessee Valley Authority and others supporting private power. The New England–New York Inter-Agency Committee spent $5 million and several years on its study and finally issued its conclusions in 1955—a forty-five-volume report weighing two hundred pounds.[82] (The National Archives regional facility in Waltham, MA reports that the final report along with correspondence, data files and committee work has a volume of twenty-one cubic feet!)[83] The commission sidestepped one big issue when it assumed water power sites in the region would be privately financed, rather than the publicly supported government authorities used elsewhere. New Englanders were opposed to public power development. The change was attributed to the changes at the highest level of government. The *Providence Journal* reported:

> With the change in administration in 1953, and the disappearance from the committee of its ardent power advocates, the climate within the federal-state group rapidly changed for the better. The project is ending with the most cordial relations between state and federal personnel, and a gratifying meeting of minds on practically all matters. Because of this change of climate, an interesting postscript to Operation NENYIAC seems likely to be written. The state representatives have signified their desire that the federal government cooperate in maintaining a continuing body, which, in cooperation with the state members, will stimulate and record action on the innumerable recommendations contained in the report.[84]

The commission studied each of the twenty-six river basins in the region and made recommendations on flood control, water supply, pollution control, hydroelectric power development, land and forest management, fish and wildlife conservation, development of mineral and recreational resources — even insect control. The report's overall tone was optimistic. It called New England a place in which "to live, work, and play."[85] It said the region has just about everything to make life worthwhile: an abundance of water, a "climate conducive to hard work," and fish and wildlife resources among the best in the nation. Three years later Congress was still awaiting legislation to implement the commission's recommendations. The widespread damage caused by Hurricane Diane in 1955 highlighted the need for additional flood control projects. The Army Corps of Engineers completed seven projects at a cost of $10 million, and still had twelve more costing $104 million under construction. It was also planning six more, including the Fox Point Hurricane Barrier in Providence.[86]

Meanwhile, in Rhode Island in 1952, Democrats were making the environment a significant issue in the statewide election campaign. Governor Dennis J. Roberts, in a broadcast from radio station WEAN, said he supported a program to develop the state's natural resources.[87] He criticized the Republican Senate for killing a bill that would have given fish and game wardens the power to search buildings and trucks without having to obtain warrants, saying it was a thoughtless act that killed a measure that would have reduced the "menace of bootlegging in undersized and polluted shellfish."[88] Roberts also pledged more efforts to support agriculture, expansion of game preserves, and hiring biologists to work on striped bass.

Massachusetts was the first state to enact legislation allowing communities to establish their own conservation commissions to save natural resources. In 1960, Rhode Island approved similar legislation.[89] It authorized communities to create commissions that would be allowed to receive gifts of money, land, buildings and other properties with local council approval.[90] Within two years, fifteen Rhode Island communities used the new law to create their own conservation commissions.[91] Charles H.W. Foster, commissioner of conservation in Massachusetts, said of efforts in his state at the time, "So far our groups have been hard at work at education. They may seem slow moving; but they are moving. Once everyone understands that to conserve means to save, we shouldn't have much trouble."[92]

In the 1960s, Governor John H. Chafee (later to become one of Rhode Island's U.S. senators) put his strong interests in the environment to work. He created Governor Chafee's Task Force on Natural Resources to study the state's conservation problems and recommend strategies for going forward. In collecting information, the task force got an earful. Bruce A. Blackwell, secretary of the East Providence Conservation Commission, released a statement calling Narragansett Bay, the state's largest natural resource, a "filthy, stinking cesspool for thousands of Rhode Islanders."[93] He said it was time to create one state agency to focus on cleaning up bay pollution. At that time, responsibilities lay with several agencies, but so narrowly that the "burden of Bay pollution falls upon no one."[94]

The task force found that 86 percent of the state's population could be classified as urban, without ready access to natural resources and outdoor recreation.[95] It incorrectly predicted the state's population would grow from 860,000 to 1,200,000 by 1985. (Rhode Island's population is little more than one million as it enters the twenty-first century.) Applying guidelines from the National Park Service, it recommended that

Commissioned in 2002, the seven-foot statue of former governor and long-time U.S. Senator John H. Chafee, the ranking member for more than a decade on the Clean Water, Fisheries, and Wildlife Subcommittee of the Senate's Environment & Public Works Committee, depicting him typically striding off to yet another mission, could have been placed anywhere in New England, for Chafee was characterized by the EPA's regional director as the legislative leader most involved in every significant fight for the environment of our generation. Yet, the site ultimately chosen for the statue was at the grass-roots level—one overlooking Rhode Island's Narragansett Bay at 460-acre Colt State Park, the place where Chafee as governor demonstrated his first, significant commitment to land conservation.
(PROVIDENCE JOURNAL)

there be set aside sixty-five acres of public recreation land for every 1,000 of population. By those standards, it found the 33,700 acres then available fell far short of the 55,900 acres that should have been set aside. And if the twenty-one-year growth prediction held true, it found that 68,000 acres should be set aside. "In one short generation, then, Rhode Island can be choking to death in an unrelieved pattern of urban slums and suburban sprawl, or it can be a model of metropolitan civilization," concluded The *Providence Journal* in an editorial.[96] The task force recommended a $5 million bond issue to acquire land, legislation to strengthen and encourage local conservation efforts, creation of a new Department of Natural Resources, and tax reforms to encourage private conservation. The report also called for a balanced natural resource program in Rhode Island based on four elements: the coastline, the metropolitan area, the interior woodlands, and the state's unique natural sites.[97] It called for setting aside open space, creating a wide variety of outdoor recreation facilities, modest development of areas prone to flooding, and building roads that provide attractive vistas rather than destroy scenic attractions. The state published a map, showing locations where open space should be acquired and preserved. The locations circled the state, from Block Island to Newport, Bristol, Prudence Island, Cumberland, Burrillville, Exeter, and Charlestown.

Governor Chafee proposed his $5 million Green Acres bond issue in the fall of 1964. It called for six new state parks, including parks at Colt Farm in Bristol, the Snake Den section of Johnston, Prudence and Patience Islands, Brenton Point in Newport, Diamond Hill in Cumberland, and East Matunuck State Beach. He borrowed the term, "Green Acres," from a similar, though much larger, program in New Jersey. His Rhode Island plan was based on a report by Julie O'Brien of the Rhode Island Development Council.[98] Opposition came primarily from residents in Charlestown, Westerly, and West Greenwich, who feared the loss of property tax income. The program was endorsed by both political parties and the Chamber of Commerce.[99] Chafee campaigned around the state. The program passed by nearly a two to one margin. The only towns to vote against it were Charlestown, Hopkinton, Richmond, and West Greenwich. Almost every potential park identified in the program was eventually developed. Green Acres was the perfect match of a strong, articulate leader, and a public willing to follow. Every open space bond issue proposed by the state in the following decades passed by wide margins.

Also in 1964, a Massachusetts judge issued a ruling that attracted attention in Rhode Island. The judged ruled that the state had the authority to prevent a private landowner from action that would harm the marine life in an adjoining marsh. The developer wanted to cut a channel through forty-nine acres of marsh on Onset Bay so he could develop a marina on nearby land. The state's commissioner of natural resources, Charles H.W. Foster, had no objections to the marina. But he was opposed to plans to dump dredged material in the marsh. He invoked a new state law that he said gave him the power to protect shellfish or marine life from dredging projects. This was the first court test of that law. Judge Horace T. Cahill ruled that protecting natural resources is an overriding interest to the developer's property rights. The *Providence Journal* observed that at that point, there was no similar law in Rhode Island and the state's chief of harbors and rivers, Henry Ise, told the *Providence Journal* he felt it was not very democratic for the state to try to tell a private landowner what he can do with his marsh.[100]

Alfred L. Hawkes was director of the Audubon Society of Rhode Island for thirty-five years and constantly educated the public about conservation in speeches, columns, and letters to the editor. He recalled during this period teaming up with an Audubon board member, Aram Berberian, a brash lawyer, to enact legislation protecting salt marshes and authorizing the creation of local conservation commissions. They had tried to stop the town of Middletown from allowing the town dump to overwhelm a salt marsh behind Second Beach. The first time they went to court, they lost because there was no law.[101] In those days, he recalled, if you wanted to get legislation passed, you worked quietly. "You got it introduced by somebody that wouldn't be questioned much. He would introduce it, and it would just kind of float through. They didn't know what they were doing. I mean they didn't understand it or why, but so-and-so wanted it, so it was all right. We got the pesticide legislation in the same way." Hawkes was always ready with a pointed comment when reporters called. In the early 1960s, he said of the state of conservation in Rhode Island: It "is very, very poor." He said there were big problems with towns creating dumps in marshlands, developers scraping away top soil, and mosquito spraying in swamps.[102] He observed that after the conservation commission legislation passed, nothing had happened. So he and Berberian talked to every Town Council in Rhode Island, and one by one, they started creating conservation commissions.[103] The first statewide conference of conservation commissions drew representatives from seven communities to a meeting place in Newport. Charles H.W. Foster, director of the Massachusetts Department of Natural Resources, was guest speaker. He reported on conservation commission activities in his state.[104]

Hawkes also was a member of an unofficial environmental think tank known as the Natural Resources Group and organized by Bradford Kenyon, a retired business executive. Members included, among others, C. Robert Frederikson, the environmental writer at *The Providence Journal*; Calvin Dunwoody, chief of the state division of forestry; and Stuart O. Hale, a former newpaperman who later became a key aide at the University of Rhode Island's Graduate School of Oceanography. Hawkes said the group would meet for about an afternoon a month at Kenyon's house, where they were well supplied with food and liquor.[105] The group would study local environmental problems and recommend solutions. It was their idea to consolidate various state agencies into one Department of Environmental Management. The reorganization did not work out entirely as they had planned, but at least it happened. They were unable to merge the state Water Resources Board because it had more clout than they did. They were successful in instigating a state wetlands law, similar to one in place in Massachusetts.

Hawkes said he felt one of the Audubon Society's biggest accomplishments during his tenure was buying and preserving open space for wildlife. The society started with a few small donations and by 2007 owned nearly 7,000 acres in refuges across the state, with another 1,800 acres protected by conservation easements. The society is one of the largest landowners in the state. Hawkes recalled buying some of the salt marsh in Barrington's Hundred Acre Cove at a tax sale. He got much of the rest by simply knocking on the door of the landowner. "She said, you know I've been wondering what the heck to do about that land. Well, eventually she gave it to us. Took her a long time, but eventually she gave us the whole thing."[106] Hawkes added: "You have to constantly give people opportunities to do things 'cause sometimes they won't do it on their own, but you give them an opportunity, you just lay it out, they will."

During this period, as new laws were put in place to protect certain fragile areas, Rhode Island continued to lose others. In 1965, the U.S. Department of the Interior published a report chronicling the loss of coastal wetlands in Rhode Island during the preceding decade. The state lost one hundred twenty-three acres of coastal wetlands in 1954.[107]

It lost another seventy-one acres from 1954 to 1959, and it lost fifty-two acres more from 1959 to 1964. Most of the losses were for road and bridge construction.

In the early 1970s, spurred on by the Natural Resources Group and local scientists, Rhode Island led the nation with new efforts to protect its coastline. Faced with a surge of development on its beaches, the state proposed creating a Coastal Resources Management Council. Details were drawn up by Statewide Planning staff person Susan Morrison and a nine-member commission.[108] Legislation failed in the first year, but passed in the second. The council got off to a rocky start. Daniel Varin, a longtime state planner and a member of the Natural Resources Group, recalled one problem: "The first council and particularly the first chairman did not get along at all with the Department of Environmental Management. I went to the first meeting of the council and in less than an hour, they were screaming at each other. That went on for several years."[109]

Varin said the new agency in its early years made decisions on an ad hoc basis, with very little input from professionals. "It was a matter of years getting that program reoriented to making decisions based on something other than who the applicant was and whatever came off the top of their heads at the time, and I had real fears that it would never happen, but it did."[110]

In fact, a year after Rhode Island launched its coastal program, Congress passed a Coastal Zone Management Act that led to the creation of similar programs in coastal states around the country. During the period that Rhode Island created its coastal agency, it also founded a think tank at the University of Rhode Island called the Coastal Resources Center. In the last three decades, scientists at the Coastal Resources Center have helped advise and guide coastal managers throughout the United States and, under the aegis of the U.S. Agency for International Development, throughout the world.[111]

The bicentennial of 1976 brought more help to the conservation movement. Peggy Sharpe, a longtime environmental advocate in Rhode Island, recalled that garden clubs, once little more than social centers for housewives, decided to mark the bicentennial by working on the American Land Trust Program. Their goal was to get every garden club in the country to work on a local land preservation project.[112] Cathy Douglas, the wife of Supreme Court Justice William O. Douglas, was spokeswoman for the campaign. Sharpe recalls making the rounds of garden clubs in Rhode Island and getting a donation from every one for land conservation. Sharpe went on to work with The Nature Conservancy in its early years of land preservation in Rhode Island in the 1970s and 1980s.

The TNC sent one person, Eve Endicott, to Rhode Island to look for land to save. One of her first projects was the now famous Lewis-Dickens Farm on Block Island. Sharpe used her contacts to bring in partners. One key partner was the Champlin Foundations, a charitable foundation endowed by a longtime Rhode Island businessman. For years Champlin focused on giving to causes such as schools, hospitals, and libraries that benefited children. But at that time it also began donating to land conservation. Since then, it has donated tens of millions of dollars. Sharpe said it was also important at that time that Robert L. Bendick Jr., the head of the state Department of Environmental Management and a planner by training, was very interested in land conservation. The DEM, The Nature Conservancy, and Champlin worked together for years on many successful conservation efforts, Sharpe recalled.[113] In 2002 The Nature Conservancy offices completed a major fund-raising campaign. The total take in little Rhode Island, Sharpe said, was $56 million. Sharpe said she thinks there was simply a turning point in peoples' awareness in the 1970s that brought the environmental movement to the forefront. Her husband, Henry D. Sharpe, ran a large machine-tool business in Rhode Island for many years. She recalls that he, as so many others, abruptly changed his attitude. He later was at his wife's side for many environmental causes in Rhode Island. She said:

> It takes a lot of time for people to become aware. You know, I remember my husband, who was in business for a lifetime, saying that some of the issues we're concerned about now never crossed their minds. You know, the fact that you had to worry about polluting the water, polluting the air, or polluting the land. Never, we just never thought about it. As I said earlier, the best things in life, which are air, water, and natural resources, were taken totally for granted. So that in those early years it was really important to really work with the lay public,

> to try to help them understand what we must do. Of course, in the seventies that was a moment when people like Sen. John H. Chafee were doing, you know, developing laws to take care of our earth. You know, the Clean Water Act, the Clean Air Act and all these things. And that was a very important time. And there was more funding for environmental protection for a while than there later was.[114]

In 1978, the Natural Resources Group helped promote legislation that created the Agricultural Lands Preservation Commission, charged with inventorying farmland in Rhode Island and empowered to buy the development rights to farms.[115] Varin, the state planner, was placed in charge of the commission in its early years. He recalls getting a two million dollar grant from the Environmental Protection Agency and using some of that money to pay the U.S. Soil Conservation Service to do a complete soil survey of the state.[116] The survey provided clear answers of where soils types could support vibrant agricultural activities, and where they couldn't.

Also in 1978, one family, the Laphams, stepped forward with an act of generosity that played a key role in the landmark conservation effort on Block Island. For years, Captain Rob Lewis, and later his sons, Keith and David Lewis, led conservation efforts on the island, starting with a campaign to save the island's Rodman's Hollow from New York developers. The state Department of Environmental Management made its own contributions by preparing a top ten list of island landmarks to be saved. At that point, all of the island's most scenic spots were still in private hands. Just before the end of the year, F. David and Elise Lapham stepped forward with an historic contribution. They gave The Nature Conservancy an easement to some 145 acres they owned on the island's northeast shore. The easement allowed the public complete access to the land and its trails. More importantly, the state DEM was to use the easement as a match to obtain more than $1 million in federal funds, which it used to buy many of the other landmarks on Block Island. The partnerships among the Block Island Land Trust, the Block Island Conservancy, the town, the DEM, The Nature Conservancy, and the Champlin Foundations continues to this day and have resulted in Block Island protecting more than 41 percent of its landscape.

RHODE ISLAND DISCOVERED—THE AUTOMOTIVE ERA

As the industrial era ground down in Rhode Island, state planners like to say another era arrived in the 1980s—the Automotive Era. The completion of new roads, better cars, and growing desire by people to escape cities and older suburbs finally caused an eruption in the 1980s—all of Rhode Island's coastal and rural places were now in commuting range. Commercial construction jumped from $44 million in 1981 to $95 million in 1987 and the number of new housing permits rose from 2,400 in 1982 to more than 7,000 in 1987.[117] State planners wrote:

> History will record the 1980s as a watershed for Rhode Island's landscape. Rhode Island was, as its auto tag slogan invited for decades, "discovered" in the '80s: by Boston-bound commuters, by tourists, by out-of-state developers, by national retailers, by global industries. While the decade's stratospheric growth rates are unlikely to be repeated soon, the pressure of growth on the land will surely return as prosperity takes root.[118]

Rhode Islanders felt threatened by that building boom. It violated their sense of place. Farms turned into subdivisions. Villages were overwhelmed with commercial development. Woodlots were cut down. Rhode Islanders responded by approving more than $100 million in bond issues to buy open space land and to preserve farms. Voters supported three statewide planning and zoning laws that required communities to better plan for the future, improve conservation and steer development to urban areas.[119] While local leaders had begun planning for parks and open spaces nearly one hundred years earlier with the Metropolitan Park Commission, the need to plan and do more was intensified by the boom. In 1984, DEM Director Robert L. Bendick Jr., whose background was in community planning, issued another challenge to Rhode Islanders that reflected his vision for the state's future.[120] He said a linear park should be created along the Blackstone River and its canals, an area of abandoned mills and old housing. U.S. Senator John H. Chafee persuaded Congress to provide the authorization and funding. Some twenty-three years later, the John H. Chafee Blackstone National Heritage Corridor is considered a model for public-private heritage partnerships around the country. An evaluation by the National Park System Advisory Board in 2006 found the:

> ... corridor fostered restoration of dozens of historic buildings for private and public use, annual cleanup efforts, regular water quality testing, and improved water access. The commission's work has generated thousands of volunteers and new recreation enthusiasts. Residents, businesses, and local governments are reconnecting with the Blackstone River, generating new economic vitality, valued at twenty-two times the National Park Service investment of $24 million over the past eighteen years."[121]

Bendick's vision included one bike path built along the river and connected to the Providence waterfront and another bike path from Providence to Colt State Park. Bendick's suggestions also included further development of the Bay Island Park system on Prudence Island and other small islands in Narragansett Bay, expansion and improvement of Scarborough Beach in Narragansett, and preservation of more land on Block Island. Almost every goal set by Bendick has been achieved, while work remains on the few that are still incomplete.

In 2004, more than 2,100 conservationists and environmentalists from around the country traveled to Providence, Rhode Island to attend the four-day National Land Trust Conference. It was the group's highest attendance ever, and many came to see the beautiful lands that Rhode Island had protected and preserved. There were sold-out trips to Block Island, Prudence Island, Little Compton, Tiverton, South County grasslands, and the Pawcatuck Borderlands—Rhode Island's largely undeveloped border with Connecticut and the only remaining expanse between Boston and Washington, D.C. that is not illuminated at night. Rhode Islanders were scheduled to vote on a $69.7 million environmental bond issue that fall with proceeds targeted toward open space acquisition, recreation facilities, and reducing pollution of Narragansett Bay.

Rand Wentworth, a former land developer who said he had an epiphany when he watched one of his own projects destroy 150 acres of beeches, wildflowers, and a spring near Atlanta, was the president of the Land Trust Alliance, the conference sponsor, and gave the plenary address.[122] He described New England as the birthplace of land conservation in the United States:

> It is appropriate that we are gathered here in New England, the place that formed so many of our country's patriots — toughened by its rocky soil, long winters, and strong ale. For perhaps the same reasons, New England also served as the birthplace for land conservation in America. In 1634, just fourteen years after the Pilgrims landed, the Boston Bay Colony set aside the Boston Common and, for the first time in American history, people voted to tax themselves to protect greenspace with the creation of the Boston Gardens. It was 150 years ago that Henry David Thoreau first published Walden, which planted seeds in the American consciousness that later blossomed into the guiding philosophy of the conservation movement. Just thirty-seven years after Walden was published, citizens of Massachusetts created America's first land trust, the Trustees of Reservations. And it was here in New England that land trusts gathered in 1982 to form the Land Trust Alliance.[123]

Wentworth talked about constant threats to natural resources caused by sprawl and new threats posed by proposed federal changes to tax shelters and charitable giving that fund much land conservation work. But he also told two uplifting stories. He described Mary McFadden, who founded the Wareham Land Trust in a blue-collar town with the lowest income in southeastern Massachusetts. Wentworth said when McFadden asked city officials if they would support a conservation bond issue, they scoffed and said that was a "yuppie issue" that would never pass. Friends told her to give up. But she appealed to the town's school children and invited them to create stories and pictures of what they loved about their town and what was changing. The children showed their work to their parents. And when the vote was tallied, Wareham passed the conservation measure with a 75 percent plurality — the largest of any town in the state.

Wentworth also told a Rhode Island story. He described Louis Escobar, who inherited his farm on Aquidneck Island from his parents, Portuguese immigrants. Escobar could have sold his farm for a fortune, Wentworth said, but he loves his farm and the children who come to his cornfield maze each fall. So he granted a conservation easement to the Aquidneck Island Land Trust that will hopefully allow him to keep milking his cows as long as he likes.

In the face of growing challenges, Wentworth said Rhode Island had forty-three land trusts that helped protect some of its most cherished landscapes. He added:

> There is a great story at work here: Conservation is moving out of federal government and into state and local hands. This is not because the federal agencies are unwilling — indeed, the federal government has been the major champion for land conservation over the past century — and we certainly need their leadership in the years ahead. But federal deficits and political opposition have driven federal funding to a new low. Fortunately, voters at the local and state level have stepped in to fill the gap, approving $24 billion in funding for conservation over the past six years. And, in just five years, local land trusts doubled the acres they have protected.

In 2006, a handful of people in the rural town of West Greenwich, R.I., proved they weren't done with conserving Rhode Island. The people in this small town voted 632–16 to borrow $8 million to help The Nature Conservancy buy 1,566 acres of land, and to block a developer from erecting 168 new houses that the town didn't want.[124] The decision to borrow the money will cost the owner of an average house in West Greenwich $150 in extra taxes annually during the next twenty years. But the vote also meant that the land will forever be set aside for hiking, hunting, and fishing. The vote took place at a town meeting, without any discussion. Town residents did not feel it was necessary. They simply lined up at the voting booths and cast their ballots. To honor their decision, the Environment Council of Rhode Island in 2007 bestowed its annual John H. Chafee Conservation Award to The Nature Conservancy, and the entire town of West Greenwich.

LOOKING AHEAD IN RHODE ISLAND

In 2007 state officials estimated that about 20 percent of Rhode Island's landscape was protected from sprawl and development.[125] The state controls about 76,000 acres through outright ownership and conservation agreements. Local governments and nongovernmental organizations owned another 43,000 acres and protected thousands more with deed restrictions or plans for further actions. But state officials were not satisfied with their efforts. A new, twenty-year statewide plan released in 2006 noted that while 80 percent of Rhode Islanders still lived in urbanized areas along the state's bay and rivers, there is a continued trend of people moving out to the state's rural areas. There was a tendency towards single-family home ownership, utilizing much larger lots than in the past. The amount of developed land was increasing nine times faster than the population and about 30 percent of the state's land is developed.[126] At that rate of growth, the planners predicted another 109,000 acres could be developed by 2025. The planners recommended continued efforts to buy and protect rural lands, and new efforts to concentrate new state and private development within existing villages and urban areas. The planners pushed for new greenways and bike paths around the state.

Will Rhode Island meet the challenge of conserving its land? There are some positives to consider. Rhode Island is institutionally more prepared than ever to act. The State Department of Environmental Management has a good working relationship with The Nature Conservancy[127] that provides the expertise to identify good open-space and agricultural prospects. The Champlin Foundation and state bond issues provide the funding to acquire them. Towns are better prepared than ever. A state that had no land trusts before 1972 now has forty-three. The public seems supportive—it has yet to turn down an open-space bond issue. A buy-local-food initiative is growing rapidly, which should provide more support for the farms that remain. Finally, the economy may help. The price of energy is soaring so high it may prompt people to reduce their commutes.

But there are still reasons for concern. Rhode Island appears strapped with chronic state budget deficits that limit the state's ability to launch new initiatives that cost money. There are concerns that the state or some other entity could find ways to develop properties that were considered set aside for conservation. In 2005, Governor Donald Carcieri proposed using a corner of the rural, Big River Reservoir property for a new state police headquarters. Environmentalists were furious because legislation had prohibited development there. The governor backed off, surprised at the hostile reception.[128] At the same time, there seems to be a consensus that taxes are too high already, so it will be difficult to generate new funding through increased taxes. And Rhode Island—with its spectacular bay, widely forested areas, and scenic shoreline—continues to look attractive to wealthy people from out of state looking for a nice quality of life. One Philadelphia developer alone confirmed plans in 2007 to invest $1 billion on Rhode Island's Aquidneck Island. He is constructing a twenty-two-story tower of condominiums that will range in cost up to $7 million, the largest marina in New England, and a "village" with 1,000 new residences.[129]

Rhode Islanders have shown through the years the propensity to do what it takes to thrive economically. So if tourism is going to continue to be a major part of the state's economy, one might expect Rhode Islanders will do what it takes to ensure their state continues to be an attractive place to visit. But, if history is any guide, the job won't be done quickly, and it won't happen without controversy and disagreement.

One final story may offer a glimpse of the future. In August 2003, a perfect combination of rainy weather, hot days, and little wind created low oxygen conditions in Narragansett Bay that caused the biggest fish kill that anyone had ever seen. The public and the state's political leaders reacted powerfully. Several study commissions were created. The next session of the General Assembly passed six new bills implementing ecosystem based management strategies so far reaching that famed oceanographer Bob Ballard toured the State House one night and congratulated legislators for enacting some of the most modern environmental legislation in the country. Further action was anything but swift. The Republican governor and Democratic legislature did not agree on who should lead a new bay management team. Legislators cut budget proposals to do more monitoring to support the ambitious management goals. But improvements did not grind to a halt. The state ordered communities to make changes at their sewage treatment plants to reduce nitrogen inputs into the bay, and they did so. The state Department of

Environmental Management cobbled money from other sources to do at least some of the monitoring it believes necessary. The state's scientific community got together to compare studies of the bay and to publish a very comprehensive book of bay science. A new management plan was scheduled to be presented in early 2008. There is no single, persuasive leader such as Rhode Island enjoyed in the past — no John Chafee, no Bob Bendick or Rob Lewis. But there is general agreement that something needs to be done, and people are finding a way to do it.

NOTES

1. "Farms for Sale or Lease in Rhode Island," R.I. State Conservation Commission, Bulletin No. 1, March 1911
2. Rhode Island Tourism Division, Rhode Island Economic Development Corp., http://www.riedc.com/riedc/industry_clusters/18/, (accessed Nov. 12, 2007).
3. Carl Bialik, "How Big is Rhode Island Anyway? *The Wall Street Journal Online*, Oct. 30, 2007, http://blogs.wsj.com/numbersguy/how-big-is-rhode-island-anyway-214/
4. William G. McLoughlin, *Rhode Island — A History*, W.W. Norton & Co., New York, 1986, p. 9.
5. Ibid.
6. Ibid.
7. Ibid.
8. Ibid. p. 115.
9. Ibid. p. 117.
10. Ibid. p. 118–119.
11. Ibid. p. 146.
12. Ibid. p. 148.
13. "Learning from the Past — Rhode Island Forest History," R.I. Department of Environmental Management.
14. Ibid.
15. Ibid.
16. "Historical Statistic of the United States, Colonial Times to 1970," U.S. Dept. of Commerce, Bureau of the Census.
17. Ibid.
18. Ibid.
19. Ibid.
20. "Farms for Sale or Lease in Rhode Island," R.I. State Conservation Commission, Bulletin No. 1, March 1911.
21. "The Forests of Rhode Island," U.S. Department of Agriculture, Sept. 2002.

22. "Providence Industrial Sites," Statewide Historical Preservation Report p. 6. Rhode Island Historical Preservation Commission, July 1981.

23 John Hutchins Cady, "Civic and Architectural Development of Providence 1636–1950," p. 184.

24. Ibid. p. 185.

25. Eighteenth Annual Report of the State Board of Health of the State of Rhode Island, for the Year Ending Dec. 31, 1895. E. L. Freeman & Sons, State Printers. 1897. P. 59

26. Ibid. p. 2.

27. See note 11, p. 146.

28. See note 11, p. 198.

29. Patrick T. Conley and Paul R. Campbell, "Providence, a Pictorial History," Publishing Co., 1982, p. 99.

30. Ibid.

31. Marilyn Massaro, museum curator, conversation with author, May 8, 2007.

32. Ibid.

33. Ibid.

34. Ibid.

35. Marilyn Massaro, "All Things Connected—Native American Creations," Museum of Natural History, Roger Williams Park, Providence, R.I. p. 8.

36. Mark V. Barrow, Jr., "The Specimen Dealer: Entrepreneurial Natural History in America's Gilded Age," *Journal of the History of Biology*, 33, 493–534, 2000.

37. Ibid.

38. Ibid.

39. "All about the Christmas Bird Count," Audubon Society, www.audubon.org/bird/cbc/history.html.

40. Ken Weber, "A Century of Dedication—The First 100 Years of the Audubon Society of Rhode Island," The Audubon Society of Rhode Island. P. 7.

41. Ibid. p. 8.

42. Ibid. p. 8.

43. Ibid. p. 8.

44. Ibid. p. 11.

45. Ibid. p. 13.

46. Ibid. p. 13.

47. See n. 1.

48. "A Metropolitan Park System: Report Upon a System of Public Reservations for the Metropolitan District of Providence Plantations to the General Assembly at the January session, 1905." Rhode Island State Library.

49. Ibid. p. 10.

50. Ibid. p. 19.

51. Ibid, p. 20.

52. Ibid. p. 43.

53. Ibid. p. 60.

54. Ibid. p. 61.

55. "A Metropolitan Park System Report Upon a System of Public Reservations for the Metropolitan District of Providence Plantations to the General Assembly at the January Session, 1908," Rhode Island State Library.

56. "Burlingame State Park — Campground History." Rhode Island Department of Environmental Management Web site, www.riparks.com/burlingamehistory.htm.

57. "A Metropolitan Park System: Report Upon a System of Public Reservations for the Metropolitan District of Providence Plantations to the General Assembly at the January session, 1935." Rhode Island State Library.

58. The Providence Journal, Feb. 11, 1961.

59. "History of Providence Water and the Scituate Reservoir," Providence Water Supply Board, www.provwater.com/site_ndx.htm, (Accessed Nov. 13, 2007).

60. See n. 8.

61. "Founding Legislation and History of the Forest Service's Traditional Role," U.S. Forest Service, www.fs.fed.us/plan/par/2003/final/html/fs_glance/founding_legislation.shtml.

62. Ibid.

63. Joseph J. Kirby, chief of division of Fish and Game for the Department of Agriculture and Conservation, Part V of the Annual Report of 1938.

64. Mathias P. Harpin. Aug. 6, 1944, *Providence Journal.*

65. Providence Journal, May 7, 1950.

66. Eleanor Johnson, "Miss Gilmore . . . Support for Conservation," The Providence Journal, Oct 25, 1949.

67. Zinn collection, Special Collections Unit, University of Rhode Island Library, Kingston campus.

68. Letter from Gilmore to Donald J. Zinn, Special Collections Unit, University of Rhode Island Kingston campus.

69. Ibid.

70. Letter from Ruth M. Gilmore to John H. Hinman, president of International Paper Co., April 17, 1951, Special Collections Unit, University of Rhode Island Library, Kingston, R.I.

71. "Donald J. Zinn; marine ecologist, professor of zoology at URI," *The Providence Journal*, Sept. 27, 1996.

72. See note 32.

73. Ibid.

74. Zinn, Box 2.

75. Zinn, Box 7.

76. Harpin, *The Providence Journal*, July 30, 1944.

77. "Resources Study Urged, New England and New York Senators Propose Commission," *The Providence Journal*, June 7, 1950.

78. "President Orders Resources of N.E. and N.Y. Surveyed — Instructs Six U.S. Agencies to Make Study Congress had Refused to Authorize," The Providence Journal, Oct. 12, 1950.

79. Ibid.

80. "Funds Sought for N.E., N.Y.—$350,000 Proposed for Natural Resources and Hydro Power Survey," *The Providence Journal*, Jan. 21, 1952.

81. "Engineer Made Water Power Group Consultant," *The Evening Bulletin*, June 27, 1951.

82. Dudley Harmon, "Public Hearing Set for Thursday on Voluminous Regional Report," *The Providence Journal*, March 7, 1955.

83. Guide to Archival Holdings at NARA's Northeastern Region, Waltham (Boston), www.archives.gov/northeast/boston/holdings/rg-050-099.html

84. Ibid.

85. Ibid.

86. Dale R. Taft, "Bills to Conserve, Develop N.E. Resources Planned," The Providence Journal, Nov. 22, 1958.

87."Roberts Gives Resource Plan," *The Providence Journal*, Oct. 10, 1952.

88. Ibid.

89. Paul G. Martasian, "Mass. Conservation Effort Makes Gains," *The Providence Journal*, April 27, 1960.

90. Conservation Commissions, sec. 45-35-1, State of Rhode Island General Laws.

91. Dickinson, Lewis E., and Hawkes, Alfred L., "Some Thoughts on Conservation," *The Providence Journal*, Jan. 22, 1962.

92. Ibid.

93. Edward F. Leydon, "Angry Man Cites Bay History: Beauty Spot to Vast Cesspool." *The Providence Journal*, Feb. 23, 1964.

94. Ibid.

95. "A Plan to Meet the 'Quiet Crisis' in R.I.," *The Providence Journa*l, March 24, 1964.

96. Ibid.

97. "Task Force Puts Stress on Coastline," The Providence Journal, March 20, 1964.

98. Daniel Varin, in e-mail to author.

99. Stuart O. Hale, *The Providence Journal*, Oct. 25, 1964.

100. Stuart O. Hale, "The marshes win—A Massachusetts Judge Rules the Owner May Not Destroy Resources People Need," *The Providence Journal*, April 14, 1964.

101. Alfred Hawkes, in oral history taken by Linda Wood, Oct. 19, 2005.

102. *The Providence Journa*l, Nov. 12, 1961.

103. Ibid.

104. *The Providence Journal*, Sept. 15, 1961.

105. Ibid.

106. Ibid.

107. Supplementary Report on the Coastal Wetlands Inventory of Rhode Island, June 1965, Folder 19, Box 2, of Special Collections Unit of the University of Rhode Island, Kingston.

108. Varin, in email to author.

109. Varin, as told to Ninian Stein, oral history.

110. Ibid.

111. Coastal Resources Center at the University of Rhode Island, About the Coastal Resources Center, www.crc.uri.edu/about.php?about_id=4, (accessed Nov. 13, 2007).

112. Peggy Sharpe, as told to Ninian Stein, oral history. P. 1.

113. Ibid. p. 2.

114. Ibid. p. 5.

115. Memo to Natural Resources Group from Calvin B. Dunwoody, R.I. Department of Natural Resources, March 15, 1978.

116.Varin. p. 16.

117. "A Greener Path—Greenspace and Greenways for Rhode Island's Future, Rhode Island Statewide Planning Program, http://www.planning.state.ri.us/greenways/title.htm, (accessed Nov. 13, 2007).

118. Ibid.

119. Varin, in e-mail to author.

120. Robert L. Bendick, Jr. "Seeking a New Point of View for Rhode Island," *The Providence Journal*, Jan. 29, 1984.

121. "Charting a Future for National Heritage Areas—A Report by the National Park System Advisory Board," 2006. Available at www.nps.gov/policy/NHAreport.htm.

122. Scott M. Lowe, Jr., "Ex-developer now favors land trusts," *The Providence Journal*, Oct. 29, 2004, B-01.

123. Rand Wentworth, "Plenary Address, 2004 Land Trust Alliance Rally, Providence, R.I.," www.lta.org/training/rally/plenary_wentworth_2004.htm.

124. Benjamin N. Gedan, "Voters OK $8-million open-space bond issue," *The Providence Journal*, April 27, 2006.

125. Paul Jordan (Supervising Geographic Information System Specialist for Rhode Island Department of Environmental Management, in discussion with author, November 8, 2007.

126. "Land Use 2025: Rhode Island State Land Use Policies and Plan," a Powerpoint presentation by state planning director Kevin Flynn at the Power of Place Summit, May 12, 2006, http://www.planning.state.ri.us/landuse/presentation.pdf.

127. Rhode Island Land Trust Council, www.rilandtrust.org/, (Accessed Nov. 13, 2007).

128. "Carcieri Drops Plans for State Police on Reservoir Site," *The Providence Journal*, June 28, 2005.

129. Meaghan Wims, "His Empire by the Bay," *The Providence Journal*, Nov. 24, 2007.

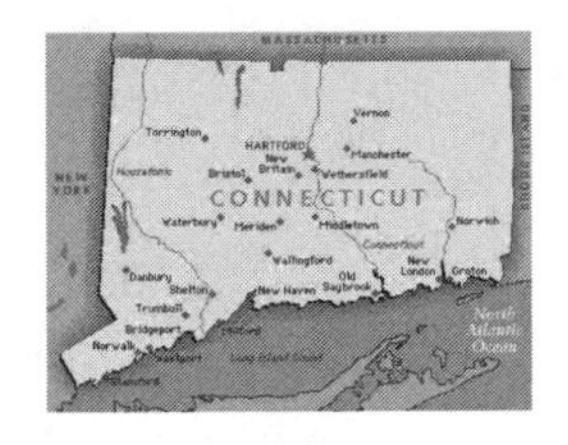

RESCUING CONNECTICUT

A STORY OF LAND-SAVING ACTIONS

RUSSELL L. BRENNEMAN[1]

This world of ours has need of those who deeply care.
There's work for us to do this very day.
And joy attends this enterprise we share
Together. The apple has been plucked. We may not stay.[2]

"For Esther,"
Richard Hale Goodwin
December 14, 1920–July 7, 2007

It would be fascinating to have been present when a small group of friends met in 1895 at the home of Rev. Horace Winslow in the village of Weatogue in Simsbury, Connecticut, where the valley of the Farmington River broadens out in fertile farmland in the lea of Talcott Mountain. For at that meeting, it could be said, Connecticut's land conservation had its beginning. But saying that would be just a convenience in a sense because beginnings in Connecticut have a way of

1. For their thoughtful review the author is indebted to John E. Hibbard, David K. Leff, Adam R. Moore, David Platt, Robert M. Ricard, David Sutherland, and Hugo F. Thomas and to Margaret Miner for reviewing the material on watershed land protection.

2. Richard H. Goodwin, *A Botanist's Window on the Twentieth Century* (Petersham, MA: Harvard Forest, 2002) p. 211.

going back and back and it is hard to say just exactly when something got started.[3] To the thousands of Native Americans who had lived in what became Connecticut for perhaps more than 6,000 years before the forebears of the friends who met in Weatogue arrived, the idea of land conservation would have seemed astonishing, if only because their communal culture seems to have provided a model of a sustainable relationship with the natural resources on which it depended without anything as self-conscious as conservation. Even to the European settlers, who brought with them their particular notions of proper governance and private property (along with their technology, religion, and diseases)[4] the abstract idea of conservation might have seemed quite odd. While a prudent farmer carefully managed his *own* crops, woodlot, wetlands, and pastures, that was really a personal matter, and only a fool would neglect good care of what belonged to him. However, the commons and town greens in many towns and cities do provide early examples of setting aside from private entitlement some property for the general benefit of the inhabitants of the newly founded communities, as they set about creating a *new* England. Remarkably, the churchmen/merchants who laid out New Haven in 1638 placed at the very center of the city they planned the New Haven Green that is still there today.

But, generally speaking, the idea of land conservation as such would have been unfamiliar. The abundance the Almighty had so (deservedly) bestowed upon the intruders was to be *used*. There was such plenty! Fish, birds, and game abounded. Every kind of tree the newcomers could imagine flourished hugely in dark forests that stretched into the boundless West of the newfound continent. Trees for homes and barns and ships and lumber and fuel and things to be fashioned from them. But change was soon afoot with the arrival of the Europeans. In the seventeenth century, much earlier than you might think, Connecticut's valleys were carpeted with farms, despite the drawback of the numberless

3. Among earlier events that can reasonably be argued as starting points would be the advocacy in the eighteen sixties by Theodore S. Gold, secretary of the State Board of Agriculture, for forest and farm conservation that anticipated many of the arguments of later foresters and ecologists. W. Storrs Lee, *The Yankees of Connecticut* (New York: Holt, 1957) pp. 192–195.

4. See generally, William Cronon, *Changes in the Land: Indians, Colonists, and the Ecology of New England* (New York: Farrar, Straus & Giroux. 1983).

stones the glacier had so generously brought down from the north and dumped in the fields. And if one tired of toiling with stones there was always easier land in the West. If resources are unlimited, what is the sense of conservation?

Another thing the Europeans brought with them in order to assure private benefits and orderly governance according to the time-honored patterns to which they were accustomed was the notion of boundaries. It only makes good sense, does it not? Samuel owns Blackacre and Josiah owns Whiteacre. This space is Pomfret and that is Ashford. This is Connecticut and that is Massachusetts. Legal rights and the powers of government are spatially defined by boundaries. The law of Connecticut is not the law of Massachusetts, nor is the government of Beacon Falls quite the same as the government of Bozrah. Connecticut is a quilt of towns, 169 of them, some of which existed nearly 150 years before Connecticut became a state. Boundaries are important to good order.

While it is obvious, it bears repeating: Nature is unfamiliar with these boundaries. Nature's "quilt" is different. The Housatonic, Connecticut, Thames, and their cousins arise and flow where ancient geological events determined they would. The creatures move within and between the types of forests and habitats that best suit their kind. The birds fly where they please, the fish come and go as their habits dictate, and the waters arise from under the land and subside into it, wherever that may occur. All this without reference to the Johnny-come-lately boundaries of our species. Looked at from above, one kind of map of Connecticut might look like a grid of lines, all the way from the boundaries dividing Connecticut from its neighboring states down to the level of counties (which are unimportant in Connecticut), towns (which are *very* important) and the lines that define the farms of Samuel and Josiah. But if we were to look at other kinds of maps, for example where rivers rise and where they flow or the range of a red fox or the sky paths of orioles or the mysterious journeys of eels, there would be very different patterns indeed, and our manmade lines would not only be irrelevant but would collide with the natural patterns on these other maps. The puzzling, altogether recent, interdiction of these primeval flows of life by mere lines on maps would be only curious were their consequences not so profound. The conservationist might want to draw from Frost's observation that "Something there is that does not love a wall" and say instead: Something there is in nature that does not love lines on maps.

There is also the matter of the towns. They comprise the political bedrock of the state. Historically, they defined the community. They are independent, distinctive, self-sufficient, proud, and in former days before relatively modern highways, often rather isolated.[5] In Connecticut particularly, the authority of each town within its own limits is so ancient and jealously guarded that it invokes the label "home rule." (While technically towns now receive their powers from the state government, many of them existed for many decades before the state as such came into being, and they find it hard to countenance such a technicality.) Times have changed, but many of these elemental characteristics of towns remain unchanged. DeTocqueville idolized them.[6] Modern ecologists find it easy to find fault with them. Why? First, because their traditional power over the use of land within their borders can conflict with any notion of super-town governance over land use even when it is appropriate for resource management or achieving sensible regional, state or national goals. Second, Connecticut's emphasis on their individual economic self-sufficiency drives towns toward development in order to enhance their tax base, putting them into economic competition with each other, regardless of the effect on others or the effect on resources. Yet these are the rules of the game in Connecticut and have been for a very long time.

How can conservation have been achieved in the context of such history, customs, and circumstances? The genius of Connecticut land conservation is that it has happened at all—that conservation has found its place within the meager 5,000 square miles or so that comprise the state. But it *has*. Why? Because another central aspect of the Connecticut experience also seems to have been a growing awareness on the part of many of the responsibility to be frugally as mindful of tomorrow as we are of today. Resources are to be used for now, but what of later? Resources are ours to enjoy today but what about the needs of others who will follow? Tomorrow has a claim that farmer Josiah would

5. It was not out of place to call them "little republics," as did Odell Shepard. *Connecticut Past and Present*, (New York: Knopf, 1939) ch. 12.

6. Alexis de Tocqueville, *Democracy in America*, vol 1, ch. 5, (New York: Schoken ed., 1961).

understand. Increasingly, the farm for many has become, at the very least, Connecticut itself. At a meeting of the Connecticut Forestry Association in 1909 Yale President Arthur Hadley set the tone for forestry in words that resonate with conservationists today. Forestry, he said, "gives even more to the future than it does to the present. It is typical of what we are all anxious to do; typical of the kind of life we shall leave for our children in the way of resources than what we shall use up and enjoy ourselves."[7] Starting with forestry, land conservation in Connecticut has become a multifaceted effort to save what a contemporary advocacy group has called no less than the "face" of Connecticut, contending that what identifies the essence of the state, its villages, farms, forests, and countryside, can be lost and is being lost in increments of change until what was once precious will be no longer present before it has been truly appreciated or understood. That summarizes the evolving story of Connecticut land conservation over the years.

Often leading the way, Connecticut has been a fertile seedbed for conservation elsewhere. To choose a few examples: The small community-based and citizen-led land trusts that proliferated in the state as early as the 1960s have provided a model that has become widely copied. The legal underpinnings for conservation easements were established by studies in Connecticut, which ultimately led to their use throughout the United States. The Nature Conservancy (first incorporated as a nonprofit in 1951 as an outgrowth of the Ecologists Union) was nurtured from its earliest days by Dr. Richard H. Goodwin, who began assembling one of its first preserves around his home property in East Haddam. The Yale Forestry School, now the Yale School of Forestry and Environmental Studies, was founded in 1900 by no less than Gifford Pinchot and his friend, Henry S. Graves, and thenceforth has pioneered professional forestry, research and education and produced legions of foresters who have served virtually everywhere that forests may be found. The Connecticut Forest and Park Association, established in 1895, was the first statewide advocacy group for land conservation, forestry and outdoor recreation, leading the way for the formation of like organizations in other states. Established in 1875, the

7. Unpublished manuscript by Connecticut state forester Austin F. Hawes, archives of the Connecticut Forest and Park Association, Middlefield, CT.

Connecticut Agricultural Experiment Station, with its forestry program that was a nursery for Yale's, was the first of its kind in the U.S.[8]; it was soon replicated elsewhere. And so on.

But to return to beginnings, to the gathering in Simsbury: The men (and one woman that we know of) who met were not concerned with philosophical abstractions. They were concerned with rescuing the Connecticut forest. Some of them, perhaps most, would have known Pinchot, who considered Simsbury his second home. Some were probably familiar with the prophetic insights of George Perkins Marsh set forth in *Man and Nature*.[9] They certainly would have known about the founding of the Connecticut Agricultural Experiment Station and its forestry program. Some may have studied (as had Pinchot) under Professor William H. Brewer at the Sheffield Scientific School of Yale College, who emphasized forestry in the courses he taught on agriculture. On the national stage, they would have read of the political struggles to establish Yosemite and Yellowstone as national parks, each an unprecedented reservation of Federal land from the onslaught of exploitation that seems much less dramatic to us today than it did at that time. As educated New Englanders, they would have been familiar with the ideas of Thoreau and Emerson. They would have known of the new parks in Hartford, which had become by then an intellectually enlightened and politically progressive city, and the work of Rev. Horace Bushnell and Hartford native Frederick Law Olmsted on city parks and landscapes.[10] All of these influences were at work within the small group that met at Reverend Winslow's home. But their focus was far more immediate and close to home. Something needed to be done to save and restore Connecticut's forests. They decided to form the Connecticut Forestry Association.

8. Originally located at Wesleyan University in Middletown, the station moved in 1882 to its present home in New Haven.

9. George Perkins Marsh, *Man and Nature: or Physical Geography as Modified by Human Action* (Scribner, 1864); Cambridge: Harvard University Press, 1965; Seattle: University of Washington Press, 2003, David Lowenthal ed.).

10. When in 1854 Hartford voters approved the expenditure of taxpayer funds to purchase a degraded industrial site to establish a park that would be "an opening in the heart of the city," in Rev. Bushnell's words, Bushnell Park became the "first municipal park in the nation to be conceived, built and paid for by citizens through a popular vote." http://www.bushnellpark.org/Content/Park_History.asp.

Viewed from the air today much of Connecticut seems covered by trees. You might well ask why in 1895 the forest needed to be restored. But Connecticut looked entirely different then than it appears today. Two-thirds of the state had been cleared for farms, dwelling places, villages, roads, canals, factories, railroads, and cities. The forest had been harvested for lumber, charcoal for iron-making, fuel for heat and manufacturing, railroad ties (the railroads had come as early as 1830 and construction of tracks boomed after the Civil War), humble boxes and hungry sawmills. Fires started by locomotives along rights of way were taking a toll, and there was no effective system for their suppression. Woodlots were neglected as farming declined from its earlier prosperity. No check on the continuing decline of the forest was in sight.

There was growing alarm among a relatively small group of concerned professionals and informed citizens in the 1880s and even before. Prophetically, experts wrote and spoke of the effects on soils and watersheds when forests are removed (this was long before ecology!).

Once heavily forested, Connecticut's forest lands had been decimated by the end of the nineteenth century through heedless harvesting, carelessness, and fires. Here members of the Connecticut Forestry Association tramp along a degraded "woodland" watercourse in Portland, Connecticut in 1905.
(Connecticut Forest and Park Association)

Others were dismayed by the diminishing of familiar species and the loss of shade trees. Concern was building in other states as well. Rampaging forest fires in Michigan, Minnesota, and other states and in Canada, commanded national attention. Forest reserves on public land had been authorized by Congress in 1891, and an embryo Bureau of Forestry had been started within the federal government. Anxiety there was on the part of the knowledgeable, but how widely was it shared? Was there a broad enough base for effective public action? And what exactly were the steps that were needed to halt the decline and begin restoration?

These were the topics that the small group that met in Reverend Winslow's home discussed. The pattern of their deliberations and their conclusions were to be repeated again and again by other groups addressing other conservation issues in Connecticut, and elsewhere to this day. What was the sequence? Assemble credible information, disseminate it broadly, identify strategies, organize constituencies, engage those able to influence change. Unknowingly, the little group in Weatogue was providing a model for future conservationists. The constitution forming the Connecticut Forestry Association that was adopted on Arbor Day (April 30) the following year, 1896, was signed by thirty members, including such notables as the governor of Connecticut, O. Vincent Coffin, and many other people of influence. There were five primary objectives: developing public appreciation of the value of forests, their protection and proper management; disseminating information; engaging the state in enacting new laws to preserve, maintain, and increase forests in Connecticut and elsewhere; establishing state and national parks and forestry reservations; and encouraging the study of "forestry and kindred topics" in schools. A more comprehensive agenda for conservation actions can scarcely be imagined more than a century later!

Among the early achievements of the new association were the passage by the General Assembly of a law creating a state fire marshal and establishing a town-based system to combat wildfires under his direction (one of the first of its kind in the nation) and a 1901 Act "Concerning the Reforestation of Barren Lands." The latter law authorized the appointment of a state forester, who was authorized to purchase land to be managed by the Connecticut Agricultural Experiment Station on behalf of the state. According to an authoritative history of the association, "Connecticut was thus the first state in the nation to have a state forester."[11] When the Portland State Forest was established in 1903 Connecticut became the first state in New England to have a state forest. The seeds of what was to become the system of state forests and parks were planted in this way.

One of the striking aspects of this earliest conservation action in Connecticut is that many of the people and families involved in it either knew each other, knew of each other, or were influenced by each other. While their numbers were few and their backgrounds distinctive, they tended to be comfortably situated, well educated, socially conscious, influential, and often politically powerful men and their families. There are many examples. Pinchot acknowledged the encouragement of Frederick Law Olmsted,[12] who himself received inspiration from Horace Bushnell. John McLean, a signer of the Connecticut Forestry Association constitution and its first secretary, was the brother of George Payne McLean, later a governor and United States senator, who was to create one of the state's first wildlife reserves (the McLean Game Refuge in Simsbury and Granby). Pinchot encouraged a fellow Yale man, Henry S. Graves, to study forestry, and it was Graves and Pinchot who were to establish the Yale School of Forestry in 1900, and Graves who became its first dean. Starling W. Childs and Frederick C. Walcott (later a United States senator from Connecticut) became acquainted with Pinchot and each other at Yale and in 1909 they began assembling what is now known as the 6,400 acre Great Mountain Forest in Norfolk and Canaan (in the midst of which is now located the seven-acre Yale Forestry Camp). George Myers, who received a master's degree from the first class to graduate from the forestry school in 1902, soon thereafter began assembling the land in Union and adjoin-

11. George McLean Milne, *Connecticut Woodlands* (Connecticut Forest and Park Association, Inc., 1995) p. 8.

12. Olmsted met Pinchot in 1891, when Pinchot was appointed forester of the Biltmore Estate and "took charge of the first large-scale example of practical forest management in the United States." Laura Wood Roper, *FLO: A Biography of Frederick Law Olmsted* (Baltimore: Johns Hopkins, 1973) p. 419. In 1905, Pinchot was appointed Chief Forester of the United States. See generally Gifford Pinchot, *Breaking New Ground* (Island Press ed. 1998; first published 1947). Graves succeeded Pinchot as Chief Forester in 1910. See also, Char Miller, *Gifford Pinchot and the Making of Modern Environmentalism* (Washington: Island Press, 2001).

ing Connecticut towns that is now the Yale-Myers Forest of Yale University; it has been a research and instructional site since 1917. The beginnings of conservation in Connecticut thus arose from the shared concerns of a relatively small group of influential men and their families. A phenomenon of land conservation in the state over the years has been a transition from a focus on forestry to a larger focus on conservation generally, and also a transition from the concerns of relatively few to a mainstream issue, a democratization of conservation advocacy, precisely, it is easy to imagine, as hoped for by conservation's pioneers.

From its start, the Connecticut Forestry Association urged a greater involvement in forest conservation and the development of parks by the state government. Influencing state action and funding was integral to the association's strategy. But it was not until 1921 that a Park and Forest Commission was created by the legislature and not until then that responsibility for state forest work was transferred from the Agricultural Experiment Station to the new commission. Still, in 1921, there were only three small state forests totaling in area little more than 4,000 acres. Once more, personal leadership and private generosity came to the fore to jump start the creation of the tens of thousands of acres that are owned and managed by the state today. An early example: a unique synergy of personal generosity and public involvement was the creative idea of Alain C. White of Litchfield, who in 1922 proposed a solicitation of gifts from the general public to raise money to buy a state forest that then would be donated to the state. The successful campaign culminated in the gift to the state of the core 400 acres of the appropriately named Peoples State Forest in Barkhamsted that was purchased entirely with contributed funds.[13] In 1913, the same Alain C. White and his sister, May, had enabled the creation of the White Memorial Foundation on 4,200 acres in Litchfield.

On the public side, leadership was also emerging, most influentially in the person of Austin F. Hawes, a dominant figure in forest conservation in Connecticut throughout the first half of the twentieth century. Hawes served as state forester from 1904 to 1909 and again from 1921 until 1945. In the intervening years he was state forester of Vermont. Hawes entered the forestry school at Yale soon after its founding and re-

13.The forest now encompasses 3,000 acres. For its history see http://www.stonemuseum.org/history_of_peoples.htm.

ceived his master's degree from that school. He saw his primary and ongoing tasks as twofold: First, develop a thorough and science-grounded information base about the condition of the forest and its needs; second, establish a system to disseminate information to foresters, forest owners, policy-making officials, and a public that was at the outset of his service relatively uninformed. Surveys undertaken under Hawes's direction and published as early as 1909 were claimed to be the first of their kind ever made in New England.

In addition to its forestry agenda, developing state parks was also a project of the Connecticut Forestry Association and its members. While a State Park Commission had existed since 1909, its budget was miniscule. Access to the water by the public over the more than six hundred miles of dry shoreline on Long Island Sound was entirely owned either privately or controlled by possessive municipalities. Today, state beaches at Sherwood Island (1927),[14] Hammonasset (1920), and Rocky Neck (1931), and a coastal reserve at Bluff Point in Groton (1975) exist largely through the advocacy and occasional funding by the association and its members, in the course of which the association changed its name to the Connecticut Forest and Park Association. Many of the state's inland parks were also established through the support of the association.

For many years it was the association that kept the lamp of land conservation burning in Connecticut through its public advocacy, meetings, publications, and the generosity of its members. Its directors tended to be well connected politically as well as in the business and professional communities. The culture of the organization consistently favored consensual and cooperative outcomes over stridency and confrontation. It was served by a succession of distinguished leaders, among its presidents Roger W. Eddy, of Newington, James L. Goodwin, Rev. George M. Milne, and David M. Smith, a Yale professor of silviculture who became director of the Yale Forests in 1954. Beginning in 1963 the leadership of secretary-forester John E. Hibbard was critical in the formative years that began with the administration of Governor John

14. Selected as a desirable location by the State Park and Forest Commission, purchases of land began in 1914, but it was not until 1937 that the state park was opened after a memorable struggle to overcome local opposition. http://friendsofsherwoodisland.org/Pages/History/H_acquisition.htm.

Beginning with the work of a small-trails committee in 1929, a network of about eight hundred miles of "blue-blazed" hiking trails across public and private land is maintained by volunteers through a program of the Connecticut Forest and Park Association. Connecticut Woodlands is the association's publication. (CONNECTICUT FOREST AND PARK ASSOCIATION)

Dempsey. While other advocacy groups arose to address their particular concerns, the association maintained consistent focus on the comprehensive issue of land conservation, outdoor recreation, and forestry, in part by means of its influential magazine, *Connecticut Woodlands*.

The pattern of extraordinary generosity of which the pioneers were exemplars continued to build the backbone of the state's forests and parks. To choose from a few examples: In 1964 James L. Goodwin donated more than two thousands acres in Chaplin and Hampton, which became the state forest that now bears his name. George Dudley Seymour, a passionate defender of New Haven's historic buildings, ambience and its Green, gave the George Dudley Seymour State Park. Seymour's generosity had by 1973 resulted in the creation of more than a dozen Connecticut forests and parks, according to Adam R. Moore, executive director of the association. In addition to the White Memorial Foundation, Alain White's philanthropy resulted in six major state forests and parks. Frances Osborne Kellogg in 1958 devised to the state 350 acres in the hills of the Naugatuck Valley. Curtis J. Veeder gave the 800 acres of Penwood State Park in Avon in 1944. In contrast with Massachusetts, where such gifts tended to go to the Trustees of Reservations, in Connecticut many have been donated directly to the state. A noteworthy exception was the gift by Katherine Ordway of 1,750 acres in Weston and Redding that now comprise The Nature Conservancy's Lucius Pond Ordway Devil's Den Preserve. At least 20,000 acres of the state's forests and parks have directly resulted from private philanthropy.

Important members of the state's land-saving community have been organizations conserving particular habitats and those whose advocacy focuses on a particular resource. The Connecticut Audubon Society, founded in 1898, protects 2,500 acres of habitat in connection with its nineteen sanctuaries and its Coastal Center at Milford Point. The National Audubon Society, manifested in Connecticut by Audubon Connecticut, manages 4,000 acres at four sanctuaries. The 103 fish and game clubs in the state own more than 12,000 acres, including the 2,000 acres owned by the Hammonasset Fishing Association on the Hammonasset River in Killingworth and Madison. Regional advocacy organizations include several watershed associations, such as the Housatonic Valley Association and the Connecticut River Watershed Council, both of which were formed in the 1940s in response to issues that arose in the multi-state watersheds of those rivers, and the Farmington River Watershed Association, organized in 1953. Umbrella resource advocacy groups such as Rivers Alliance of Connecticut have played a significant role in land protection, particularly for watershed land. The Friends of Connecticut State Parks were organized as guardians of parks. Broad associations such as the Endangered Lands Coalition have attracted large and diverse supporters of a comprehensive conservation strategy.

Assembling and managing conservation land at the level of the state government until 1969 was carried on primarily under the direction of three boards and commissions: the Connecticut Park and Forest Commission, the Connecticut Board of Fisheries and Game, and the Water Resources Commission. The combination of forestry and parks under the same commission in 1921, rather than combining forestry with agriculture, was a pivotal choice by the legislature that was hailed as a "milestone."[15] In 1959 Governor Abraham Ribicoff collected all three agencies under a commissioner of the new Department of Agriculture and Natural Resources, Joseph N. Gill.

Characteristically for Connecticut, the idea of creating a state trail system came from a few friends who enjoyed the out-of-doors, who enlisted the help of the Connecticut Forest and Park Association in 1929. A Trails Committee was established that has overseen the development and maintenance of the Connecticut "Blue-Blazed" Hiking Trail System. The system consists of approximately 800 miles of hiking trails that are located both on public land and private land.[16] The trails are looked after entirely by volunteers under the supervision of the Trails Committee. The value of such linear parks was revisited in 1992 when Governor Lowell P. Weicker, Jr., appointed a study committee whose recommendations for a state greenway program,[17] resulted in the formation of the Connecticut Greenways Council charged with develop-

15. Austin F. Hawes, *History of Forestry in Connecticut* (unpublished manuscript, archives of the Connecticut Forest and Park Association, Rockfall, CT) vol. I, p 84.

16. The trails are described in *The Connecticut Walk Book East* and *West*, 19th ed. (Connecticut Forest and Park Association, 2007).

17. See *Greenways for Connecticut—A Report to the Governor from the Connecticut Greenways Committee* (December, 1994). This document can be found in the Connecticut State Library, Hartford, CT.

ing greenways[18] to "link places we live to places we love," in the words of Greenways Committee co-chair Susan Merrow. While the recreational aspects of greenways have predominated, greenway advocates have also stressed using them to maintain wildlife habitat corridors, protect ridgelines, and preserve noteworthy vistas. Greenways also fit into the program to stimulate outdoor experiences, particularly for families and children, embodied in the "No Child Left Inside"[19] campaign announced by Governor M. Jodi Rell in 2006, the brainchild of Connecticut Department of Environmental Protection Commissioner Gina McCarthy.[20]

As the pivotal environmental decade of the 1960s began, Connecticut conservationists focused on the nexus where land-saving battles are often won or lost in Connecticut—the decisions of local agencies such as the planning and zoning commissions in each town. Decisions on the development of private and public property, they contended, were being made without due consideration of the resources affected by them. Legislation authorizing municipal conservation commissions emerged from the demands of citizens for a formal representation of conservation values within town government. Prominent leaders in this effort were Deborah Eddy—wife of state senator Roger W. Eddy—of Newington, and Marion Richardson of Glastonbury. Legislation in 1961 authorized the formation of conservation commissions by towns.[21] While their role is advisory, their vision is intended to augment the local planning and zoning process so as to assure the recognition of habitat, wetlands, farm and forest land, and other open space of community significance. Guidance was made available by the Department of Agriculture and Natural Resources until that department's role was supplanted by the new Department of Environmental Protection. A small grant program initiated by the Ford Foundation for the states of Connecticut, New York, and New Jersey provided support at the outset for innovative projects. Exchange of information among

18. Connecticut Public Act 95-335, Connecticut General Statutes sec. 23-100-102.

19. http://www.nochildleftinside.org/.

20. See generally, Richard Luov, *Last Child in the Woods: Saving Our Children from Nature Deficit Disorder* (Chapel Hill, NC: Algonquin, 2005).

21. Connecticut Public Act 61-310, Connecticut General Statutes sec. 7-131a.

commissions was facilitated by the formation of the Connecticut Association of Conservation Commissions, now the Connecticut Association of Conservation and Inland Wetlands Commissions, and their periodic newsletter.[22] By 2006, 149 municipal conservation commissions had been established in the state's 169 towns.

As was true of the rest of the United States as the modern environmental movement blossomed, the decade of the 1960s witnessed dramatic changes in the direction of public policies in Connecticut, defining new responsibilities for government and establishing new institutions to carry out those responsibilities. These changes moved land conservation in the state onto a different plane. Political leadership came from the very top, Governor John Dempsey. A cluster of initiatives emerged from the 1962 report to Governor Dempsey by the distinguished writer and activist William H. ("Holly") Whyte, a former editor of *Fortune* magazine, acclaimed land-use commentator, and author of *The Organization Man.* (Six years ahead lay the publication of Whyte's prophetic book, *The Last Landscape*.) In commissioning Whyte, Dempsey attached particular urgency to saving the state from helter-skelter growth. The governor noted as a reason to get moving the likelihood of a federal open-space funding program when the recommendations of the national Outdoor Recreation Resources Review Commission established in 1958 were adopted by Congress, as in fact they were with the creation of the Land and Water Conservation Fund in 1964.[23] Dempsey also noted Title VII of the 1961 Housing Act that authorized federal open space grants to local bodies for a portion of acquisition costs. Leveraging federal money was to become an important aspect of Whyte's recommendations.

The wide-ranging report by Whyte[24] contained a comprehensive snapshot of what was happening in Connecticut and a dour forecast of what lay on the horizon in the absence of action. Noting that for some locales (in 1962!) it was "already too late," the report in dramatic lan-

22. See generally, Connecticut Association of Conservation Commissions, *The History of Conservation Commissions in Connecticut* (2002).

23. Public Law 88-578, September 3, 1964, 87 Stat. 897, 16 United States Code sec, 460l-4-11.

24. William H. Whyte, Jr., *Connecticut's Natural Resources–A Proposal for Action* (1962). This document may be found in the Connecticut State Library, Hartford, CT.

guage emphasized the importance of landscape to the quality of life in the state and argued that forests and wetlands, parks and streams, traditional vistas, farms, and natural habitats all can be lost, bit by bit, without mindful policies of the state and municipalities to save them. Whyte emphasized the practical economic benefits of an integrated conservation program. The heart of Whyte's recommendations was *public investment* in conservation through aggressive open space grant programs. He urged Connecticut to join the "great tide of conservation that was then sweeping the country."

His recommendations were succinct: save the wetlands, save the rivers, save the flood plains, clean up the water, develop state parks, save the ridge tops, expand recreation, retain farmland, plan comprehensively. To accomplish this: open space grants for towns, comprehensive planning at the town, regional, and state levels,[25] tie-ins with federal funding programs, cluster development, acquisition of ridgelines, expansion of the forests and parks through state funding, preferential taxation for farms, a natural area preserves program, and protection of the floodplains and river valleys. At the end, as if looking into the future, Whyte wondered whether there should be a "Reappraisal of Riparian Rights Concept – What are the obstacles to proper stream flow regulation?" All in thirty-two pages!

In a preface to a reprinting of the report in 1970, Whyte was able to report that much progress in the open-space grant program had been accomplished in the years following 1962. Conservation commissions and local land trusts had burgeoned. Major water-pollution legislation had been enacted in 1967, with significant state bond funding to match federal funds. Public Act 490 had been enacted providing for "preferential" assessment of farm, forest, and open space land for purposes of the municipal tax on real property.[26] However, land acquisition, in Whyte's opinion, had lagged. Nothing had been done about the ridgelines. Regional and state planning was toothless. Economic forces were

25. In 1971 the Connecticut General Assembly passed a joint resolution calling for a state plan of conservation and development, which became official policy for the executive branch in 1974. Legislation was enacted in 1976 establishing a formal process for adopting the plan and updating it every five years. Connecticut Public Act 76-130, Connecticut General Statutes sec. 24-33, as amended.

26. Connecticut Public Act 63-490, Connecticut General Statutes sec. 12-107a-f., as amended.

continuing to overwhelm the countryside. The battle, he argued, needed to be continued if Connecticut were to be rescued from mindless alteration.

In 1959 Holly Whyte had made another contribution that was important not only to Connecticut but to land savers everywhere. He understood that very often the conservation of a parcel of land whose present use is compatible with a conservation goal can be accomplished without acquiring outright ownership of it. He recognized that government restrictions, however strict, are always subject to change. Could there be some form of restraint in the nature of a property interest that would be more permanent? His trail-blazing monograph for the Urban Land Institute in 1959 about a new and somewhat awkwardly named device he called a "conservation easement"[27] broke new ground. Such an easement, Whyte contended, could be used to strip away the rights that an owner otherwise would possess to change the use of property in unwanted ways while leaving in place all other aspects of ownership. However, the legal underpinnings for such an unfamiliar technique were unclear. Whyte's monograph caught the attention of Dr. Richard H. Goodwin and the small foundation that he had established, the Conservation Research Foundation, which promptly funded a study of the legal aspects of conservation easements. The published report proved to be the first of its kind.[28] The study concluded that while there was little legal precedent there also seemed to be no reason why such easements could not be used to protect land. Despite their awkward name, the report pointed out, such easements are best thought of as a way to sequester development rights by transferring those rights to a holder that is obligated not to use them. There followed the enactment of a Connecticut statute firmly establishing the new technique for conservation and preservation purposes.[29] This led in turn to the development of a "uniform act" governing conservation easements[30]

27. William H. Whyte, Jr. *Securing Open Space for Urban America: Conservation Easements*, Urban Land Institute Technical Bulletin 36 (1959): 1-67.

28. Russell L. Brenneman, *Private Approaches to the Preservation of Open Land* (Conservation and Research Foundation, 1967).

29. Connecticut, Public Act 1971-173, Connecticut General Statutes sec 47-42a-c. Such an "easement" in Connecticut is termed a "restriction."

30. National Conference of Commissioners on Uniform State Laws, *Uniform Conservation Easement Act* (1981).

and the passage of that or similar laws in many states. Conservation easements are now widely used to protect wetlands, forest, farmland, and other sorts of conservation lands throughout the country.

Farmland in Connecticut, as Whyte's report emphasized, was under challenge. That was still true in 2007. Of the 22,000 farms in Connecticut at the close of World War II, 4,200 remain in 2007 (far fewer than in Whyte's time), and of the approximately 360,000 acres of farmland remaining, hundreds of acres are being lost to development each year, according to the Connecticut Farmland Trust. The pressure for development is acute as habitation spreads into the countryside, particularly into the same broad valleys that attracted the state's earliest farmers. Once rural communities are being engulfed by sprawl. Fertile soils are being covered by macadam. However much a family wants to hang onto its farm, as land values rise along with taxes it can make less and less sense. Farmland preservation is a conservation issue in Connecticut not merely to preserve a valuable economic resource but also because the preservation of farms aids in the conservation of wetlands, habitat, valuable soil, watersheds, and other natural resources. A primary tool in the state's farmland-protection program arises from the same conceptual root as conservation easements: the acquisition of development rights. As early as 1978, the legislature authorized the purchase by the state of development rights in selected farmland,[31] but the program lagged because of meager funding and administrative holdups. Advocacy of a more aggressive program emerged when the Working Lands Alliance was formed by the Hartford Food System, the American Farmland Trust, and other agriculture groups. In 2007 Governor M. Jodi Rell reaffirmed a goal of saving, through the farmland preservation program 130,000 acres, including 85,000 acres of cropland. According to the Connecticut Department of Agriculture, as of January 2007, 24 percent of the goal had been achieved through the purchase of development rights in more than 31,000 acres on 228 farms. Development rights on 830 additional acres have been acquired by the Connecticut Farmland Trust.

When you hear of a "land trust" in Connecticut, what is being talked about is not statewide giants like the Trustees of Reservations in

31. Connecticut Public Act 78-232, Connecticut General Statutes sec. 22-26aa-cc, as amended.

Massachusetts (which is probably the granddaddy of them all) or national and international land savers like The Nature Conservancy, but rather a local town-based group of volunteers supported by contributions from the community and more often than not without a paid staff. They reflect precisely the notion that each town is aware of and should protect resources that are of particular local importance. They developed because of the desire of inhabitants to have protected open spaces and trails within their own towns that do not necessarily fit into the acquisition priorities of larger organizations or the state. Although trusts for conservation purposes had certainly been created by private individuals before then,[32] the probable first such *community* conservation land trust in Connecticut was organized in 1964 in the Town of Madison as the Madison Land Conservation Trust. It was soon followed by others, often with their own distinctive names, like Joshua's Tract Conservation and Historic Trust, named after a Native American, The Weantinoge Heritage Land Trust in western Connecticut, and the tri-town Great Meadows Conservation Trust, which is focused on the expansive meadow and flood plain of the Connecticut River below Hartford. As the trusts proliferated, the Connecticut Chapter of The Nature Conservancy organized the Connecticut Land Trust Service Bureau to provide support for them. There are now about 128 land trusts in the state. They own in excess of 51,000 acres "in fee" and protect more than 21,000 additional acres through easements. Connecticut land trusts have often been the recipients of land that has been set aside by developers as part of the local approval process for subdivisions. The land-trust movement has swept westward so that as of 2007 there were more than 1,500 land trusts in the country, according to the Land Trust Alliance.[33] Remarkably, 8 percent of these trusts are in Connecticut, the third smallest state in the Union.

Very often in Connecticut it has taken a dramatic crisis to stimulate dramatic land-saving actions. In the 1960s the proposal of the Rykar Development Corporation to develop 277 acres of the Great Meadows Marsh was such an event. The Great Meadows Marsh is

32. For example, the Pond Mountain trust in Kent, the Peace Sanctuary in Mystic, and Aton Forest in Norfolk.

33. See generally Richard Brewer, *Conservancy: The Land Trust Movement in America*, (Hanover, NH: University Press of New England, 2003).

The nearly three-hundred-acre Great Meadows Marsh at the mouth of the Housatonic River was saved by citizen action in the 1960s, which paved the way for protective measures for all of Connecticut's coastline. It is now part of the Stewart B. McKinney National Wildlife Refuge.
(U.S. Fish and Wildlife Service)

where the majestic Housatonic River, having been joined a bit upstream by the Naugatuck, flows into Long Island Sound between the towns of Stratford and Milford. In the 1920s Roger Tory Peterson, Alan Crinkshaw, and other prominent ornithologists would take the train out from New York in order to explore wildlife in the meadows, which are said to comprise one of the largest contiguous ecosystems of its kind in New England. In its hundreds of acres of broad tidal marsh live an abundance of fish, crustaceans, birds, mammals, and plants, including some rare and at-risk species, and the marsh serves the vital filtering and flood buffering role common to such wetlands. Located on the Atlantic flyway, it is an important stopover for migratory birds. Such marshes were once common all along the Connecticut shore, but by the time of the Rykar proposal, roughly half of the coastal wetlands that existed in 1900 had been destroyed through dredging, filling, and ditching. While important fragments of habitat had been acquired for wildlife conservation purposes, the vast majority of shoreline lay unprotected at the time of the Rykar project, except in some towns by meager and often problematic local zoning regulations.

While Rykar had to get a permit from the state Water Resources Commission to dredge or fill the land under tidal waters because it is owned by the state,[34] there was at that time no requirement for any kind of formal environmental review or impact analysis. The commission, which had been established as the Water Commission in 1925, had enjoyed distinguished leadership and membership and had made important contributions to water policy, but its regulatory powers over coastal wetlands were limited. Attempts to bolster the regulatory regime by the legislature had been unsuccessful because the Board of Fisheries and Game preferred an acquisition program. However, funding for that had been grossly insufficient. Thus, the Rykar project was subject only to a rather antique state "dredge and fill" law that contained no declaration whatsoever that it is the policy of the state to protect and preserve tidal wetlands.

Conservationists were enraged by Rykar's audacious notion to build a port facility in the pristine wetland. The fuss was sufficient to cause the chairman of the Water Resources Commission, John J. Curry, who presided over the permit hearing, to listen to what they had to say. The hearing became a strident cry for the commission to consider the environmental impacts of the Rykar proposal on the ecosystem of the Great Meadows Marsh. Essentially, the groundwork was being laid not only for a hoped for denial of the permit but also for a new law to protect tidal wetlands. The Connecticut Conservation Association, a nonprofit advocacy group, called the most important witness, Dr. William A. Niering, a distinguished professor of botany at Connecticut College, who was to become a world recognized authority on coastal wetlands. Professor Niering was the first ecologist (an uncommon word in those days) in Connecticut to forego the safety of the classroom to plunge into the abrasive fray of public testimony. His scientifically meticulous testimony persuasively spelled out the environmental and societal benefits of pre-

34. That is, the soil seaward of the high-water mark of navigable tidal waters.

serving coastal wetlands, a view that resonated forward in Connecticut and beyond. The permit was denied.

While the Great Meadows Marsh was saved largely through Dr. Niering's testimony, of even greater importance to saving Connecticut's tidal wetlands was his going on to help draft a tidal wetlands protection law that was enacted by the Connecticut General Assembly.[35] It was unique in at least two ways. It set up the direct regulation by the state of proposed development in tidal wetlands, thus running counter to the time-honored tradition of exclusive town home rule over land use. Secondly, it resolved the pesky issue of describing the land that was subject to regulation by referring not only to its elevation but also by reference to plants which "may grow or be capable of growing" there, which are identified in a detailed list of specific plants, a technique that has since been used elsewhere. The Rykar event and Dr. Niering's education of the broader community also laid the basis for Connecticut's later comprehensive approach to the regulation of inland wetlands.[36]

Just as significantly, the Rykar challenge brought together a coalition of governmental and nongovernmental organizations that successfully lobbied for more than twelve million dollars in federal funds to enable the U.S. Fish and Wildlife Service to acquire 450 acres in the Great Meadows Marsh. It is one of the ten wildlife areas comprising the Stewart B. McKinney National Wildlife Refuge.[37] Partners in its protection include The Nature Conservancy, Connecticut Audubon Society (with its adjacent Coastal Center), the Connecticut Department of Environmental Protection and even, to a limited extent, the Stratford Development Company, Rykar's successor corporation. The Great Meadows Marsh is a keystone for the protection of a much larger complex at or near the mouth of the Housatonic. Typically for Con-

35. Connecticut Public Act 69-695, Connecticut General Statutes sec. 22a-28-35.

36. Connecticut Public Act 72-155, Connecticut General Statutes sec. 22a-36-44.

37. While Federal ownership of conservation land in Connecticut is limited, an unusual holding is the Weir Farm National Historic Site of the National Park Service in Ridgefield and Wilton, which is maintained to preserve the house, buildings and landscape where J. Alden Weir and other notable American artists worked and painted beginning in the nineteenth century.

necticut, citizen response to the Rykar challenge resulted not only in saving a particular resource but stimulated land-saving actions on a much larger scale.[38]

A challenge of a different sort confronted communities in the lower Connecticut River valley in the 1960s. Anyone with the good luck to have visited the river below Middletown in a boat is astonished by its beauty and relatively unspoiled appearance. In short, it is now and for decades has been one of the most beautiful vistas in New England. It is also a natural area of much biological importance. But by the 1960s the traditional character of the region was being threatened. Aesthetically, a proposal to construct a high-voltage power line across the river at East Haddam was seen as an eyesore in a landscape that theretofore had had no such reminders of encroaching technology. The construction of a nuclear power generating station at Haddam Neck offended many. Were these the harbingers of more to come? How could the lower valley protect itself when the only restraints were the inadequate planning and zoning regulations of each individual town? As these questions grew, Senator (and former Connecticut governor) Abraham Ribicoff came up with a riveting idea: He requested the National Park Service to study the feasibility of turning the Connecticut River Valley into a National Recreation Area.

It would be wrong to report that the senator's suggestion was received with widespread enthusiasm. Particularly at that time and in that place, even the idea of *studying* the *possibility* of a *federal* presence on the imagined scale was distasteful to those who all too easily conjured up images of losing local control and being overwhelmed by hordes from parts unknown. Possible benefits from such an arrangement vanished in the presence of these quite typical Connecticut worries. Yet the alarm bell had sounded. Clearly something had to be done to save the lower valley from harmful change. And clearly no one town could accomplish it on its own. Yet the towns were not used to cooperating with each other, particularly when situated on opposite sides of the wide river. Once again in response to a unique challenge a unique land saving solution emerged.

With the patient guidance of Gregory G. Curtis of the University

38. On Long Island Sound issues see generally, Tom Anderson, *This Fine Piece of Water* (New Haven: Yale University Press, 2002).

of Connecticut Cooperative Extension System,[39] the Gateway Advisory Committee examined the Ribicoff proposal. Citizens and officials of the several towns began to work together as never before to save the landscape that was precious to all of them. Their discussions led to the establishment of the Lower Connecticut River Conservation Zone by state statute in 1973.[40] After declaring the public interest in preventing the "deterioration of the natural and traditional river way scene," the new law created a process whereby a representative commission would adopt minimum standards for local zoning regulations that would uniformly be adopted by the towns in a zone that included parts of the towns of Old Saybrook, Essex, Deep River, Chester, Haddam, East Haddam, Lyme, and Old Lyme. The process would be overseen by a new Connecticut River Gateway Commission. Each town would retain its customary role with respect to land use, subject only to the minimum requirements adopted to protect the zone. While its complexity is not suggested by this brief description and while the administration of the arrangement has not been problem free, this creative blending of authoritative regional standards with traditional town regulatory processes was unique in Connecticut. It is testimony to the acknowledged preciousness of the lower Connecticut River as well as the shock effect of Senator Ribicoff's modest proposal.

The later designation of the lower river by the Connecticut Chapter of The Nature Conservancy as one of the "Last Great Places" in 1993, and the recognition of its tidal wetlands as "wetlands of international importance" pursuant to the 1994 Ramsar Convention underscores the ecological importance of the region. Through a program that began in 1960, The Nature Conservancy protects 4,000 acres along the lower river and its tributaries. Notably, in 1991 the entire Connecticut River watershed was identified within the still somewhat indistinct Silvio O. Conte National Fish and Wildlife Refuge. The challenge of land savers in this watershed, including The Nature Conservancy and the U.S. Fish and Wildlife Service, will be to design conservation programs that work in the midst of mature systems of ownership and local land-use regulation, much the same problem faced by the creators of the

39. Formerly "Service."

40. Connecticut Public Act 73-349, Connecticut General Statutes sec. 25-102a et seq., as amended.

Lower Connecticut River Conservation Zone, recognizing that protection of what is valuable need not and cannot entail ownership of all resources and understanding that there are many tools other than acquisition that can influence how land is used consistently with accomplishing conservation objectives. This is especially true in Connecticut, with its strong traditions of home rule and local control of land-use policies.[41]

Dolbia Hill, now heavily wooded, is a modest rise in the highlands easterly of the Connecticut River along the border between East Haddam and Salem above North Plain. Burnham Brook[42] is born there and after being joined by Fawn Brook continues westward before flowing into the pristine Eight Mile River,[43] which in its turn joins the Connecticut at Hamburg Cove. It was at Dolbia Hill that Richard Hale Goodwin, then a young botany professor at Connecticut College, chose to buy a 170-acre farm together with a friend in 1956, where he was to build a home for his wife, Esther, and their family. Dr. Goodwin had been involved with the Ecologists Union, which changed its name to The Nature Conservancy in 1950. In 1960 he and friends donated to the fledgling conservancy forty-eight acres of land as the core property of what was to become the Burnham Brook Preserve. It was one of the first acquisitions of the conservancy. Today the Burnham Brook Preserve comprises more than 1,100 acres, most of which were acquired through Dr. Goodwin's efforts. Maintained as a natural area, it contains more than 100 types of trees, shrubs, and vines, all carefully identified, as well as 400 species of flowering plants and 180 species of birds, Dr. Goodwin told the *New York Times* in 1991. The Burnham Brook ex-

41. A Federal approach to the conservation of river corridors having both cultural and natural significance was the designation by Congress in 1994 of the Quinebaug and Shetucket Rivers Valley Natural Heritage Corridor in northeastern Connecticut and south-central Massachusetts.

42. Named for Jacob Burnham, who owned the top of Dolbia Hill before 1744. Richard H. Goodwin, *Burnham Brook Preserve of The Nature Conservancy, 1960 to 2005* (Privately published: Richard H. Goodwin, 2006) p. 17.

43. The Eight Mile was added to the National Wild and Scenic Rivers System by the Consolidated Natural Resources Act of 2008, signed into law by President George W. Bush May 8, 2008. It thus joined fourteen miles of the West Branch of the Farmington River as the second river in Connecticut to receive that designation. Designating the remainder of the Farmington River is under study.

ample once again points out that land conservation in Connecticut often has begun with the generosity of a single individual.

Personal volunteerism has also been a hallmark of Connecticut conservation. In addition to his role as the on-site steward of the preserve he established, Dr. Goodwin could be considered the midwife of The Nature Conservancy itself. Were it not for his early support and his skillful leadership at critical times during two different tenures as its president and his negotiation of the earliest conservancy acquisitions, the conservancy might never have survived its infancy to attain its present form.[44] Today, operating in 50 states and more than 30 countries, the million-member conservancy has protected 117 million acres of land and five thousand miles of rivers around the world. As Burnham Brook might tell us, if she could speak, and as the Weatogue group might affirm, from small beginnings in Connecticut great things often have come.[45]

Following the 1962 Whyte Report, Governor Dempsey prophetically moved Connecticut into the vanguard among the states that were engaging in environmental issues at that time by establishing, successively, blue-ribbon task forces on water quality and air pollution. Their reports, together with the Whyte report, made clear the interconnectedness of air, water, and land-use issues and the need for more overarching policies. Accordingly, Dempsey in 1970 appointed the Governor's Committee on Environmental Policy under the chairmanship of Dr. James G. Horsfall, director of the Connecticut Agricultural Experiment Station. Its fifty-four members included a broad spectrum of business, academic, conservation, young people, and citizen activists. Their report[46] influenced land conservation and environmental protection thenceforward.

Policy recommendations included supporting continued preferential taxation of farm, forest, and open-space land at the town level; reducing dependence on the real property tax; encouraging the acquisition of less-than-fee interests in land as a conservation tool;

44. See generally, Richard H. Goodwin, *A Botanist's Window on the Twentieth Century* (Petersham, MA: Harvard Forest, 2002)

45. Dr. Goodwin's urging was also the driving influence behind the 1971 legislation establishing the Connecticut Natural Areas Preserve System. Connecticut Public Act 71-727, Connecticut General Statutes sec. 23-5a-e, as amended.

46. *An Environmental Policy for Connecticut, Report of the Governor's Committee on Environmental Policy* (1970). Available at the Connecticut State Library, Hartford, CT.

preparing an official state plan of conservation and development, a study of employing state or regional zoning for projects having statewide significance; granting additional powers for municipal land-use authorities to augment conservation efforts; and studying additional ways to save farmland in addition to Public Act 490. The report recommended further investment by the state and municipalities in land acquisition, with the goal of acquiring 200,000 acres of forestlands, 50,000 to 80,000 acres of ridgetops, 50,000 acres of parkland and 25,000 acres of coastal wetland by 1980. The committee encouraged new laws to provide further tax advantages to landowners of private land who wished to maintain existing open space.

The governor's committee addressed concerns that had been raised about state level governance of conservation and environmental matters, responsibility for which had been distributed among various separate boards, commissions, and agencies. It urged the creation of a Council on Environmental Quality, to some extent echoing a recent step by the federal government. Controversially, among its other functions the recommended council was to be empowered to evaluate and approve (and by inference disapprove) proposals for "developments which because of their unusual size and nature are capable of causing significant damage to the environment." Equally controversially, the Committee recommended a study of "the possibility of a limited use of state or regional zoning" for flood plains, major highway intersections, "conservation lands" and "major commercial and industrial installations." As it turned out, the legislature did authorize a Council for Environmental Quality to report annually on environmental conditions and provide a forum for citizens' concerns, but gave the council few additional powers.[47] However, with respect to the governance of environmental and conservation matters, the 1971 session of the Connecticut General Assembly took a far bolder step than the committee had in mind.

The legislature created an entirely new agency, the Department of Environmental Protection.[48] The task forces of the sixties and the gov-

47. Connecticut Public Act 71-872, Connecticut General Statutes sec. 22a-11-13.

48. Connecticut Public Act 71-872 Connecticut General Statutes sec. 22a-2. The statute establishing the Department also enacted a "little NEPA" and granted standing to certain parties to enforce environmental laws under some circumstances.

An Environmental Policy For Connecticut

Report of the Governor's Committee on Environmental Policy

New institutions of governance, such as the Department of Environmental Protection and the Council on Environmental Quality, and comprehensive new policies emerged from the Report of the Governor's Committee on Environmental Policy appointed by Governor John Dempsey in 1970.

ernor's committee report had expressed concerns about the absence of effective processes at the gubernatorial level to resolve conflicts among resource agencies. However, under the political leadership of state senator Stanley J. Pac, Senate chairman of the newly created Joint Standing Committee on the Environment,[49] the legislature was not content, as the committee had been, with simply adding another layer of government on top of existing boards and commissions in the form of the Council on Environmental Quality. Although, as has been noted, many of these agencies, for example the Park and Forest Commission, had distinguished records of service over many years and had been part of the foundation of conservation in the state, the legislature chose simply to abolish them. Gone at a stroke of the governor's pen were the Park and Forest Commission, the Board of Fisheries, and Game and the Water Resources Commission, among many other agencies, and their responsibilities and authorities were vested in a single Commissioner of Environmental Protection. The rationale was to end wrangling, provide for more efficient and coordinated acquisition, management and regulatory policies, and improve integrated information collection, organization and dissemination. Had not Barry Commoner said that in ecology "everything is connected to everything else?"[50] Why, then, should not government agencies dealing with ecological issues also be connected?

The new department was assigned responsibilities heretofore scattered among a number of different boards, commissions, and committees of the executive branch, notably responsibilities for parks and forests, habitat and wildlife, regulation of coastal wetlands and regulation of air, water and waste disposal, together with a myriad of other authorities and duties. The staffs of the eliminated boards and commissions were largely absorbed into the new department. Notably, while the foundations for the remarkable new approach had been laid during the incumbency of Democratic governor Dempsey and the legislation establishing the department enacted by a General Assembly with Democratic majorities in both Houses, it fell to Republican governor Thomas J. Meskill to appoint the first Commissioner of Environmental Protection, Dan W. Lufkin.

49. Later Connecticut Department of Environmental Protection commissioner.

50. Barry Commoner, *The Closing Circle* (New York: Knopf, 1971) p. 33.

The momentum of initiatives begun in the 1960s carried forward immediately into the next decade under the more centralized governance, with ample support from both Democrats and Republicans in the state legislature. Concern about wetlands was extended to inland wetlands by the 1972 Inland Wetlands and Watercourses Act.[51] In contrast with tidal wetlands, regulatory authority over inland wetlands was granted to towns through establishing inland wetlands agencies that adopted regulations meeting the standards of the department. These agencies could be combined with the municipal conservation commission[52] if the municipality chose to do so. The Connecticut Association of Conservation Commissions reconstituted itself in 1974 as the Connecticut Association of Conservation and Inland Wetlands Commissions [sic]. Because a determinative permitting function often involving technical details had been assigned at the local level to frequently inexpert laypersons, the sensitivity of the land-use decision-making system to resource issues depended significantly on educating the volunteer members of the town commissions on resource issues. The department instituted programs to assist them as well as all town-level officials making resource sensitive land–use decisions.

More particularly, from the outset the new Department of Environmental Protection placed a high priority on establishing systems to gather a more complete and cross-disciplinary information base for land conservation and other purposes than had been conveniently achievable previously when natural resource data and knowledge had been scattered among different agencies. Not only decisions about land conservation but many other policies as well needed to be based on sound planning, as Whyte had pointed out ten years earlier. And sound resource planning necessarily must rest on reliable information that is not bottled up in divided agencies (so the thinking went) or confined to particular levels of the bureaucracy. The integration of gathering and assembling information and its transmission within an evolving land-use policy framework that included not only state agencies but also municipal decision-making boards and commissions (conservation, inland

51. Connecticut Public Act 72-155, Connecticut General Statutes sec. 22a-37.

52. As has been noted, municipal conservation commissions were authorized in 1961. Public Act 61-310, Connecticut General Statutes sec. 131a-d.

wetlands, and planning and zoning particularly) was the goal in order to assure that resource-conscious decisions could be enabled. This effort was particularly important in Connecticut, where so much authority over land use and resource conservation resides in the municipalities.

Under the leadership of Dr. Hugo F. Thomas, an interdisciplinary group produced a keystone document, *The Use of Natural Resource Data in Land and Water Planning.*[53] This established a methodology to standardize, integrate, and apply resource information for environmental and conservation decision making. The next step was to begin an ongoing joint effort by the department and the University of Connecticut Cooperative Extension Service to go to local communities to provide workshops and field experiences to help cognizant boards and commissions use resource information most effectively. A third step was creating multidisciplinary environmental review teams composed of professionals from federal, state, regional, and local agencies representatives, whose role was to assist municipalities when called upon, without cost to them, to evaluate sites proposed for development and provide natural resource inventories for planning purposes. These activities were particularly important for land conservation to reduce the likelihood of inadvertent decisions by inexpert members of town boards and commissions.

The value to land savers of having a centralized, integrated natural resource database was soon recognized. Expanding, coordinating, and integrating the database was the job of a new Natural Resources Center led by Dr. Thomas, which ultimately transferred data to a Geographic Information System that was made available to all users through the Internet. The new center became a broadly available clearinghouse for information on earth sciences and biology, developing methods to apply resource information to decisions of all levels of government and by the private sector, and providing needed technical assistance, particularly to municipalities. A spin-off particularly significant to land conservation was the creation of the Natural Diversity Data Base as a central repository of information on biology and population status relevant to natural diversity, rare and endangered species, and critical

53. Hill and Thomas, *The Connecticut Agricultural Experiment Station Bulletin*, New Haven, CT, 723 (1972).

habitats at risk. Protection of the critical coastal region along Long Island Sound was aided by yet another database established by the department at the Avery Point Branch of the University of Connecticut. The Long Island Sound Research Center integrates standard library inquiry methodology with a natural resource geographical information system through an interactive program available on the Internet, to the end of improving the decision-making process and assisting in establishing land-conservation priorities for the Sound region.

A significant increase in state funding for open-space acquisition that was consistent with the recommendations in the Whyte report, the report of the governor's committee, and the urgings over many years of the Connecticut Forest and Park Association, The Nature Conservancy, and many land-saving coalitions was promised with the enactment of the Connecticut Recreation and Natural Heritage Trust Program in 1986.[54] This program directs the commissioner to identify priority sites according to their value for recreation, forestry, fisheries, and wildlife that are consistent with the state's plans for outdoor recreation, conservation, and development, are a "prime natural feature" of the landscape, or are the habitat of a species that is threatened, endangered, or is of special concern—among other criteria. The commissioner is authorized to proceed with acquisitions using state bond proceeds, as they may be made available, as well as matching funds from private sources, the federal government, municipalities, or transfers of property that meet the criteria of the program. However, the first two years of funding for the program were disappointing, averaging only three and one-half million dollars per year. Several major conservation organizations established the Land Conservation Coalition to lobby for greater support. Responding to their pleas, Governor William A. O'Neill increased the funding to fifteen million dollars annually. This program has become the primary tool for the acquisition of conservation land.

To some, the ten square mile tract of land in very quiet Easton, in Fairfield County, would seem an ideal place for a championship golf course

54. Connecticut Public Act 86-406, Connecticut General Statutes sec. 23-73-79.

bordered by expensive homes in the midst of other protected open space. To the owner of the property, Aquarion (which had acquired the land when it purchased its previous owner, the Bridgeport Hydraulic Company) it seemed an attractive idea. However, when Aquarion unveiled the proposal in 1994, to a great many others it spelled desecration of precious Trout Brook Valley. As is characteristic in Connecticut, local citizens rallied to Trout Brook Valley's defense. The controversy reopened the larger issue of what to do about the 110,000 or so acres of "water company land" that is scattered throughout the state.[55]

Although, as has been said, beginnings in Connecticut are hard to track down sometimes, the start of what became the water company land issue may have been the fight in the 1930s by residents who were determined to save the historic Valley Forge village in Weston from inundation at the hands of the Bridgeport Hydraulic Company, which planned to flood the entire valley where it was located for a water supply reservoir. The valley was so named because of a forge there that had supplied munitions to American soldiers and sailors in four wars, starting with the American Revolution. The forge is said to be the place where were made the first cast-iron plows in New England.[56] When the water company announced its plans, which included taking homes and other private property in the valley by eminent domain, militant citizens formed the Saugatuck Valley Association, a coalition that included residents, writers, artists, politicians, and other influential Fairfield County folks. They challenged the company in court. But ultimately they failed to persuade the judge, and the valley and its history now sleep beneath the waters of Saugatuck Reservoir. The historic village had been drowned, but the issue had not died with it.

It surfaced again in the early seventies when news circulated that

55. Predominately, in Connecticut water for communities has been supplied by privately owned water companies. A noteworthy exception is the Metropolitan District Commission, which was chartered by the legislature in 1929 to provide water for Hartford and a cluster of central Connecticut municipalities. Kevin Murphy, *Water for Hartford* (Wethersfield, CT: Shining Tramp Press, 2004) p. 208. The MDC owns more than 30,000 acres of watershed land, 25,000 acres of which is actively managed for forestry and a significant portion of which is open for public recreation.

56. James Lomuscio, *Village of the Dammed* (Lebanon, NH: University Press of New England, 2005) p. 16.

the New Haven Water Company, another investor-owned public utility, planned to sell some 16,000 acres of the 26,000 that it had acquired for water-supply purposes because, it said, they were no longer needed and the company needed the money from the sale to facilitate compliance with the new federal Safe Drinking Water Act.[57] Like the Bridgeport Hydraulic Company and other public service companies, the New Haven Water Company had been granted the power to take private property by eminent domain. Two issues arose immediately. First, should such a company, without public oversight, be allowed to sell any real estate it wanted to merely by a boardroom conclusion that it was no longer needed? Second, if land that had been bought by a company possessing the trump card of the power of eminent domain (whether or not it was exercised) was sold, should the private investors in the company be entitled to take home all of any profit? Both issues were important because private and public water companies possessed tens of thousands of acres of increasingly valuable open-space land strewn all over the state. The Natural Resources Council of Connecticut, a prestigious environmental group, raised these issues at a public forum, and they became the particular concern of Dorothy S. McCluskey, a state representative from North Branford, who soon secured the passage of a law that placed a two-year moratorium on such sales and authorized a new Council on Water Company Lands to study the matter and make recommendations.

As a consequence of the council's recommendations, the General Assembly adopted a law requiring water companies to notify specified government agencies and nongovernmental organizations of any proposed sales or other dispositions of land and granted to them an option for a period of time to negotiate its purchase.[58] Representative McCluskey led the charge, along with citizen activist Claire C. Bennitt, who, together with many others, rallied political support. Ten years later a system was legislated[59] that requires the classification of land

57. See Dorothy S. McCluskey and Claire C. Bennitt, *Who Wants to Buy A Water Company?: From Private to Public Control in New Haven* (Bethel, CT: Rutledge Books, 1996).

58. Connecticut General Statutes sec. 16-50b-e, as amended.

59. Connecticut Public Act 77-606, Connecticut General Statutes. sec. 25-37a-i, as amended.

owned by public and private water companies into three categories according to the importance of the land for water supply purposes, restricts the transfer or change in use of the two most critical categories, and establishes a state Department of Public Health approval process for certain land sales.[60] How sale proceeds are to be allocated as between ratepayers and shareholders is to be decided by the Department of Public Utility Control. This legislation places Connecticut in a leadership position in recognizing both that open space, healthy soils, and natural vegetation are the best purifiers of drinking waters and in establishing state processes to assure the protection of water supply land.

In the case of the New Haven Water Company, which was organized in 1849 and owned land in sixteen south-central towns, the legislature authorized the formation of the South Central Regional Water Authority,[61] which in 1980 purchased the entirety of the company's land. In that region, ownership by a service-providing public authority managed with the participation of member towns was deemed the most appropriate approach. Today, 16,700 acres of that land are committed by the Authority to "renewable and nonrenewable" resources, within which is located the Eli Whitney Forest, one of the oldest managed forests in the United States. The authority conducts extensive educational and recreational programs on land that has been "saved" from inappropriate development. Thus, for south-central Connecticut the issue was taken care of through the energetic advocacy of regional caretakers.

Another large step forward was taken in 1998, when Governor John G. Rowland, responding to the recommendations of a blue-ribbon task force that he had appointed under the chairmanship of DEP deputy commissioner David K. Leff, announced a goal of saving 21 percent of the state's land area, roughly 673,000 acres, by the year 2023. The acreage target, but not the year, was ratified by the General Assembly, and it remains Connecticut's goal. The task force noted that at the time of its report the state was about two-thirds of the way to the target. Sources of state funding would be the Recreation and Natural Heritage Trust Program established in 1996 and a new "protected open space and

60. Ongoing consideration of water supply land and related issues is the charge of the Water Planning Council that was established in 2001. Connecticut Public Act 01-177, Connecticut General Statutes. sec. 25-33o.

61. Connecticut Special Acts 77-98 and 78-14.

watershed land acquisition" matching grants program enacted in 1998 as a consequence of the task force's recommendations.[62]

The Easton golf course debacle was resolved only after the dogged intervention of citizens and nongovernmental organizations such as the Trust for Public Land, The Nature Conservancy, the Connecticut Fund for the Environment, and the Aspetuck Land Trust. Inspired by the leadership of Easton activist Patricia Falkenhagen, Gail Bromer, chair of the Easton Open Space Task Force, and Bruce LePage of the Aspetuck Land Trust, the Easton Coalition to Preserve Trout Brook Valley rallied public opinion, enlisted the support of celebrities, and got the attention of Governor Rowland, who announced in June 1998 that the state would put up six million dollars for the purchase of Trout Brook Valley, to be combined with monies from the Town of Weston (where part of the valley is located) and the land trust.

Trout Brook Valley was thus saved, but what about the remaining thousands of other acres owned by Aquarion? The Trout Brook coalition, with the encouragement of the Connecticut Fund for the Environment, The Nature Conservancy, the Trust for Public Land, and other land-saving nonprofits, broadened its outreach and enlisted the participation of even more influential people, officials, and conservation groups. The Nature Conservancy joined Deputy Commissioner Leff at the negotiating table with representatives of the new owner, the British Kelda Group (Aquarion having been acquired).[63] After weeks of complex negotiation, in the end Governor Rowland was able to announce the acquisition by the state of 15,300 acres for a purchase price of ninety million dollars, eighty million dollars from the state surplus through the Open Space and Watershed Land Acquisition Grant Program[64] and ten million dollars from The Nature Conservancy, which emerged the holder of conservation easements as a continuing safeguard. An ongo-

62. Connecticut Public Act 98-157, Connecticut General Statutes sec, 131d(a)-(c).

63. Aquarion was sold in early 2007 to Macquarie Bank of Australia, marking the transition of control over significant Connecticut regional water services from a Connecticut public-service company to a foreign financial institution.

64. As of June 2007, 480,550 acres of conservation land had been protected under this continuing program, which is administered by the Department of Environmental Protection.

ing management committee was established with members from Kelda, DEP, and The Nature Conservancy.

Success in the watershed-lands struggle that was fueled by relentless citizen activism over decades is reflected in the enactment of laws that appear to establish significant additional protections and put in place a process that protectively governs the disposition of land or changes in its use by all possessors of watershed lands, including both private and public water companies. In addition to the watershed-land classification system and regulatory protections established in the 1970s, the legislature has provided attractive economic incentives if the land disposed of is restricted for use solely for conservation or recreational purposes[65] and, importantly, has provided a program to fund the purchase of land for conservation.

The Connecticut approach to the watershed land issue thus combines two important land saving techniques: *Ownership* of land, either by a regional authority, as is the case in south-central Connecticut, or by the state, municipalities, or land-saving nonprofits, as is the case with much of the Aquarion property; and *regulation* of land that remains in the ownership of private and public water companies. With respect to watershed land that is protected solely by regulation, advocacy groups such as Rivers Alliance of Connecticut and others have noted that the effectiveness of watershed-land regulation, as well as other regulatory regimes for land protection,[66] depends upon the competence and zeal of the officials entrusted with their administration[67] and the willingness of officials to resist political and economic pressures by remaining entirely faithful to the legislative mandate establishing the protection, as well as enduring fidelity to the regulatory protection by the legislative body that has established it.[68]

65. Connecticut General Statutes, sec. 16-43.

66. This is especially true of local land-use decision making by planning, zoning, and inland wetland commissions, whose ill-considered decisions on proposals for resource-sensitive land can result in the degradation of wetlands, fragmentation of habitat, and a landscape spattered with eyesore development.

67. The Endangered Lands Coalition maintains particular surveillance over regulatory programs affecting conservation land.

68. In one notable departure, the Connecticut General Assembly voted to bypass the protection of water-supply land that the legislature itself had established. Connecticut Public Act 07-244. After an uproar, that legislation was soon repealed in a special session. Connecticut H.B. 8006 (2007).

In a responsive collaboration that is typical in Connecticut, about fifty organizations representing many thousands of people allied themselves in 2007 to pursue a campaign called the Face of Connecticut, borrowing its name from an important book that takes an integrated view of "people, geology and the land," to quote its subtitle.[69] The purpose of the campaign is to increase reliable funding for land conservation, farmland protection, historic preservation, comprehensive planning, watershed protection and stewardship, and to support responsible growth strategies. The wide array of sponsoring organizations reflects how interrelated the particular agendas of the sponsors are seen by them to have become, a perception that has evolved from experience. Conservation of forests is not in many respects much different from conservation of wetlands or protection of farmland or safeguarding habitat or defending landscapes or enhancing the amenities of cities and towns. It is not odd, then, to find among the sponsors of the same campaign the Connecticut Conference of Municipalities, the Connecticut Farm Bureau, the Connecticut League of Conservation Voters, the Garden Club of New Haven, and the Connecticut Trust for Historic Preservation, to list only a few of the participants. The Face coalition reflects this learning pointedly: while each element of community preservation has understandably emerged from a discrete concern that has attracted its own special advocacy, in land conservation, as in ecology, the perception is emerging that here also everything seems to be connected to everything else.

A gubernatorial Green Plan to guide acquisition and land protection by the state was first promulgated in 2001, and an updated version for the years 2007–2012 was unveiled by Governor M. Jodi Rell in 2007. It knits together strategies to satisfy the objectives of a variety of state plans pertaining to conservation and development, wildlife conservation, recreation, and climate change, and reaffirms the land-conservation goals set forth in commitments by prior governors and the legislature. Governor Rell's updated Green Plan carries out her Executive Order No. 15, which created an Office of Responsible Growth "to coordinate state initiatives to control rampant, ill-conceived development that threatens Connecticut's special character."[70] To put it

69. Michael Bell, *The Face of Connecticut* (Hartford, CT: State Geological and Natural History Survey, 1985).

70. Executive Order No. 15, 2006. The Office of Governor M. Jodi Rell.

another way: to save the face of Connecticut from ill-considered change.

It would be enlightening if the little group that met in Weatogue in 1895 were able to return for a day—like Emily in *Our Town* and her onlookers from their starry resting place—so that we could ask them what they think of what has happened to their legacy and charge. None of what has been recounted here had yet happened then, of course. Of what would they approve? What has been neglected? About what would they be dismayed? It would be an interesting conversation.

Reminding them that Connecticut started with a mere 12,000 acres of state forests in 1921, we could tell them that today there are 32 state forests of almost 170,000 acres and one hundred state parks. Much of the state is once again blanketed with trees. They would learn that the state has acquired a total of 251,000 acres for its system of parks, forests, and natural resource management areas. They could be told that state and municipal governments have been organized so as to attend methodically to conservation programs. They would be happy to learn that natural resource information is being accumulated systematically and made available to conservation decision makers. It would be pleasing to tell them that much of this has happened through the efforts of informed citizens and a galaxy of organizations that have sprung up, much in the model of their Connecticut Forestry Association, to address myriad aspects of the conservation effort. While they would be saddened by the dwindling of farmland, they would be happy to discover that something is being done to save those farms that remain. They would be excited that the No Child Left Inside initiative is so successful in enticing children and families to have fun in the state parks the founders helped create.

They would probably be astonished by the population growth,[71] disturbed to see the countryside being gobbled up by sprawl,[72] and worried about the decline in village and city amenities, including some urban parks. They would be upset by the frequent elevation of private claims over community entitlements in the local zoning process. They

71. In 2007 more than three million people on three million-plus acres.

72. As well as the social consequences. See http://www.oua-adh.org/centerEdge_project.

would wish that the same spirit of hope, generosity, civic responsibility, and volunteerism that motivated them would be abundantly carried forward by *all* who call Connecticut home today and tomorrow. They would argue sternly that the state forests and parks that exist largely because of what they started should be supported by the regular appropriation of sufficient public monies to assure their proper maintenance and appropriate use, and they would be offended by past neglect. They would cry shame at the continuing loss of privately owned forests through development's often undiscriminating axe. From their removed yet not indifferent vantage point, they might observe that while our century just past has witnessed challenges and turmoil and the present is not free of uncertainty, their time, too, had not lacked its own tribulation; and they might remind us that through it all the need to rescue from indifference the too conveniently ignored intimacy of man and nature has not changed. Whatever the circumstances, they might tell us, the metaphorical forest always will have need of our defense and tending.

As the stories are examined, they would conclude that land conservation in Connecticut over the last century and into this one has been characterized by distinctive qualities. Important among them have been collaboration, responsiveness, invention, individual initiative, personal volunteerism, generosity, adaptability, and creative synergy between the role of government and the role of citizens and their organizations. As the Simsbury group would know well, traditionally the rights and responsibilities of individuals have been the keystone of the political compact that from the beginning has defined Connecticut; the individual has the right to participate in public life and the responsibility to do so. Nowhere has this been more evident than in conservation efforts in the state. Very often it has been individual citizen leaders and the collaborations they have formed that have been the catalysts for action.

The culture of the towns puts citizens in close contact with their own government and their responsibility for local action, so it would come as no surprise to the Weatogue group to learn that Connecticut has been the seedbed of town-based land trusts. One of the success stories has been the bipartisan support of a succession of governors, legislators, state administrators, and town officials over the years and the creative relationships that have been forged between government at all levels and the proliferating nongovernmental organizations. Collaborative problem solving has been a hallmark of Connecticut conservation. Inventiveness also has been an important aspect of conservation in the state. Early on people in Connecticut understood that the newfangled notion of a conservation easement could be used to determine what happens to a parcel of land when we are less concerned about its present use than we are about what might happen later on. The model of regional land-use cooperation in the lower Connecticut Valley is another example of innovation. So also is how the water supply lands issue is variously being addressed.

An instance of a need for still further inventiveness might be addressing as a work in progress defining the appropriate roles for municipal, state, and federal governments and their powers and resources in a state that treasures home rule, nurtures personal responsibility, profoundly respects private property, and wishes to hold onto all the good in those things while yet deeply aware that the mindless exercise of these time-honored privileges may work against the sustainable functioning of healthy natural systems and the appropriately broad enjoyment of resources and amenities. Connecticut understands that there is no pat answer to achieving the desired balance. The same Connecticut "tinkerer" mentality that brought forth the cotton gin, the sewing machine and the can opener will be needed for new applications as the state continues to struggle, in the words of President John F. Kennedy, "to develop new instruments of foresight and protection and nurture in order to recover the relationship between man and nature and to make sure that the national estate we pass on to our multiplying descendants is green and flourishing." In less grandiose words, perhaps, that was the task the Weatogue group set for themselves that has resonated forward into all that has followed and into this very time, thus defining the "work" to which Dr. Goodwin's words refer.

FEDERAL LAND CONSERVATION IN NEW ENGLAND

CRISIS, RESPONSE, AND ADAPTATION

Robert W. McIntosh, Rolf Diamant, and Nora J. Mitchell

INTRODUCTION

To understand the evolution of federal programs addressing land conservation in New England, it is first necessary to look more carefully at the region's environmental history and cycle of environmental crises that have transformed the face of New England — the deforestation in the nineteenth century, industrial pollution and suburbanization of the mid-twentieth century, and increasing habitat fragmentation of the twenty-first century. In response to each crisis, the national government, mostly in conjunction with local initiative, used tools that had been tried and tested elsewhere, however, in many cases, they were adapted in ways that were strongly influenced by regional conditions and the unique environmental and social characteristics of New England. The national government's ever-changing role in the conservation of the New England landscape has always reflected a number of national trends and policies that have been uniquely interpreted and applied to the region. Model programs and designations that appeared to work in other sections of the country would require a reinterpretation if not a more substantial reinvention in the independent, small town, home-rule environment of the region that first introduced America to the concept of town meetings.

In the White and Green Mountain National Forests in New Hampshire and Vermont, for example, relatively late additions to the nation's forest system, new precedents were introduced for buying and conserving cutover private forest lands, a programmatic emphasis on scenery and recreation, and a grassroots network of public support and advocacy. The implementation of the 1911 Weeks Act forever changed the U.S. Forest Service. As part of the original Wilderness Act, the first wilderness designation in New England, the Great Gulf Wilderness Area on the northeast slopes of Mount Washington would preface a fundamental reinterpretation of wilderness values. Championed by Vermont's U.S. Senator George Aiken, the 1975 Eastern Wilderness Act established a substantially more inclusive vision of federally protected wild lands. Musing on his 4,000-acre backyard — Bristol Cliffs wilderness in the Green Mountains — writer John Elder explains in the book *Wilderness Comes Home: Re-wilding the Northeast*, "Gates of the Artic they're not. Stone walls break through the ferns and jewelweed of these slopes, broken choker cables lie half-buried beside trails that were logging roads not along ago, and cellar holes collect and compost weeds . . . These tracts of third-growth forest were not included under the original 1964 Wilderness Act, being neither 'primeval nor "untrammeled' . . . These were lands 'protected because of their beauty and their biological significance.' "[1]

In some instances, the adaptations themselves have proven so successful that they have been "exported" back out of New England and used in other parts of the United States. The first National Heritage Area was established in Illinois in the 1980s, but today's growing system of thirty-seven national heritage areas, from Gullah Geechee on the Carolina coast to Arizona's Yuma Crossing, has been largely influenced by the experiences and adaptations of the National Heritage Area along the Blackstone River and Canal in Massachusetts and Rhode Island. The Blackstone Valley is a venerable artifact of centuries old New England mill towns surrounded by equally old farm and forest landscapes, and the heritage corridor commission works to support a balance between heritage preservation, tourism, and diversified economic development. Similarly, for years the National Wild and Scenic River

Program was generally perceived as applicable to western rivers exclusively on federal lands and managed by federal agencies. There were exceptions in New England however. The policy debates of the 1960s and '70s over the protection of the public interest along Maine's Allagash River were resolved by establishing a category of state administered rivers in the National Wild and Scenic River System. Later, in the 1980s, the program underwent a further metamorphosis with the designation of New Hampshire's Wildcat River. The Wildcat was the first of several rivers largely outside the federal estate that were nationally designated but protected primarily through private land-conservation activities and comanaged with local communities. This model has now been replicated in Massachusetts and Connecticut and several other states.

When attempts have been made to rigidly impose a federal model for parks and conservation, regional resistance has been intense. The unhappy fate of the proposed Green Mountain Parkway (1930s), the Connecticut River National Recreation Area (1960s), and the Housatonic National Wild and Scenic River (1970s) are a testament to the limits of any top-down solution. It is not that New Englanders did not wish to conserve land or to work with federal agencies on conservation or receive assistance — it was just that conservation in New England had to be much more locally driven, collaborative, and flexible. It is therefore not surprising that today there is a model "partnership" national wildlife refuge on the Connecticut River and a national heritage area in the upper Housatonic Valley. Vermont finally has a national park — Marsh-Billings-Rockefeller — but a national park with a commitment to continuing a long legacy of responsible stewardship, active forest management, and community engagement. The long standing proposal for a 3.2 million acre "Maine Woods" national park has limited popular support and has not been introduced in Congress; however, the U.S. Forest Service's Forest Legacy Program has provided financial assistance for forestland protection on a willing seller basis in Maine and throughout New England. In northern Maine, New Hampshire, and Vermont the program has assisted in the establishment of some of the largest forestland conservation easements in the United States.

This essay is an overview, and one no doubt shaded by the authors' close association with the National Park Service, though contributions from others more familiar with the U.S. Forest Service and U.S. Fish and Wildlife Service have broadened its perspective. The narrative is not intended in any way to be a comprehensive history. Rather this chapter draws on a handful of examples to illustrate how the federal government fits into the land-conservation mosaic of New England in the past and how it might in the future.

BEGINNING THE TWENTIETH CENTURY

At the beginning of the twentieth century, the forested mountains of central Vermont and northern New Hampshire were largely clear cut, scarred with slash fires, and stream sedimentation was clogging local rivers as well as posing a threat to the water-driven mills downstream. On Mount Desert Island, coastal Maine summer residents were focused on protecting the island landscape from the undesirable effects of the growing community of Bar Harbor and the increasing number of "summer cottages." The land-use conditions and trends established in the nineteenth century were the subject of debates at the local, state, and national levels. In response, the U. S. Forest Service in the northern woods and the National Park Service on the Maine coast, would play a part in shaping the twentieth-century land use in these areas. Importantly, both federal agencies were in their early formative years and their early New England experiences would become very much a part of institutional development.

NEW ENGLAND'S FIRST NATIONAL PARK

The effort to establish a national park on Maine's Mount Desert Island was led by George Dorr, a member of a family of long-term summer residents from Massachusetts.[2] Dorr enlisted the help of others, including Charles Eliot, then president of Harvard University. With others, they created the Hancock County Trustees of Public Reservations, a nonprofit corporation, and once established, they immediately began to purchase properties they considered vulnerable to development. Concurrently Dorr and his associates were using their Washington connections to establish a national park on the island. In 1916, using the authority of the 1906 Antiquities Act, President Wilson proclaimed the establishment of Sieur de Monts National Monument (to become congressionally designated as Lafayette National Park in 1919 and

Acadia National Park in 1929). The properties included within the original Sieur de Monts National Monument were acquired privately and then donated to the United States. Not only was Sieur de Monts the first national park in New England, but it was also the first national park east of the Mississippi River.

The use of a presidential proclamation was aggressively pursued by Dorr. Members of the Maine legislature were opposed to a nonprofit organization buying land for public purposes and were seeking legislation to extinguish the Hancock County Trustee's corporate charter. Dorr wanted immediate, permanent protection and judged that he could not wait for a lengthy and uncertain congressional process. This early and unprecedented approach demonstrated both the tension and the

Members of the "Path Committee" of concerned local citizens, under the leadership of George Dorr (far right), meet in September of 1923 to plan for the future of Jordan Pond, a part of what will later become Acadia National Park. Those attending (from left to right) included Joseph Allen of Seal Harbor, Walter H. Buell of Southwest Harbor, Fred D. Weeks of Bar Harbor, Professor C.H. Grandgent of Southwest Harbor, William J. Turner of Northeast Harbor, and T.A. MacEntire of Seal Harbor.
(National Park Service)

opportunity inherent in private nonprofit and federal partnerships in conservation, characteristics evident even today.

NEW ENGLAND'S FIRST NATIONAL FORESTS

At the same time, debate over the appropriate use of the extensive forests of New Hampshire, Vermont, and the southern Appalachian Mountain Range resulted in the passage of the Weeks Act in 1911. This act provided the authority to establish and purchase lands for national forests in the Appalachian Mountain region and authorized $11 million initially for such purposes. In 1918, the U.S. Forest Service established the White Mountain National Forest in New Hampshire and Maine and, in the 1930s, Vermont's Green Mountain National Forest.

Dr. Perry R. Hagenstein, writing about the establishment of the White (New Hampshire) and Green (Vermont) Mountain National Forests, notes that the creation of a system of national forests was a landmark conservation achievement of the Progressive Movement of the late nineteenth and early twentieth centuries.[3] The forest reserves, later known as national forests, established between 1891 and 1910 were reserved from the federal public domain west of the Mississippi River. The initial reserves were in response to concerns with water supplies affected by mining and agricultural activities in this mainly semi-arid region. Assuring future timber supplies, along with various other multiple uses, were added later. Support for establishing the national forests was heightened by the devastation that had commonly accompanied logging in eastern forests

The success of the western experience with the new idea of federal conservation reserves in the late nineteenth century quickly led to calls to extend the basic idea to the still forested, although often ravaged forests, of the eastern states. For example, heavy logging in northern New England following the Civil War supported the economic growth of the region, but it was also seen as a threat to the tourist industry, which was dependent on the natural scenery provided by the forests and mountains. As early as 1847, George Perkins Marsh warned of the ecological and social impacts of logging in Vermont.[4] Concerns also mounted over the effects of logging and associated erosion on the streams and rivers whose headwaters were in the forests of the North Country. But conditions in New England differed from those in the West.

The western forests that were set aside for conservation were already in the federal public domain and the immediate costs of reserving them were low. The forests in New England were privately owned and the states were generally strapped for funds. Access to federal funds was an attractive idea, but there were doubts in the U.S. Congress on its authority to vote for funds for buying forests. Changing the congressional view of its authority and gaining its willingness to spend money to buy forest lands would take a major effort, one that would ultimately involve many people and many interests. A few individuals and interests stand out but, without the others, the main effort may well have failed. The efforts that led to creation of the White Mountain and Green Mountain National Forests and their place in New England's conservation history were built on tenacity, individual and organizational leadership, a commitment to civic engagement, and a pragmatic view of conservation that recognized the breadth of the public's support of conservation.

Campaigns were started even before 1900 to tap the federal coffers for money to buy cutover forest lands. These campaigns led to building coalitions among private groups who already had some history of interest in the forests along the Appalachians in New England and in other states. In New England, the Appalachian Mountain Club, formed in 1875 to pursue hiking and mountaineering activities in the White Mountains and elsewhere in the Northeast, was one leader. Its interests in protecting the forests of the northeastern region of the Appalachians coincided broadly with those of groups in southern sections of the Appalachians, especially the Appalachian National Park Association, which pushed for creation of a national park in the southern Appalachians. The American Forestry Association, a national citizen's conservation group in Washington, D. C., helped to build support for protection of forest lands among a broader national constituency.

To bring together a coalition of diverse elements to, among other things, support a federal forest reserve in New England, the Society for the Protection of New Hampshire Forests (SPNHF or the Forest Society) was formed in 1901. Philip Ayres, the Forest Society's first forester, was convinced that getting congressional support for federal funding had to be based on a pragmatic appeal to economics. At the same time, Ayres recognized that aesthetics were also an important rationale. According to New Hampshire Governor Sherman Adams,

Ayres put together a coalition of "diverse elements" that included "loggers and pulp manufactures, nature lovers, hotel owners, political leaders, literary figures . . . who could see the economic and environmental advantages to saving the White Mountains."[5]

Marshall Hapgood, a lumberman in Vermont, was an early proponent of public ownership of forest lands. In 1905, he offered to sell a portion of his forest land to either the federal or state government. His lumber business depended on a supply of large timber and he worried about logging methods that were reducing such supplies. After hearing Gifford Pinchot's talks in Vermont about the advantages of proper forest management, Hapgood became convinced of the advantages of public ownership. His view was that public reservations would provide for "combined watershed, game, scenic, and lumber uses," a combination that presaged the list of uses (except for range) in the 1960 Multiple Use Act.[6]

With the federal government lacking authority and funds for buying forest lands, the state of Vermont began to explore initiating a state park system to take advantage of Hapgood's offer. But Vermont was not in a position to buy land for parks. The federal purse was the obvious potential source of land acquisition funds, but even after passage of the Weeks Law, Vermont had difficulty in getting a place in line for funds and gaining approval of the National Forest Reservation Commission and of the state for a national forest in Vermont. Some of this difficulty was opposition by the U.S. Forest Service, which wanted acquisition funds to go to the White Mountain National Forest.

Similar to building support over time for passage of the Weeks Law, a grassroots effort was needed. Some members of the Green Mountain Club, which had been established in 1910, began discussing a federal forest reserve in 1919, and they were helped by Congressman Frank Greene, who later also represented Vermont in the U.S. Senate. He pushed to have the earlier offers of land by Hapgood and Joseph Batell considered for possible Weeks Law purchase. The U.S. Forest Service came around on this and pushed for a recommendation by the National Forest Reservation Commission (NFRC) for establishment of the Green Mountain National Forest. The NFRC turned down the recommendation, arguing that other parts of the country were in more need of the acquisition funds. Perhaps reflecting only lukewarm support on the part of the Forest Service, W. B. Greeley, who was chief of the Forest Service at the time, in response to Senator Greene's prodding, said, "Thanks to the natural conditions, the common sense, and conservative temper of the people, and the effective work of the state department of forestry, Vermont is in splendid shape in comparison with other states."[7] These commendations flew in the face of a continued decline in condition of Vermont's forests, which became increasingly recognized within the state. A major flood in the state in 1927 added to pressure for favorable action by the NFRC, which decided late in 1928 that a National Forest in the Green Mountains was justified. Even then, it took until 1932 for the Green Mountain National Forest to be established.

NEW ENGLAND'S FIRST NATIONAL WILDLIFE REFUGE

The Moosehorn National Wildlife Refuge in Washington County, Maine, New England's first national refuge, came into being in the 1930s. The first National Wildlife Refuge was established in Florida in 1903. In 1937, the Congress authorized the Federal Aid in Wildlife Restoration Program (Pittman-Robertson) which, over the years, has played a key role in the conservation of New England's fish and wildlife resources. This U.S. Fish & Wildlife Service program provides financial assistance for the protection, restoration, rehabilitation, and improvement of land and water areas adaptable as feeding, resting, or breeding places for wildlife. Since the inception of the program, funds exceeding $255 million have been provided to New England.

THE NEW DEAL ERA

Changing Perceptions: The Era of Emergency Conservation and Setting the Groundwork for Future Collaboration

The Roosevelt Administration's response to the Great Depression fundamentally reshaped federally led conservation efforts in New England and the rest of the United States. Emergency Conservation programs like the Civilian Conservation Corps (CCC) greatly broadened the traditional role of federal land-management agencies, in particular, the National Park Service (NPS). Nationwide, the CCC, primarily managed and led by the NPS professional staff, established more than seven hundred state parks undertaking reforestation projects along with the construction of roads, campgrounds, picnic areas, lodges, and recreational facilities. In 1935, for example, the NPS was running 117 CCC camps in national park areas and 457 camps in state parks. In New England, federal responsibilities were expanded far beyond the boundaries of a handful of existing national parks and forests to organize and lead public works conservation projects in every corner of the region. As Ronald Foresta noted in his book *National Parks and Their Keepers*, "The reality beneath the image is that neither the national parks nor their keepers stand apart from our times; they are very much subject to the problems and dilemmas of modern American life."[8] This would not be the last time that the administration and the Congress would employ the capabilities and resources of federal conservation skills and resources in pursuit of a larger social agenda.

However, not all these national efforts were successfully replicated in the New England states. For example, the Blue Ridge Parkway in Virginia and the Natchez Trace Parkway in Tennessee and Mississippi were undertaken by the Roosevelt administration as national models for scenic recreational motoring and at the same time public relief projects particularly targeting some of the poorest rural communities in the United States. In 1933, the National Park Service was directed to survey and study a proposed 260-mile parkway modeled on the Blue Ridge project, to be built along the ridgetops of Vermont's Green Mountains. This proposal was developed in concert with other federal efforts led by the Resettlement Administration to relocate Vermonters from "submarginal" hill farms to new communities and add thousands of reclaimed acres to new public forests and parks. Local opposition ultimately killed both efforts.[9] In the case of the proposed Green Mountain Parkway, once again a national public land management and development model from "outside" did not fit with local approaches to recreation, tourism development, and conservation. This was particularly true for the Green Mountain Club, which vigorously fought the parkway to preserve the integrity and isolation of the recently completed Long Trail built along the spine of the Green Mountains.

Federal relief and conservation efforts that better meshed with community needs and aspirations and the desire for local control, however, were received very differently. State park systems throughout New England got an enormous shot in the arm from the CCC and emergency

At the time of the early Weeks Law deliberations (1914-19), New England faced extensive areas impacted by debris from logging operations, such as this scene on Mt. Carrigain in the White Mountains. Individual leaders such as (from left to right) *Chief Forester Henry Graves of the U.S. Forest Service, White Mountain Forest Supervisor J. J. Fritz, District Forester Franklin Reed, NH Forest Protection Society Forester Philip Ayres,* Boston Transcript *writer and Appalachian Mountain Club director Allen Chamberlain, and District Ranger C.B. Shiffer joined forces to plan ways to remedy the damage.*
(White Mountain National Forest)

conservation activities. In some instances, as in the case of Vermont's Elmore State Park, land was purchased by the federal government and developed into a park by a resident CCC camp and then turned over to state administration. In other instances, existing state properties were greatly expanded and improved, as in the case of Massachusetts's Mount Greylock Reservation in the Berkshire Mountains where the CCC built the rustic Bascom Lodge.

The National Park Service administered CCC Camps in all six New England states; one in Connecticut, eight in Maine, thirty in Massachusetts, five in New Hampshire, three in Rhode Island, and nine in Vermont. In Maine, for example, two of these camps were in Acadia National Park; three were in Recreation Demonstration Areas and the rest were in state parks. The creation of Recreation Demonstration Areas in many cases involved reclaiming "sub-marginal" agricultural lands originally acquired by the Resettlement Administration. The demonstration areas provided both relief for impoverished rural landowners and jobs for the unemployed. The demonstration areas were also envisioned to offer recreational benefits to an unusual range of users "embracing children's camps, family camps, and industrial and social organization camps, offering opportunities for low-income groups of populous urban and rural sections, public and semipublic organizations, and others to enjoy low-cost vacations of outdoor life for short periods."[10] Upon completion by the National Park Service and the CCC, these areas, such as Camden Hills in Maine, Bear Brook in New Hampshire, and Beach Pond in Rhode Island, were turned over to their respective state park agencies to administer. In Massachusetts, a total of seventy CCC camps were established in what would become future state parks and forests. Even in independent-minded Vermont, no less than twenty-four CCC camps were established in state parks and forests employing more than 11,000 young men between 1933 and 1942.[11] Programs of the Works Progress Administration and the Farm Security Administration that buttressed rural life and heritage were also popular throughout New England.

Much of the iconic, rustic park and forest development done by the CCC, that opened these public lands for broad public use and enjoyment, has stood the test of time and, in many cases, placed on the National Register of Historic Places. Taken as whole, the CCC work continues to invoke a lasting nostalgia for a time when there was national consensus—and federal support—accomplishing significant conservation endeavors and investments, and when public park and forest facilities were built with craftsmanship and pride. The emergency conservation work of the thirties would also establish a precedent for future state and federal cooperative conservation efforts such as the Land and Water Conservation Fund and the Forest Legacy Program. During this period both the U.S. Forest Service and National Park Service were also directed to undertake national recreational studies that would eventually lead to the establishment of a new generation of near-urban units including national lakeshores and seashores like Cape Cod. The formula of federal recreation and heritage linked with technical assistance and funding combined with a requirement for strong state and local community participation and influence — so important to New Englanders — would also presage future development of a National Heritage Area movement which, not coincidently, would begin to take shape in the New England states at the close of the twentieth century.

POST-WORLD WAR II — A NEW ERA OF CONSERVATION AND RECREATION

A decade after World War II, the nation became increasingly aware of environmental and public health issues and made increasing demands for outdoor recreation opportunities. A newly conceived Interstate Highway System was starting to lay its web across the nation. Post World War II income levels were enabling families to see America, and the number one recreational activity was "driving for pleasure." In response, President Eisenhower and the Congress, with bipartisan support, established the Outdoor Recreation Resources Review Commission (ORRRC) in 1958 to determine the nation's needs and to make recommendations for the future. Once again, this important national initiative would have deep New England roots.

The commission was chaired by Laurance S. Rockefeller whose family had been central to the establishment and growth of several of the early national parks, including Acadia National Park in Maine. Laurance and his wife, Mary, would later donate the Marsh-Billings-Rockefeller property in Vermont to the National Park Service as a National Historical Park. The Commission's executive director was Francis W. Sargent, a Massachusetts conservation official who would be elected the commonwealth's governor in 1970. State contact officers in New England included Park and Forest Director Donald Matthews in Connecticut, Director of State Parks Lawrence Stuart in Maine, Natural Resources Commissioner Charles H.W. Foster in Massachusetts, Attorney General Maurice J. Murphy, Jr. in New Hampshire, Development Council Administrative Chief Henry C. Gagnon in Rhode Island, and Forest and Parks Director Perry Merrill in Vermont.

The major recommendations of the ORRRC Report, published in 1962, were adopted as national policy, and many of the programs recommended were quickly established—among them a national outdoor recreation plan, a land and water conservation fund, a national trails system, a national wild and scenic river system, and a national wilderness system. In addition, the commission called for the states, with federal assistance, to plan, acquire land, and provide for outdoor recreation through investments of their own.

NATIONAL WILDERNESS SYSTEM

As Dr. Hagenstein has described, the Wilderness Act of 1964 provided a way to resolve decades of conflict about one aspect of whether, where, and how the nation's public lands and forests would be managed.[12] The Act provided instant wilderness status for some nine million acres of federal forest land nationally and set up a system for future wilderness designations by Congress. This legislation and the National Forest Management Act of 1976 would set the stage for an expanded role by citizens that continues to this very day. Both laws responded to issues that were not particularly New England in nature, but by providing for a new level of concern and citizen involvement in federal land management, both would materially affect the future of the White and Green Mountain National Forests in the region.

The 1964 act created problems for New England by describing wilderness as areas where "the earth and its community of life are untrammeled by man."[13] Strictly defined, there were few such areas in New England, and certainly none of any significant size. Some cultural artifacts, such as evidence of logging, railroads, and camps, would have to be allowed. Thus, the only area in New England that qualified for immediate protection as a wilderness area was the Great Gulf Wilderness (5,400 acres) in New Hampshire, a portion of the White Mountain National Forest that had been classified administratively as a "wild area" for some time.

Support for what later became the Eastern Wilderness Act of 1975 built slowly over several years. But, in the end, the Act became law. It gave immediate wilderness designation to the Presidential Range-Dry River Wilderness (20, 380 acres) in the White Mountain National Forest and the Lye Brook Wilderness (14,300 acres) in the Green Mountain National Forest. It also led to an often acrimonious process of building support for new wilderness areas that continues today. As of

2007, within the Forest Service and Fish and Wildlife Services' thirty-six established federal management areas in New England, there are seventeen units of the national wilderness system covering more than 260,000 acres of land.

Land and Water Conservation Fund

The Land and Water Conservation Fund (LAWCON) was established in 1965. The funds for this program were derived from a share of federal offshore leasing revenues and their availability has fueled many initiatives by federal, state, and local agencies over four decades. As of June 2007, $110 million in LAWCON funds had helped stimulate 868 separate land conservation projects in New England covering 241,300 acres—most during the early years of the program. The first major and largest acquisition came in 1967 with the protection of 26,000 acres along Maine's newly established Allagash Wilderness Waterway at a federal cost of $1.2 million. Since then, Connecticut and New Hampshire have recorded the largest number of state LAWCON projects over the years (198 and 184, respectively)—Vermont has recorded the largest amount of acreage protected (70,900). Despite its size, Rhode Island has participated actively (seventy-seven projects covering more than 14,000 acres using a total of $7.9 million in federal funds).

In recent years, however, Congress has been reluctant to fully fund the Land and Water Conservation Program, mostly limiting LAWCON appropriations to federal agency projects. The lack of LAWCON funding for state and local projects and the increased concern for forest land protection has stimulated new thinking about possible partnership conservation models applicable to the states. Needless to say, New England, with its reputation for innovation, has become a leader in this effort.

National Wild and Scenic Rivers System

The National Wild and Scenic Rivers System was established through the enactment of Public Law 90-542 in October 1968. Like the earliest national forests and national parks, the system was conceived on the basis that units of the system would be largely on federal lands. Yet, it was a visionary piece of legislation, laying the framework for a national system of protected rivers at the federal level, as well as prompting states and local river protection efforts with federal assistance and incentives.[14]

For New England the legislation only included the Allagash River of northern Maine. A national and local debate over the protection of Maine's Allagash River prompted the creation of an additional category, a state administered river within the national system. For the Allagash and the citizens of Maine, a compromise — the river would enjoy national recognition and yet be protected and managed by the state. The Allagash was included as the inaugural component of a class of "state-administered" Wild and Scenic Rivers under Section 2(a)(ii) of the Act (pending anticipated application by Maine's Governor). And, the governor did apply, after the legislature established the Allagash Wilderness Waterway and the state purchased an average five hundred-foot wide buffer along the ninety-two-mile waterway. As noted before, this 26,000-acre buffer was purchased with federal matching funds from the newly established Land and Water Conservation Fund.

Absent the unique Allagash resolution, none of the original components of the system were found in New England — not surprising given the relative lack of federal lands, the density of the population, and the region's prevalence of communities based around their rivers. And yet, the Act clearly anticipated that rivers without adjacent public lands and with some limited degree of adjacent development should be considered and included in the system. For these river segments specific provisions limit land acquisition authority on rivers where communities had enacted "compatible" zoning (Section 6c); and encourage local and state participation in administration and management (Sections 10 and 11).

The state administered river concept has been copied across the country and the concept of nonfederal management also provided a bridge to New England's creation of another new model, the Partnership Rivers of the Wild and Scenic River System. The partnership model grew out of a traditional National Wild and Scenic Study on Connecticut's Housatonic River and was first implemented on the Wildcat Brook in Jackson, New Hampshire.

The following excerpt from the 1978 Housatonic (CT) Wild and Scenic River Study (Summary of Findings and Recommendations) exemplifies the early consideration of National Wild and Scenic River designations in New England.

> The forty-one mile segment of the Housatonic River . . . is eligible for inclusion in the National Wild and Scenic River System upon completion of an acceptable management plan through local action. . . . If National Wild and Scenic River designation is desired, the completed management plan should be presented to the local towns for approval, and then to the state legislature for recognition as a state scenic river and for legislation officially recognizing the managing agency. Finally, the governor should submit the plan to the Secretary of the Interior with a request for National Wild and Scenic designation as a state-designated unit, as provided for under Section 2(a)(ii) of the Wild and Scenic Rivers Act.

This quotation from the Housatonic study crystallized the issues inherent to making a federal Wild and Scenic designation work in a private lands dominated, New England setting. As with the Allagash, local control and management were critical, but in the Housatonic Valley large-scale public ownership was not desired or affordable. The study's proposed Management Plan and the Management Structure became the key. The Management Plan—How do you protect identified "outstandingly remarkable" values of a river when they are not on public lands? Everyone needed to know the answer to this question before a designation would be acceptable to the landowners, the communities, the state agencies, and the National Park Service. Without the plan, there was no basis of common understanding. The management structure—How will local, state, and federal jurisdictions coordinate? What is the role of landowners? Who is in charge? How will coordination occur? Who has the funding responsibility? How do you limit the federal role and the unwelcome federal condemnation of lands? These questions as well needed to be answered prior to designation.

This is the issue embedded behind the Housatonic's recommendation to pursue designation under Section 2(a)(ii) of the Act, which leaves the state in control, and prohibits federal administration or acquisition. The citizens of the Housatonic did not agree, however, two pioneering efforts soon picked up the challenge, and in different ways, have laid the groundwork for solving the challenges described above.

The common element—complete the plan and work out the details as a part of the study process prior to designation.

Wildcat Brook (approximately nine miles in the White Mountain National Forest and six miles in the Town of Jackson, New Hampshire) has been designated as a component of the National Wild and Scenic River System administered by the secretary of agriculture through a cooperative agreement between the U.S. Forest Service, the state of New Hampshire, and the town of Jackson.[15] Spurred by the threat of an unwanted hydroelectric development, the town of Jackson, New Hampshire successfully partnered with members of Congress, the White Mountain National Forest, and the National Park Service to authorize in 1984 a new kind of Wild and Scenic River Study—one that would answer the questions that thwarted the Housatonic by developing and implementing, as the centerpiece of the study process, a successful river conservation plan.

The plan, developed by the town with support of the National Park Service and a specially formed local advisory committee, identified and implemented local zoning, conservation easements, and riverfront restoration elements necessary to protect the river's special values. In 1988, the Wildcat Brook Conservation Plan became the basis of federal legislation designating the Wildcat Brook as a component of the national system. This designation enjoyed the support of landowners, local and state officials, and the federal government. The level of local participation and support for the project was unprecedented; Jackson Town Meeting voted unanimously in favor of the town pursuing congressional National Wild and Scenic River designation.

In the Wildcat Study, limited resources for technical assistance were carefully targeted to leverage the efforts of others and sustain momentum. The National Park Service was particularly effective in building credibility and public support by being perceived as a fair, objective, and reliable partner in the conservation process, not out in front, but always close at hand. On several occasions this enabled the National Park Service to mediate and resolve conflict between different interests and help define a cooperative conservation agenda. Success required the mobilization of a broad conservation coalition and responsiveness to a variety of issues and interests. The local community's stake in the project had to be clearly defined. In Jackson, the commitment and motivation of the study advisory committee laid the cornerstone for every accomplishment. Like a political campaign, success also required consistent leadership, organization, good communication, and a sense of where opportunities lay and when to seize them.

The Westfield River in Massachusetts utilized a similar approach, but one that took advantages of the built-in mechanisms of Section 2(a)(ii) of the Act to limit and define the federal role. The critical element still was to complete the plan in partnership with local communities and landowners prior to designation. For the Westfield, this was accomplished through the assistance of the National Park Service acting through the Rivers and Trails Conservation Assistance Program (rather than under a congressionally authorized study), and through state planning grants. Chris Curtis of the Pioneer Valley Planning Commission initiated this process in 1984, also choosing to form a locally based advisory committee to assist in developing the conservation plan. In 1992, the completed Greenway Plan was submitted by Governor William Weld to the Secretary of the Interior as the basis of designation in 1993 as a "state-administered" component of the Wild and Scenic Rivers System. This submission had the support of local communities, landowners, and state and federal officials.

The pioneering works of the Wildcat Brook and Westfield River have continued to bear fruit and evolve. Today, the term "Partnership Wild and Scenic Rivers" is used by the National Park Service and Congress to describe a distinct subset of the Wild and Scenic Rivers System based on the principles of the Wildcat and Westfield. In 2007, three New England areas (New Hampshire's Lamprey River, Massachusetts' Sudbury-Assabet-Concord Rivers, and Connecticut's West Branch Farmington River) are currently authorized as "Partnership Wild and Scenic Rivers." Three additional rivers have been designated by Congress for study (Connecticut's Lower Farmington and Eight Mile Rivers and Massachusetts's Taunton). Two of the three (the Eight Mile and the Taunton) have bills authorizing designation pending before Congress. All of these rivers share the following common characteristics:

- Federal Land Ownership Not Authorized;
- Land Use Governed by State and Local Statutes;
- Management Coordinated Through Local-State-Federal Advisory Councils;
- No Federal Superintendent or Federal Law Enforcement;

- Management Costs and Responsibilities Shared among Partners; and
- Management Directed According to Partnership-Based Management Plans Developed and Endorsed Locally Prior to Designation.

Under this umbrella, the process to achieve National Wild and Scenic River designation has proven an excellent tool to bring together local, state, federal, and nongovernmental partners together with a

This pristine scene of a portion of Connecticut's Eight Mile River, a tributary of the lower Connecticut River, illustrates a variant of the National Wild and Scenic River System pioneered by New Englanders. Called "Partnership Wild and Scenic Rivers", these designations share the following characteristics: no federal ownership, land use governed by state and local statutes, partnership plans developed and endorsed locally prior to federal designation, management coordinated through local-state-federal advisory councils, and management costs and responsibilities shared among the partners.

single vision for the future of outstanding free-flowing rivers. The Partnership Wild and Scenic Rivers have established a model for successful adaptation of the Wild and Scenic Rivers Act to a community based, private lands setting. In 2007, the Ash Institute for Democratic Governance and Innovation at Harvard University's John F. Kennedy School of Government named the Partnership Rivers to its list of the top fifty government innovations linking citizens with important public services.

National Trail System

The National Trail System Act, authorizing a system of national historic, scenic, and recreation trails, also became law in 1968. With it came National Scenic Trail status for the famed Appalachian Trail, incubated in the 1920s by Massachusetts's Benton MacKaye, assisted in its construction by the Civilian Conservation Corps, and ultimately maintained by a network of local, volunteer trails enthusiasts. The Appalachian Trail has served as a model for the new additions to the National Trail System.

Predicated upon McKaye's familiarity with New England's Appalachian and Green Mountain Clubs, and their artful and strategic ability to sustain volunteers to establish and maintain trails on both private and public lands, the Appalachian Trail has also set the standard for cooperative resource management for more than eighty years. A second long-existing New England footpath, the Metacomet-Monadnock-Mattabesett Trail, running from the Connecticut coast through Massachusetts to the New Hampshire border, is currently under consideration by Congress for designation as the New England National Scenic Trail.

Rivers and trails designations have been greatly facilitated by the NPS's national Rivers, Trails and Conservation Assistance (RTCA) program authorized to carry out the mandated federal responsibility to provide assistance to states, other political subdivisions, private organizations, and volunteers. Not surprisingly, this program has roots deep in New England. Starting in the Northeast, the RTCA broke the traditional governmental role of land manager, regulator, grants provider, and information clearinghouse and instead began operating as facilitators and catalysts for on-the-ground projects.[16] RTCA staff helped local groups identify goals, focus on achievable objectives, find ways to broaden local support, and lead community and nonprofit groups to sources of funding. The key to their success has been prominent participation in highly visible projects in every state.

NEW MANAGEMENT MODELS FOR THE NATIONAL PARK SYSTEM

Between 1916 and 1959, only two additional units of the National Park System were established in New England: the Salem Maritime National Historic Site (1938) and the Adams Mansion National Historical Site (1946), both in Massachusetts. By the 1950s, in part, in response to the same national consciousness that spurred the creation of ORRRC, the post-World War II increased suburban development, mobility, and pressures for recreation opportunities, a locally driven momentum resulted in the establishment of two new national parks in Massachusetts: the Minute Man National Historical Park (1959) and the Cape Cod National Seashore (1961).[17] Both of these parks with their relatively large acreage and recreation purposes, were the first, outside of the National Capital area, to be established in near-urban settings and on lands that had been previously settled and largely in private ownership. According to historian Ronald Foresta, ". . . views within the Park Service on what the agency should be preserving changed and expanded in the 1960s and 1970s. When agency leadership started thinking of cultural as well as historical and natural preservation, and when whole living landscapes came under its consideration, there evolved a new preservation synthesis which allowed the agency's mission to be viewed in entirely new terms."[18] For designation of these types of sites, a new approach was needed that integrated national park designation with long-established local communities. It was in this environment that the ideas of using purchase of land with federal funds while retaining within the park boundary large permanent private land inholdings under public regulation were introduced.

Cape Cod National Seashore

In 1959, Massachusetts Senators John F. Kennedy and Leverett Saltonstall and Congressman Hastings Keith developed a creative legislative proposal that fit the somewhat unique circumstances of Cape Cod. Under this legislation, Cape Cod National Seashore, like Minute Man National Historical Park, would be authorized to acquire private lands with federal funds, an authority the first of its kind after World War II.[19] This authority was necessary because both of these areas, unlike earlier national parks, were not created either from the public domain or from private gifts, but largely from land purchased from individual owners. On the Cape the park boundaries overlaid portions of six communities. Developed residential lots existing prior to the park's establishment were exempt from acquisition as long as the owners complied with local zoning ordinances modified to comply with the zoning standards established by the secretary of the interior for the seashore. It was also specified in the legislation that should the zoning standards be violated, the privately held parcel was subject to federal acquisition and by condemnation if necessary.[20] Finally, the act created a Citizens Advisory Committee to guide the park's management and development including review of plans for recreational development and the issuance of any federal permits.

This so-called Cape Cod formula created a new framework for establishing national parks that was used in other settled portions of the United States, particularly on the East Coast. Despite ongoing management challenges, it is generally acknowledged that this framework has been very effective in retaining public access to the coast and also has greatly limited, with the seashore owning more than 27,000 acres, the impact of intense development pressures across the broad landscape of the outer Cape—both of these objectives were part of the original intent of citizen supporters and environmental visionaries of the 1930s.[21]

Connecticut River and the Silvio O. Conte National Fish and Wildlife Refuge

In 1966, buoyed by the successful designation of Cape Cod National Seashore, Connecticut Senator Abraham Ribicoff led the New England congressional delegation in authorizing a study of the four-state main stem of the Connecticut River. The study report released in 1968 proposed three federally acquired recreation areas in the headwater Connecticut Lakes region, the Mount Holyoke/Mount Tom segment in Massachusetts, and the Connecticut River Gateway region at the mouth of the river near Essex, Old Lyme, and Old Saybrook in Connecticut. Citizens of the valley were concerned about federal land acquisition and the fear of condemnation; and their concerns were fueled by stories of heavy-handed federal land acquisition at Minute Man National Historical Park and Cape Cod National Seashore. In the Connecticut Valley citizens were being asked to support a very broad and general concept; their many questions had no specific answers. Resting the future in the hands of a federal agency and somewhat undefined planning process was not acceptable. Consequently, this initiative faltered and by 1972 the Bureau of Outdoor Recreation, the National Park Service, and the elected officials withdrew the proposal from consideration.

This debate did, however, focus local attention on these outstandingly remarkable resources and while the federal proposal was defeated many citizens recognized the need for stewardship and began to undertake that responsibility.[22] Each of the three areas responded in their own but different ways. In time, and in some cases with federal participation, state, local, and private efforts have taken major steps to protect these areas. In 1973, the Connecticut General Assembly established the Gateway Commission, a state and eight-town local compact for the protection of the lower river. Comprehensive and evolutionary in its approach the local towns together with the help of state and federal agencies and nonprofits have used zoning, acquisition, and education to protect this area.[23] Massachusetts undertook its own initiative in line with the federal report recommendation. Starting in 1977 through 2002, the commonwealth has acquired more than 17,700 acres in the vicinity of Mount Holyoke.[24] One of the commonwealth's newest state parks is the Connecticut River Greenway encompassing twelve miles of permanently protected shoreline.[25] In the Connecticut Lakes region of northern Vermont and New Hampshire, just recently, a joint public and private effort has resulted in the permanent protection of 171,000 acres.[26]

The continuing public interest in the future of the entire watershed eventually resulted in the establishment of the Silvio O. Conte National Fish and Wildlife Refuge in 1991, a refuge that embraces the entire 7.2 million acre Connecticut River basin.[27] Representing several innovations to the protected area model, it is the only refuge in the national system established to encompass a large-scale watershed and a multitude of heavily populated areas in four states. The citizens of the region continue to insist upon an open planning process for the new refuge, maintenance of local controls, and a program to maximize federal-state-local cooperation and minimize the direct acquisition of land by the federal government. In response, under a concept of "No Ordinary Refuge," the conservation for this large landscape recognizes the importance of working with the people of the watershed, especially landowners and land managers, in environmental education and cooperative management projects and minimizing conventional land acquisition by the federal government. The result has been an accepted vision statement making clear the refuge's intent to forge a collaborative model that may well influence other national refuges of the future.

Lowell National Historical Park

Since the authorization of Cape Cod National Seashore in 1961, thirteen new units of the National Park System have been established in New England to recognize people and places that have played a role in the history of the United States.[28] Each of these national parks is engaged in partnerships with local communities, other organizations, or other government agencies, and often these partnerships are specified in the park's enabling legislation. Lowell National Historical Park and its principle partner, the legislatively authorized Lowell Preservation Commission, established in 1978, proved to be one of the early examples of this partnership model. Located at specific sites in downtown Lowell, Massachusetts, the park's primary purpose is to preserve and interpret the city's social, industrial, and economic history from the earliest days of America's industrial revolution. The commission had a broader community-wide mandate and through its planning and grant programs fueled Main Street revitalization, historic neighborhood improvements, and culturally based programs. Given both the park and the commission's role, it has been said that it is hard know where the park ends and the city begins or vice versa. The sponsors of the legislation, Massachusetts Congressman Paul Tsongas and Senators Edward M. Kennedy and Edward Brooke, saw the establishment of this national park as a way to celebrate Lowell's history, renew the citizens' pride in their city, introduce tourism to the city's economy, and attract a broader spectrum of economic development. One can say for certain that the vision of thirty years ago has become a reality today. The DNA from the Lowell model can be found in several park areas in New England including Boston Harbor Islands National Recreation Area and New Bedford Whaling National Historical Park. Most notable is Lowell's contribution to the National Heritage Area concept (see discussion below). The Lowell Historic Preservation Commission proved that a community based, federal agency with a local mandate could plan and execute heritage conservation and economic development programs.

Weir Farm National Historic Site

The designation of many of the more recent national parks is the result of the leadership of local individuals and organizations seeking to protect nationally significant sites in their communities and, in many cases, continue their stewardship role as a partner to the National Park Service after designation. For example, at Weir Farm National Historic Site, the partnerships with private organizations, individuals, and government at the local, state, and federal levels were instrumental in creating and in managing this park.[29] A thirty-year grassroots effort of local residents, the Trust for Public Land, The Nature Conservancy, the State of Connecticut as well as curators and art historians throughout the country created the 126-acre Weir Farm National Historic Site in southwestern Connecticut in 1990. This site was the summer home for nearly forty years of Julian Alden Weir, a leading figure in the American art world and in the development of American Impressionism. This is the first national park dedicated to an American painter, and the site preserves the farm largely intact as well as the home, studio, artwork, and grounds integral to his art.

The Trust for Public Land (TPL), in particular, played a pivotal role in acquiring the property, funding the feasibility study, and establishing and managing the park. TPL served as an interim manager prior to the park's authorization and also laid the groundwork for the Weir Farm Heritage Trust, an organization which included Weir, Young, and Andrews family members, elected officials, environmental leaders, artists and art historians, and concerned local citizens. Today the Weir Farm Art Center (formerly the Trust) is the park's cooperating association and most recently the owners of the Weir Preserve, 110 acres which abut the park, formerly owned by The Nature Conservancy.

Marsh-Billings-Rockefeller National Historical Park

Marsh-Billings-Rockefeller National Historical Park (MBRNHP), opened in 1998, tells the story of conservation history, evolution of land stewardship, and emergence of a conservation ethic. This park is named for three families that owned this property and made significant contributions to American conservation. George Perkins Marsh, the author of *Man and Nature* a landmark book published in 1864, was a major influence on the formation of the modern conservation movement. Frederick Billings and his family put Marsh's philosophy into practice with a farsighted program of progressive farming and forestry, using pioneering reforestation techniques borrowed from nineteenth-century Europe. Billings's granddaughter Mary and her husband, Laurance S. Rockefeller, inherited the property in the mid-twentieth century, sustaining and enriching it through their own conservation projects. They donated the property to the NPS in 1992 with a legislative mandate to maintain and build upon the 140-year legacy of conservation stewardship and sustainable forest management.

A major innovation at Marsh-Billings-Rockefeller involved its mandate to maintain a working landscape—in this case a managed forest emblematic of much of northern New England—and to demonstrate responsible forest stewardship, community engagement, and sustainability. The NPS explored the application of contemporary conservation ideas, such as the production of traditional value-added forest products and branding through third-party certification. In 2005, MBRNHP became the first national park or national forest in the U.S. to be third-party certified by the Forest Stewardship Council. This independent certification offered a way for the park to receive recognition for careful, responsible forest practices and promote a broader public dialogue on the future of New England woodlands and sustainability of rural communities. The park continues to explore options for labeling and co-labeling of its wood products to illustrate the connection between traditional and sustainable products and landscape stewardship. So, almost sixty years after the Green Mountain Parkway concept was defeated, Vermont has a small national park of a very different stripe, one encouraging the stewardship of working landscapes and viable, sustainable communities.

Another innovation at Marsh-Billings-Rockefeller is the Conservation Study Institute established by the National Park Service to advance collaborative conservation throughout the National Park System. This park-based institute works in partnership with academic and nonprofit organizations to serve a national audience by conducting demonstration projects, distilling and sharing lessons learned, and facilitating information exchange. Institute programs on collaborative leadership, civic engagement, partnership networks, and program evaluation maintain a dialogue on the best thinking and practice in conservation.

The mid 1980s and '90s gave life to new partnership models, networking, and public-private collaboration at a larger, landscape scale. Civic engagement, locally based political support, and integration with the local economy became central to this model.

National Heritage Areas

In 1986, Congress established the John H. Chafee Blackstone River Valley National Heritage Corridor in Rhode Island and Massachusetts as the nation's second national heritage area to preserve and interpret the nationally significant contributions of the valley to the history of industry in America.[30] The Blackstone River Valley is one of the nation's richest, best-preserved repositories of landscapes, structures, and sites attesting to the social revolution that transformed an agricultural society into an industrial giant. These two forces — agriculture and industry — shaped the patterns of settlement, land use, and growth in the valley. Congress also created the Blackstone River Valley Corridor Commission (a federal commission), as the management entity to assist the two states, the municipalities, and other partners and to develop a management plan to unify historic preservation and interpretation goals and strategies throughout the region. The management plan for the corridor had to complement state plans and required approval by both governors and the secretary of the interior. The plan's implementation strategy emphasizes "integrated, linked actions rather than single, stand-alone projects," observing further that balanced ac-

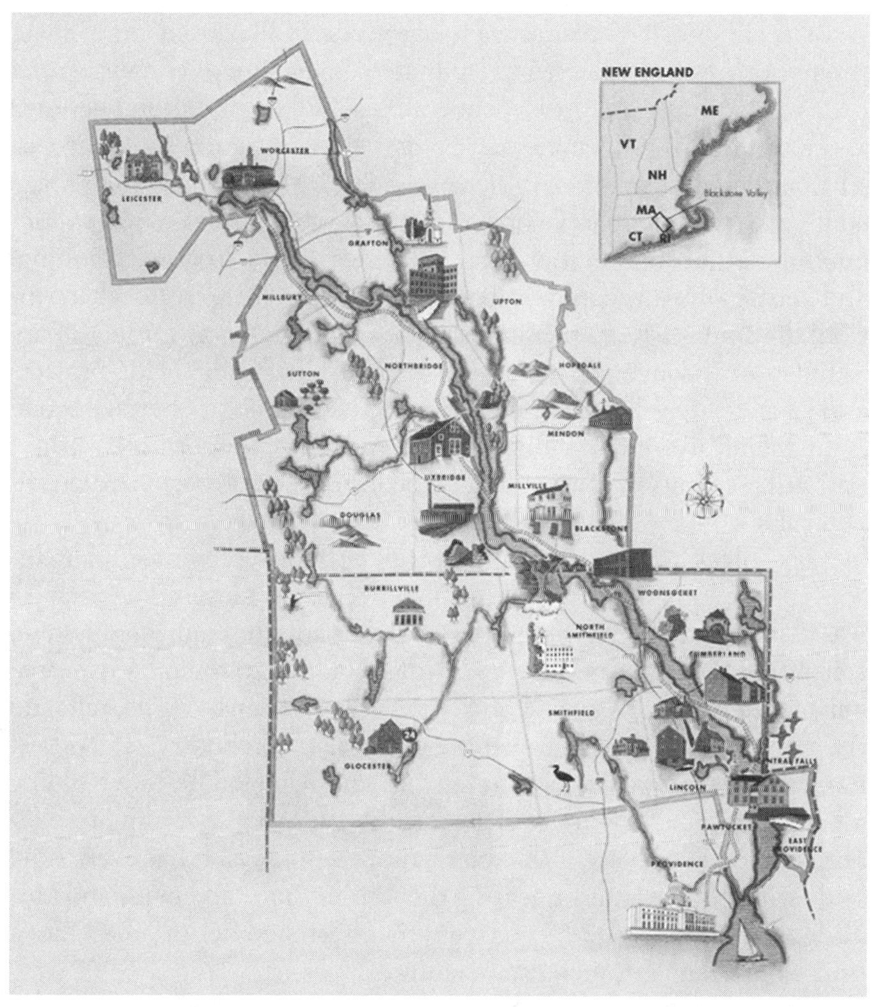

Only the second in a new concept of national heritage corridors, the federally authorized Blackstone River Valley Heritage Corridor, established in 1986, is dedicated to achieving harmony among preservation, recreation, and development interests within a forty-six-mile stretch extending from Providence, RI to Worcester, MA. As of October 2007, this unique concept of federal, state, and local partnerships formed to secure and celebrate the regional identity of natural systems, history, and culture had spread to four other New England states.
(National Park Service)

tion "is critical to achieving harmony among preservation, recreation, and development."[31]

The commission's planning process invited valley residents to share their priorities and values, and facilitated development of a vision of a cohesive region — an interdependent place linked by cultural heritage and a common set of economic, natural, and cultural resources goals. The commission set an ambitious agenda for the corridor — heritage education, recreation development, ethnic and cultural conservation, environmental conservation, historic preservation, land-use planning, and heritage-based economic development. Since the corridor's inception, the commission, staff, and partners have created an extraordinary foundation through strategic investment in key valley-wide projects and partnerships. The federal role for the National Park Service is one of financial and technical assistance and, most importantly, brings national recognition of a region's national story and associated resources and values.

The Blackstone National Heritage Corridor represented an ambitious, fresh approach to thinking about regional places. The premise was that if residents understood the valley as an interconnected system, this would engender new attitudes that would help revitalize the area as a place to visit, live, work, and invest. The commission, its staff, and the network of partners are working together to conserve important natural and cultural heritage in lived-in landscapes. They are building a partnership culture in the Blackstone Valley that is leading to conservation of an important story and unique resources. It has everything to do with people and connecting them to heritage and place and kindling a sense of stewardship. In this process, shared heritage becomes a bridge between past, present, and future.

Today, twenty-one years after designation, change is visible throughout the valley. Renovated mills and historic buildings provide housing, business space, museums, and arts facilities. Along the Blackstone River, cleanup efforts have removed shoreline trash and debris, and the river itself is beginning to recover. State and federal governments, local jurisdictions, historical societies, environmental organizations, businesses, sports groups, and private landowners collaborate to promote and care for the things that make the area special. The Corridor Commission is providing a model for how the federal government can work in partnership with others toward common goals of historic preservation, a cleaner environment, and revitalized communities.

The success of this initiative in the Blackstone National Heritage Corridor helped to launch a national movement in heritage areas. National heritage areas have become increasingly popular in the U.S. as a way of protecting important landscapes that have a regional identity formed by natural systems and shaped by history and culture. The heritage area concept is based in partnerships, engaging every level of government and the people who live there. The interest in heritage areas has been fueled by recent trends in conservation, all of which are employed in heritage area work, including the:

- emerging strength of community-based conservation, driven by local initiative;
- growth of partnerships and collaborative management;
- linking of conservation and social and economic objectives;
- rise in place-based education;
- integration of cultural and natural conservation;
- evolving concept of national parks and their purpose; and,
- effectiveness of landscape-scale conservation.[32]

As of October 2007, there are thirty-seven national heritage areas nationwide, five of which are located in New England.[33] In addition to the Blackstone, they include the Quinebaug-Shetucket Rivers Valley National Heritage Area based in northern Connecticut, the Essex National Heritage Area in northeastern Massachusetts, the Upper Housatonic National Heritage Area in northwestern Connecticut and southwestern Massachusetts and, most recently, the Lake Champlain National Heritage Partnership in the states of Vermont and New York.

The Great Bay Resource Protection Partnership is a concept similar to national heritage areas. Focused on an 8.9 square mile estuary fed by seven major freshwater rivers lies in one of the fastest-growing and most ecologically important areas of New Hampshire.[34] The area has been used by Native Americans and European settlers for at least 10,000 years. Its rivers, salt marshes, and uplands have helped serve as the sites and sources of natural resources for thriving communities between Newmarket and Exeter on the west and Portsmouth on the east.

A proposed refinery in 1973, and the threatened closure of the Strategic Air Command's Pease Air Force Base, galvanized community action to preserve the area. Since 1994, the resulting collaborative effort, involving a consortium of nine organizations and agencies, has secured more than $54.4 million in public and private grants and directly protected more than 4,800 acres around the bay, including places of the highest ecological value. This partnership effort has also contributed to the conservation of an additional 3,000 acres by individual partners using private funds and public sources. As The Nature Conservancy spokesman Eric Aldrich has observed, "Any partnership is a collaboration not simply of agencies and organizations; it's about people, too."[35] In the case of Great Bay, civic engagement has been most instrumental in making this project one of New Hampshire's most significant conservation success stories.

Federal Land Management Areas in New England in 2007

From a start one hundred years ago, the efforts described above have resulted in the prominent presence of three federal land management agencies in New England. Today, New England is host to two units of the National Forest System, nineteen units of the National Park System, and thirty-four units of the National Wildlife Refuge System. The White and Green Mountain National Forests encompass approximately 800,000 acres each. The largest National Wildlife refuges are Monomoy in Massachusetts (30,000 acres), Moosehorn in Maine (28,700 acres), and Umbagog in New Hampshire (20,500 acres). The largest units of the National Park System are Acadia in Maine (47,400 acres) and Cape Cod in Massachusetts (43,600 acres).

Forest Legacy Program

Entering the twenty-first century, another major federal contribution to New England, in addition to the federally managed areas, has been the Forest Legacy Program administered by the U.S. Forest Service. The program is state-based and involves both private for-profit and nonprofit. As of June 2007, more than 933,000 acres of forest land had been protected in all six New England states at a federal cost of $101 million. Forest Legacy has been one answer to the declining Land and

Water Conservation Fund. The roots of this program are strongly New England, stemming from the direct stimulus given Forest Legacy by the work in the 1980s of New England and New York's Northern Forest Lands Council. The program has also benefited from the region's extensive interest and support for favorable treatment of land conservation in the federal tax code as well as active participation by the states and land trusts. Especially significant for the future has been the unusual flexibility the Forest Legacy program provides for land conservation, and the encouragement of ways private landowners can participate effectively in a conservation program without losing all their rights in their land.

CONCLUDING OBSERVATIONS

As we have seen, the conservation strategies of federal agencies have evolved over the last hundred years. Federal land conservation started in the late nineteenth century in the west where large areas were first extracted from the public domain to establish national forests, national parks, and national monuments. National park areas like Yosemite, Yellowstone, and Grand Canyon National parks were, after 1916, managed by the National Park Service. These lands were located in remote, sparsely populated areas, surrounded by other public ownerships, and subject to limited state and local governmental controls. In the east, with a different political landscape, federal land management practices required substantial adaptation and evolution.

In New England, national parks, forests, and refuges were established in areas directly governed by states, counties, and local governments; and none were prepared to cede total responsibility to the federal government. Agency traditions, and directives from Washington, were not always appreciated in self-reliant New England. Moreover, municipalities and homes and businesses that found themselves surrounded by the federal property often experienced conflicting values and land-management philosophies. The federal-local dynamic was also influenced by the presence of nonprofit organizations that often predated the existence of the federal agencies. The Green Mountain and Appalachian Mountain Clubs are both examples. In Maine, George Dorr and his colleagues also predated the establishment of Acadia National Park. These individuals and organizations took the position that the federal government was being invited to be a partner in the stewardship of these natural areas.

In later years, other challenges to the federal local partnership emerged. Coincident with the generally welcome establishment of the Cape Cod and Minute Man national parks in Massachusetts came the often unwelcome use of federal land acquisition and, in some cases, condemnation of private property. In combination with the arrival of federal management policies also generated tension. The practice of career rotation of senior federal managers added a sense of inconsistency that exacerbated the uncertainties of federal agency land management. The depth of these concerns was evident when the proposed National Recreational Area encompassing the headwaters, middle, and lower Connecticut River failed to receive public support. However, New Englanders still wanted to protect their natural and cultural resources and continued to seek help from the federal government in other ways.

By the 1980s, increased recognition that the natural resource-based economy was in decline led to a growing awareness that heritage tourism could play a role in the future. At the same time, first and second home development continued to impact much of rural New England's prime scenic and ecologically valued areas. The region saw an increased willingness of its residents to commute long distances or simply telecommute, and the infusion of light manufacturing and service economies began to precipitate big-box and residential development in historically undeveloped areas. In the Northern Forest, large tracts were being sold at an accelerating rate, further increasing uncertainty within the region. It was uncertain that New England's traditional "special places" would be protected.

Over time, the federal agencies and their new neighbors began to learn to work cooperatively to find common ground. A New England model of a working federal—local partnership is seen today in a diverse array of cooperative resource stewardship examples described in this chapter. This strategy was crafted in response to three major transition points in the twentieth century. This series of crises followed, in general, the impacts of shifts in major economic drivers. First, in the early 1900s, more than a century of seemingly unlimited timber harvests in the headwaters of the region's major rivers supplied billions of board feet needed to build and expand the mill cities along those very same rivers and their tributaries. This large scale clear cutting operations in New Hampshire and Vermont and, in addition to timber harvests on Maine's Mount Desert Island, the threat of expanding second home development, influenced the establishment of federal areas managed by the National Park Service, the U.S. Forest Service and the U.S. Fish and Wildlife Service.

In mid-century, New England and the nation reacted to the post-World War II expanding population and economy and the resulting increased leisure time and mobility, suburbanization, and the most obvious environmental impacts of the industrial revolution. For the first time national policy was established to protect open space, provide for outdoor recreation opportunities, and to restore clean air and water. Each of the states responded appropriately with significant positive results including accomplishments of the Land and Water Conservation Fund, the National Wild and Scenic Rivers Program, the National Trails Program, the National Wilderness System, and a new generation of national park areas, exemplified in New England by the establishment of Cape Cod National Seashore.

Toward the end of the century New Englanders grew concerned about the overall loss of forest and farm land and the unsustainable forest management practices in northern New England and upstate New York. In addition, their continuing interest in conservation, but dislike for federal land acquisition and management led to the creation of several partnership models of conservation that include all levels of government and the private sector. The transition from a locally controlled forest product industry to one controlled by foreign-based multinational corporations and increased globalization of the industry heightened these concerns. In response, the New England congressional delegation played a leadership role in the establishment of the Forest Legacy Program to protect and maintain sustainable working forests. It should also be noted here that the national, regional, and local land trust non-governmental organizations have for more than a hundred years but especially in the last forty years, played a major role in New England land conservation, often working in cooperation with the state and federal government. In the populated parts of New England, locally based, federally assisted efforts created New England partnership parks in Lowell, Boston Harbor Islands, and New Bedford, the Partnership Wild and

Scenic Rivers and the National Heritage Areas. This collaborative model integrates best practices for natural and cultural resource stewardship with natural and cultural heritage tourism and other sustainable economic strategies.

Accelerated climate change is a major new factor that will influence conservation strategies in the twenty-first century that will continue to grow in importance in New England and around the world. Forecast, but ignored for more than a century, the environmental, public health, and economic impacts of global climate change are only now receiving the attention required. Already there is evidence of the change in floral and faunal species composition, sea-level rise, and public health issues at a global scale. Larger in scope than the clean air and water issues fifty years ago, managing a response to climate change will require global solutions and unprecedented global cooperation. Increased awareness of the need to reduce fossil fuel emissions and the need for carbon sequestration will once again reshape the strategies for land conservation.

Twenty-first-century conservation, writ large, will require new collaborative models that leverage all levels of government and all aspects of the private sector. In shaping these new models, the following are important considerations:

- Examining the cause and effect relationships between sustainable resources, sustainable economies, and livable, healthy communities;
- Analyzing effects and sustainability of traditional vs. innovative conservation finance models and identifying incentives necessary for them to work;
- Accepting that solutions may require unprecedented forms of intrastate and megaregional, national and global cooperation.

Finally, new models should build on the self-reliance and self-determination so evident in New England over the years, a tradition that can build a comprehensive shared vision for the future and a common understanding of the issues. As they have done so successfully in the past, New Englanders can be the ones to provide leadership for meeting the twenty-first century conservation challenges by effectively linking local, regional, and national strategies in a global context.

NOTES

1. John Elder, "A Conversation at the Edge of Wilderness," in Wilderness Comes Home: Rewilding the Northeast, edited by Christopher McGrory Klyza (Hanover and London: University of New England Press, 2001), p. 256.

2. Acadia National Park draft essay prepared in October 2007, photocopy on file at the NPS Conservation Study Institute, 54 Elm Street, Woodstock, VT 05091.

3. Dr. Perry Hagenstein, "White Mountain and Green Mountain National Forests" draft essay prepared in October 2007, photocopy on file at the Conservation Study Institute. The remainder of this section was primarily excerpted from this essay.

4. George Perkins Marsh, Address Delivered before the Agricultural Society Rutland County, September 30, 1847. Rutland, VT 1848.

5. Hagenstein, 2007.

6. Ibid.

7. Ibid.

8.Ronald A. Foresta, America's National Parks and Their Keepers (Washington, D.C.: Resources for the Future, Inc., 1984), p 1.

9. Sara M. Gregg, Vermont History, vol. 69 (Winter/Spring 2001): 201-221.

10. Harlan D. Unrau and G.Frank Williss, Adminstrative History: Expansion of the National Park Service in the 1930s (Denver, CO: National Park Service, 1983), see http://www.nps.gov/history/history/online_books/unrau-williss/ adhi4i .htm (December 30, 2007).

11.Vermont Historical Society see http://www.freedomandunity.org/create _image/ccc.html (December 30, 2007)

12. Hagenstein, 2007.

13. Ibid.

14. Jamie Fosburgh, draft essay on National Wild and Scenic River System prepared in October 2007, photocopy on file at the Conservation Study Institute.

15. Rolf Diamant, draft essay on the Wildcat River project prepared in October 2007, photocopy on file at the Conservation Study Institute.

16. Steve Golden, draft essay on RTCA prepared in October 2007, photocopy on file at the Conservation Study Institute.

17. The subsequent passage of the Land and Water Conservation Fund in 1965 provided a reliable funding source for land acquisition associated with both of these new parks.

18. Foresta, 1984, pp. 2389

19. Minute Man National Historical Park, established in 1959, was the first national park area in New England to be authorized to purchase private lands.

20. "Although it was argued, and probably with some justification, that the presence of towns made up of quaint clapboard and cedar-shingled buildings enhanced the national seashore, it was also true that eliminating the towns would have involved displacement and condemnation on a scale not previously seen in

connection with the establishment of national parks. The political costs of this type of mass displacement would have been enormous." Foresta, 1984, p. 238.

21. Cape Cod National Seashore, draft essay prepared in October 2007, photocopy on file at the NPS Conservation Study Institute.

22. Russell Brenneman, see the chapter on Connecticut in this volume.

23. Connecticut River Estuary Regional Planning Commission, Protecting the Character of the Lower Connecticut River, Connecticut River Gateway Commission, 2004.

24. Chronos Historical Services, "Mount Holyoke Historical Timelines," Chronos Historical Services, http://www.chronos-historical.org/ mtholyoke/index.html, October 28, 2007.

25. Massachusetts Department of Conservation, "Connecticut River Greenway State Park," Massachusetts Department of Conservation, www.mass.gov/dcr/parks/central/crgw.htm, October 28, 2007.

26. Judd Gregg, Senator Gregg Attends Ceremony Marking the Final Protection of Connecticut Lakes Headwaters Lands, new release October 10, 2003.

27. Rob Moir, Charles H.W. Foster, and Beth Goettel, "Silvio O. Conte National Wish and Wildlife Refuge" draft essay prepared in October 2007, photocopy on file at the NPS Conservation Study Institute.

28. Several of these sites, such as the Saugus Ironworks, Lowell National Historical Park, and Springfield Armory, have their story embedded in the development of the natural resources of the region. Others such as Adams National Historic Site represents the home and surrounding landscape of one of the founding families of the American Revolution. The Frederick Law Olmsted Site in Brookline, Massachusetts, preserves both the home and studio of the Olmsted Sr and successor firms.

29. Linda Cook draft essay on Weir Farm NHS prepared in October 2007, photocopy on file at the NPS Conservation Study Institute.

30. Sarah Peskin draft essay on National Heritage Areas prepared in October 2007, photocopy on file at the NPS Conservation Study Institute;. Jacquelyn L. Tuxill, Nora J. Mitchell, Phillip B. Huffman, Daniel Laven, Suzanne Copping, and Gayle Gifford. Reflecting on the Past, Looking to the Future: Sustainability Study Report, A Technical Assistance Report to the John H. Chafee Blackstone River Valley National Heritage Corridor Commission (Woodstock, VT: Conservation Study Institute), 2005. http://www.nps.gov/csi/

31. Cultural Heritage and Land Management Plan for the Blackstone River Valley National Heritage Corridor (1989). See http://www.nps.gov/blac/home.htm

32. For more information on the heritage area movement, see Brenda Barrett and Nora Mitchell, editors, "Stewardship of Heritage Areas," special issue, George Wright Forum 20, no. 2 (2003); and Brenda Barrett, guest editor, and Elizabeth Byrd Wood, editor, "Regional Heritage Areas: Connecting People to Places and History," special issue, Forum Journal 17, no. 4 (2003), published by the Center for Preservation Leadership at the National Trust for Historic Preservation. For a more in-depth treatment of the evolution of park models, see Paula A. Degen,

Branching Out: Approaches in National Park Stewardship (Fort Washington, Pennsylvania: Eastern National, 2003); Jacquelyn L. Tuxill and Nora J. Mitchell, editors, Collaboration and Conservation: Lessons Learned in Areas Managed through National Park Service Partnerships (Woodstock, VT: Conservation Study Institute, 2001); and Jacquelyn L. Tuxill, Nora J. Mitchell, and Jessica Brown, editors, Collaboration and Conservation: Lessons Learned from National Park Service Partnership Areas in the Western United States (Woodstock, VT: Conservation Study Institute, 2004). The latter two publications are available at http://www.nps.gov/csi/

33. See NPS National Heritage Areas Web site http://www.cr.nps/gov/heritageareas/

34. Eric Aldrich draft essay prepared in October 2007, photocopy on file at the NPS Conservation Study Institute.

35. Ibid.

THE PAST AS PROLOGUE

CHARLES H.W. FOSTER

To begin with, let us return to the six themes identified at the outset of this project: the New Englander's stubborn commitment to self-determination, the effective use of innovation in this region, the affinity shown for particular places; the reliance on individual leadership, the important role played by civic engagement, and the pragmatic view of active management as well as preservation of natural resources that has marked the ownership of land ever since the first settlers settled in Plymouth in 1620. We will then give you a sampling of experiences with these themes from the preceding accounts, follow with a sampling of individual views, and conclude with a set of challenges and recommendations for future land conservation in New England.

A SAMPLING OF EXPERIENCES

Self-determination

Much has been written about the legacy of the Revolutionary War in setting the stage for unremitting resistance to the imposition of dictates by outsiders, in that instance the British Crown. However, this New England trait actually derives from the nature of the first settlers who were not just land occupiers, but had faith, education, skills, experience, and convictions of their own to offer. These early colonists were committed to building a new world, not just a new England. But until this account, little has been said about the way self-determination has been manifest in conservation over the years.

For example, we have learned that even in conservation there has been a seemingly inborn resistance to having others make decisions and impose initiatives and programs arbitrarily upon the region. One example was the mid-century, New Deal proposal for federal valley authorities that would develop resources and generate public power from New England's major watersheds and waterways. These were vigorously opposed by an alliance of business, civic, and political leaders. Even the presidential decision in 1950 to establish a New England–New York Interagency Committee (NENYIAC) to simply inventory resources was deemed unacceptable until its state cooperators were given coequal standing.

Among other examples was the celebrated occasion in the early 1960s of the proposed, federally authorized Allagash River Wilderness Waterway in Maine, fiercely resisted as an undue invasion of states' rights. The unrelenting opposition by former Vermont governor and later senior Senator George Aiken at mid-century to a role for the Corps of Engineers in flood control underscored the New Englander's long memory and legendary stance of stubborn, unforgiving, and often cantankerous representation. And to illustrate that self determination has had a long and determined life, the principle of co-equality reemerged as a major issue in the creation and activities of New York–New England's Northern Forest Lands Council during the 1980s.

Nevertheless, in a pragmatic blend of principle and opportunism, New Englanders have never been averse to seeking help from the federal government — as long as it was on their own terms. The classic example was New Hampshire's pursuit of a White Mountain National Forest through passage and activation of the Weeks law of 1911, and the later (1932) Green Mountain National Forest proposed by Vermont, that began the long and painful process of rebuilding New England's ravaged forest resources. Another was the creation of the New England Regional Planning Commission in 1927 — the first in the nation — which was accepted by the New England governors only after it had withstood state and local scrutiny. In later years, the region welcomed the New England River Basins Commission (1963) and the New England Regional Commission (1965), funded primarily by the federal government, but only after the principle of coequality had yet again been sustained.

Perhaps the ultimate in these new forms of self-determined, imaginative, cooperative, federal-state-local land conservation approaches has been the Resource Protection Partnership pioneered by The Nature Conservancy for New Hampshire's Great Bay in the Durham-Portsmouth area, a project involving a NOAA National Estuarine Research Reserve designation, the active participation of two additional federal agencies, the USDA Natural Resources Conservation Service and the Environmental Protection Agency, and numerous state, local, and private conservation organizations. Testimony to the value of such alliances has been the receipt of more than $50 million in designated federal funds to directly protect more than 4,800 acres around the bay.

Equally informative has been the changing role of the federal government, and the perceptions of its leaders, as to what degree of federal participation will be received favorably within our region. The overview we include by one such federal agency, the National Park Service, will give future leaders valuable clues as to the part to be played by a potential federal partner. In the meantime, however, we must credit the specter of a federal presence with a substantial role in stimulating conservation by New Englanders themselves.

In the case of the Allagash, for example, objection to the federal proposal stimulated the state of Maine to pass Allagash Wilderness Waterway legislation of its own in 1970. And as we have seen, the proposals for valley authorities on New England's major river systems, so objectionable to the investor-owned utility industry, turned out to have a silver lining because they helped stimulate the formation of the first New England watershed association, the Connecticut River Watershed Council (CRWC), operating under independent, citizen auspices. From this early beginning in 1952 has come a now-vibrant, grassroots-based, region-wide watershed movement to further state and local actions relating to land and water conservation in all six of the New England states. And the highly successful Martha's Vineyard Commission, put in place by a coalition of state and local leaders, would never have occurred without Senator Edward Kennedy's threatening federal legislation in April of 1972 calling for a federal trust for Nantucket, Martha's Vineyard, and the Elizabeth Islands.

Innovation

New England has long been renowned as a region where inventiveness has been a way of life for centuries. Innovation has been prompted by a number of factors. A problem in practice might require a solution. The much-heralded New England stone wall, for example, was a response to the pragmatic need to mark boundaries and to find a place to put the glacial erratics after the land was cleared for farming. A societal need might also demand attention. The founding of the field of sanitary engineering, and the creation of the pioneering Lawrence (MA) Experiment Station on the Merrimack River in 1886, were directly attributable to the inadequacy of drinking water and the absence of pollution control in mill communities at the turn of the twentieth century. A business opportunity was often the stimulus for an invention. For example, New England–born Eli Whitney returned to New Haven, CT from Georgia in 1793 to perfect his invention of the cotton gin. An external event could also trigger innovation.

After the American Revolution, New England moved quickly to build and operate its fleet of coastal and clipper ships and to begin profiting from the infamous trilateral trade of timber, rum, and slaves. The War of 1812 slowed the era of shipbuilding, but released accumulated private fortunes to serve as capital for investments elsewhere. The result was the founding of the textile and mill economies of southern New England, so much a part of the Industrial Revolution, and the beginnings of the extensive family owned forest holdings in Maine. Not surprisingly, conservation has also been helped measurably by innovation. The following vignettes are simply examples of those mentioned more fully in the state overview chapters.

At the turn of the twentieth century, the concept of the land trust made its entrance, the brainchild of Massachusetts landscape architect Charles Eliot and the Trustees of (Public) Reservations he helped found. The inspiration was a mix of a societal need to improve the health of urban dwellers, and the intellectual objective of accumulating a "collection" of parks and open spaces analogous to the art and literary holdings of museums and libraries.

One of the most significant legal innovations affecting habitat conservation in the region was the 1896 action of the Supreme Court in *Greer v. Connecticut.* Justice Edward White, writing for the Court's majority, found that as successors to the British sovereign, states held the right to control and regulate wildlife "as a trust for the benefit of the people." The implications for conservation in New England and the nation would be profound. Free-ranging fish and wildlife, frequently assumed to be part of the property of a private landowner, actually were a public resource.

Vermont, Connecticut, and Massachusetts have been responsible for a number of important innovations in the field of agriculture. For example, Vermont Congressman Justin Morrill sponsored a legislative measure in 1862 establishing a national system of land-grant colleges for the scientific study of agriculture. Connecticut's Hamden-based Agricultural Experiment Station, founded in 1875, was the first of its kind in the nation under purely state auspices. A half century later, a young Massachusetts commissioner of food and agriculture, Frederick Winthrop, sought and won support from his legislature to acquire development rights on prime agricultural land, thereby providing a regional and national model for saving productive farmland from development and enabling farm families to access the capital needed to improve their operations and remain on the land.

A New England nonprofit head, Harris Reynolds, on his honeymoon in Germany in 1913, spotted a community level forest initiative that seemed tailor-made for his region. The result was the system of town forests that now occurs in all six states and has spread successfully elsewhere in the country. Four decades later, another community conservation innovation appeared—the 1957 Massachusetts authorizing legislation for the establishment of city and town conservation commissions. Now a thousand members strong, they are among the most effective advocates for state-level conservation in the entire Commonwealth of Massachusetts.

Sense of Place

It is hard to review conservation in New England without appreciating the affinity New Englanders seem to have for particular places. More times than not, it is here that their intentions and actions have been focused most effectively. To be sure, New England has also had its share of broad, general, land-conservation program initiatives.

New Hampshire's 1988 Land Conservation Investment Program is one such example, an effort that ultimately spent nearly $50 million to protect more than 100,000 acres. The Land for Maine's Future initiative, authorized in 1987 at a $35 million level, is another example of program action that has helped purchase over 192,000 acres. The initiative proved so popular that it has been replenished twice. Massachusetts's state parks expansion program, operating over a planned twenty year period at an estimated cost of $100 million, was authorized in 1957 with what was then a tremendous commitment of $1 million per year. Connecticut and Vermont have been equally supportive of substantial parks and open space acquisitions — the former declaring in 1998 a goal of keeping 21 percent of the state's entire land area as open space; the latter making effective use of its unique 1987 Vermont Housing and Conservation Trust Fund. The finest example proportionately must be tiny Rhode Island where author Peter Lord asserts that no major state or local bond issue for parks and open space has ever been turned down by the voters.

But when New Englanders rally to the cause of a particular place is where the qualities of conservation we are evaluating rise to particular prominence. A few examples of this principle are offered to make this point.

Connecticut's lower Connecticut River land conservation campaign, focused on the estuary of the river, is one that comes immediately to mind. As author Russell Brenneman has written, the effort was bottom-up rather than top-down, acting through a state-authorized, but locally inspired, Connecticut River Gateway Commission. Today, through a combination of state and local action, abetted by the private efforts of The Nature Conservancy, a place important to Connecticut and the nation has been preserved. In doing so, Connecticut has been faithful to the keystone of its state's historic compact with its citizens — their right to participate in public life and their responsibility to do so.

A similar story can be told for Block Island in Rhode Island. As author Peter Lord has described, a task force on natural resources created by then-governor John Chafee during the 1960s spotlighted eleven-square-mile Block Island as deserving of preservation, but it was not until landowners F. David and Elise Lapham stepped forward with a historic easement contribution of some 145 acres that the effort to conserve Block Island really got underway. Thanks to effective networking among the Block Island Land Trust, the Block Island Conservancy, The Nature Conservancy, the state, and the town—all abetted by philanthropic funds supplied by the Champlin Foundation—more than 41 percent of the island's landscape is now protected.

Camel's Hump in Vermont is another case where concern for a particular place stimulated a range of subsequent conservation actions. Once the property of Colonel Joseph Battell, an early Vermont conservationist, the land was given to the state in 1911 to become part of the state's forest and park system. But as authors Robert McCullough, Clare Ginger, and Michelle Baumflek have described, University of Vermont scientists Hubert Vogelmann and Richard Klein, in the 1970s, noted a decline in the condition of the red spruce on Camel's Hump. Further research attributed this condition to acid precipitation arising from power plants west of New England. This triggered a nationwide inquiry into the effects of what came to be called "acid rain" on forests, streams, and water bodies and their dependent fish and wildlife populations—all started by a concern for the conservation of a particular place.

Elsewhere in northern New England, the experience with Maine's (Mount) Katahdin inland and Bar Harbor on the coast are equally instructive. These stories, told more completely by author Thomas Urquhart, are illustrations of the extraordinary value of place connectedness. Katahdin was the life work of one individual, business leader and later governor Percival Baxter who, at the time of his death in 1969, left as his personal endowment to the state more than 200,000 acres as Baxter State Park, all acquired without a single dollar of public money. In Hancock County (ME), another place-committed individual, George Bucknam Dorr, described as a Boston boy who took a shine to Maine, virtually single-handedly engineered the creation of Sieur de Monts National Monument, renamed in 1929 as Acadia National Park, again without a dollar of public funds. Over a span of more than three decades, Dorr was able to forge a supportive alliance of local and summer people that endures to this very day.

Our selective examples of the importance of sense of place in conservation practice are rounded out by the case of the Sudbury Valley in Massachusetts, nearby to the homes of such celebrities as Ralph Waldo Emerson and Henry David Thoreau. This place so imbued with human meaning was the proving ground for one of New England's most effective, modern conservation leaders, Allen H. Morgan, executive vice president of the Massachusetts Audubon Society during the 1960s and 1970s. As author Charles Foster has written, Morgan converted his personal interest in birding to the formation of the Sudbury Valley Trustees (SVT), a nonprofit organization that subsequently compiled an outstanding record of land conservation throughout the valley. When Morgan rejoined his beloved SVT in 1980, he had built a valley-wide legacy of not only local land protection actions, but a federally designated Wild and Scenic River and a new national wildlife refuge encompassing the entire Sudbury River meadows.

Individual Leadership

Throughout this account of New England conservation echo the footsteps of individuals who have made it happen. In the introduction to the history, we talk about the qualities of leadership, and the various kinds of leaders, needed to attain a complex goal such as conservation. The individual state histories are replete with evidence of how profoundly fortunate New England has been in its reservoir of human talent over the years. In singling out some and not others, we are open to criticism that our choices are incomplete. Nevertheless, it would seem constructive to trace the manner in which individual leadership has been responsible for our story and to offer a few vignettes of those who contributed in significant ways. In the various state accounts, we meet a panoply of individuals who have been key to New England conservation. They range from the elite to the ordinary. Yet, such is the pervasive nature of conservation that places have been found for each to make an individual contribution. This makes them collectively extraordinary.

In Maine, for example, author Thomas Urquhart introduces us to such political leaders as Republican Governor Percival Baxter, the donor of Baxter State Park; House Majority leader and State Senator Harrison Richardson, the leader of conservation reform for many years in the Maine legislature; and Kenneth Curtis, reputed to be the "greenest" governor in Maine's history. On the citizen side are former Attorney General Jon Lund, publisher of the influential *Maine Sportsman*; businessman Thomas Cabot; long-time Maine Coast Heritage Trust leader Jay Espy; environmental journalists John Cole, Peter Cox, and Robert Cummings; and philanthropists Peggy and David Rockefeller and Roxanne Quimby. One of the most extraordinary illustrations of what we have called transactional leadership was that of then-Director of Public Lands Richard Barringer's negotiations, completed in the early 1980s, to convert Maine's heritage of scattered public land lots into what is now a nearly 600,000 acre, permanent, Maine Public Reserved Lands system.

In Vermont, authors Robert McCullough, Clare Ginger, and Michelle Baumflek have also given us glimpses of some extraordinary conservation leaders. At the head of the list, of course, is George Perkins Marsh, Vermont's and the country's founding conservationist. Although Marsh is renowned for his seminal work, *Man and Nature*, published in 1864, his contributions to state-level conservation included service as a Vermont commissioner of fisheries and the state's first commissioner of railroads, and a keen eye for the effects of transportation, development, deforestation, and over-farming on the state's rivers, streams, and dependent resources. More recent leaders singled out for recognition by McCullough and Ginger are contained in a list of twenty individuals from whom the University of Vermont and the Vermont Folk-life Center hope to take oral histories.

For New Hampshire we are indebted to both the overview written by James Collins and the two published histories of the Society for the Protection of New Hampshire Forests (SPNHF). The most recent of these (2001), edited by Rosemary Conroy and Richard Ober, bears the appropriate title *People and Place*. These accounts recall to mind the significance of men like Philip Wheelock Ayres, the SPNHF's first secretary-forester; the life-long contributions of former Governor, U.S. Senator, and presidential advisor Sherman Adams; the eloquence of academic environmental activist Donella Meadows; and the role of generous landowners like Martha and Bill Rotch of Milford, Philip Heald of Wilton, Mary Lyn Ray of South Danbury, Roger and Ann Sweet of Sullivan, and Contoocook landowner and woodworker Ernest Hebert who once remarked that owning and knowing a forest is much like the man who shapes a spoon being himself shaped by the spoon. We are also introduced to the large-as-life confidante of philanthropists, party leaders, and presidents, Jean Lande Hennessey, granddaughter of one of the early "horse whisperers" in Washington state, who rose to help found the New Hampshire Charitable Foundation, served as budget director for the state of New Hampshire during the Gallen administration, and rounded out her many contributions to conservation in the capacity of a U.S. commissioner of the International Joint Commission.

In his Massachusetts overview, author Charles Foster pays tribute to a similar spectrum of leaders who have given the commonwealth a reputation as a state where conservation innovation and responsibility are to be expected. Among these are the Arnold Arboretum's imperious Charles Sprague Sargent, a contemporary of Gifford Pinchot and John Muir and a member of the National Forest Commission of 1897; dour Benton MacKaye, Harvard forester and social activist and the proposer of what is now the Appalachian National Scenic Trail; physician and fish and game activist John Charles Phillips, an initiator of the New England wildlife conservation movement; modest, self-effacing Harold Jefferson Coolidge, whose recounting of the founding of the International Union for the Conservation of Nature and Natural Resources often took place under a giant Alaskan moose head mounted in his summer lodge at Squam Lake, New Hampshire; and master conservationist R. Frank Gregg, executive director of the Boston-based New England River Basins Commission from 1967 to 1978, who, in moments of stress, would reportedly take off his glasses, close his eyes, and appear to take a direct message from God.

For Rhode Island, *The Providence Journal* writer and environmental reporter Peter Lord provides us an account of twentieth-century conservation in the smallest state in the nation. It, too, is a story of significant, individual leadership validated by seven formal oral histories, including those of Alfred Hawkes, for thirty-five years the director of the Audubon Society of Rhode Island; Daniel Varin, long-time state director of planning; Robert L. Bendick, Jr., director of the Department of Environmental Management from 1984–88; and Mrs. Henry D. (Peggy) Sharpe, a citizen activist involved in the work of The Nature Conservancy and many other nonprofits. We are brought back to the nation's first managed private forest (1820) established by mill owner Zachariah Allen in Smithfield, RI. Lord then pays tribute to the 1900-era civic and political leaders who established citizen-led commissions to plan and acquire land for parks, forests, and boulevards in order to improve the health and welfare of industry workers and their families, and the modern leaders such as Governors Theodore Green and John Chafee who worked to consolidate government services and expand its conservation activities. We meet John L. Curran, lawyer and sportsman, who organized the Rhode Island Fish and Game Protective Association; Dr. Donald J. Zinn, a marine ecologist, professor of zoology at the University of Rhode Island, and an active volunteer with the National Wildlife Federation; and writer Ruth M. Gilmore, an indefatiguable citizen conservation activist. Of special interest is Lord's account of the Rhode Island Natural Resources Group, an informal think tank of activist influentials that would meet at the home of businessman Bradford Kenyon during the 1960s to study local environmental problems and recommend solutions—an approach that would seemingly work well anywhere in New England.

Rounding out the six state overviews is author Russell Brenneman's account of conservation in Connecticut over the past century. We are introduced to the now-familiar approach of a small group of friends meeting privately in 1895 to discuss the need for a state conservation movement. The result was the founding of an organizational leader, the Connecticut Forestry Association (later changed to the Connecticut Forest and Park Association), and the rapid enlistment of other luminaries. In the course of this account, we meet Gifford Pinchot, who grew up in Simsbury; George Payne McLean, the creator of the state's first wildlife reserve (the McLean Game Refuge); Henry S. Graves and George Myers of the Yale School of Forestry; Austin Hawes, who served as state forester in both Connecticut and Vermont; and ecologist Richard Hale Goodwin, the much-beloved co-founder of The Nature Conservancy. Of particular note is Brenneman's account of the recruitment of William H. (Holly) Whyte in 1962 to lay out the case for a state program of public investment in conservation. Closely associated with Laurance S. Rockefeller in New York, Whyte brought to Connecticut a strong and charismatic advocacy of a humane metropolis, a view that would help Connecticut reconcile the needs of its rural and urbanizing populations. An important additional story with profound implications for land conservation is the way Connecticut addressed the future of its unique heritage of private water company lands. Here we meet another kind of leader, State Representative Dorothy S. McCluskey, whose district included the 26,000 acre New Haven Water Company lands. In later years, we find McCluskey, now retired to Rhode Island, deeply involved in the conservation of Block Island where she now resides.

Civic Engagement

The considerations of self-determination, innovation, place connectedness, and leadership are fine attributes to bring to an issue, but they beg the question of how a situation can be resolved in practice. The answer in New England seems to rest on the degree to which citizens are engaged. As has been observed earlier, leaders must have followers if they are to succeed. Fortunately for us, the accounts of conservation in each of the states illustrate many instances when this has been so.

In the Maine overview, for example, Thomas Urquhart takes us through the evolution of the $35 million Land for Maine's Future initiative. Had the program rested solely on the leadership of government agencies, it never would have happened. As late as the 1950s, the consensus was that since all of the industry lands were open to the public, Maine did not need any of its own. By 1987, however, attitudes had changed. Citizens to Save Maine's Heritage was formed consisting of an alliance of environmental and conservation organizations plus a cross section of well-known business and political leaders. Modern campaign techniques were enlisted—legal analysis, advertising, and public-attitude polling. When the proposed bond issue came before the voters in November 1987, support was on a level of two to one. Maine land conservation had now come of age.

In New Hampshire, the campaign to save Franconia Notch that occurred not once but twice constitutes another example of the importance of civic engagement. Again, it represents an amalgam of the qualities mentioned above. The first effort in the 1920s, under the campaign slogan "Sawed or Saved?," focused on acquiring the 6,000 acres of the Notch in lumbermen's hands within seven miles of the Daniel Webster Highway. What was known as the "little Yosemite valley" included a view of the popular rock formation known as the Old Man of the Mountain. Women's clubs took up the challenge, school children contributed their nickels, and under the Forest Protection Society's skillful leadership, trees were "purchased" with donations of a dollar each. The second campaign for the notch in the 1960s involved a proposed federal-state interstate highway that would blast and clear space for a four-lane, divided, superhighway. For seven years this proposal was fought to a standstill by a coalition of environmental, business, and political leaders until, in 1978, agreement was reached on a more modest two-lane parkway at the heart of the Notch. Civic engagement had again won the day.

The story of the proposed Green Mountain Parkway in Vermont, told by Robert McCullough, Clare Ginger, and Michelle Baumflek, is another instance where citizen involvement ultimately resolved a complicated and controversial issue. In essence, the parkway was a metaphor for the fundamental ambivalence of Vermonters' long-term goals for both tourism and conservation. Proposed as a parkway during the Depression years, and modeled after the Blue Ridge Parkway in Virginia and North Carolina, the road would occupy a corridor stretching from just south of the Massachusetts state line to the Canadian bor-

1908 2008

A similar image today might find the young pine in the first chapter in need of release from competing hardwoods — evidence that to forestall unintended change, professional management must be an integral part of any modern effort to conserve land.

(2007 PEN AND INK SKETCH BY ARTIST LENITA BOFINGER OF CONCORD, NEW HAMPSHIRE) der. The route would impact more than thirty Vermont towns. After much debate, Vermonters decisively rejected the proposal on Town Meeting Day in 1936, voting to forego more than $10 million in New Deal dollars and to keep the state unspoiled. As the authors have observed, the controversy marked the beginning of a long and continuing era during which the overwhelming influence of the automobile has forced the shaping of opinion about the ways in which Vermonters should practice conservation.

Equally determined was the Massachusetts decision in 1961 to have the Cape Cod National Seashore authorization defy tradition and become a specifically designed, landmark exercise in federal, state, and local cooperation. According to Charles Foster, the result was a Seashore Act, tailor-made by the state's Congressional delegation (including newly elected President John F. Kennedy) and enacted after a series of contentious local hearings, that included the pathbreaking authorization of a standing Cape Cod National Seashore Advisory Commission. Constituted of individuals nominated by the six Seashore towns, and designees of both the state and the county, the advisory commission remains a viable and trusted instrument of civic engagement to this very day. At the twentieth anniversary ceremony held at the Salt Pond Visitor Center on August 7, 1981, chief draftsman David B. H. Martin, observed that the seashore had fulfilled his highest hopes and expectations, "enlisting a legion of participants, transforming the act into a living experience, and remaining fully in touch with social reality."

Rhode Island, too, has had its share of experiences with civic engagement. Author Peter Lord tells the story of Ruth M. Gilmore of Providence, a citizen activist with a social conscience who became a determined conservation advocate. Cofounding the Rhode Island Conservation Workshop in 1949, a cooperative environmental education project involving the Rhode Island Wildlife Federation, the Rhode Island College of Education, the state Department of Agriculture and Conservation, and Rhode Island State College, "Miss Gilmore" set about personally to make changes in state, national, and even international conservation policy. For example, she urged an Izaak Walton League convention in Cincinnati to pay more attention to conservation and even took on elements of forest industry over their impacts on the environment, asserting that "they try to make people believe that we will have abundant forests forever." Along with her good friend,

University of Rhode Island professor Donald J. Zinn, a man with the professional credentials she lacked, the pair represent an effective example of civic engagement in practice.

As author Russell Brenneman attests, Connecticut's experience with civic engagement has also been substantial. At the heart of it are the state's 169 cities and towns, some of which existed 150 years before Connecticut became a state. In addition to playing a direct role in the use of land, towns serve as an important reservoir of citizens potentially able to directly influence conservation. Their value is seen in Connecticut's 700-mile Blue Blazed Trail system, founded in 1929 and maintained today largely by local volunteers Another example is the burgeoning local land-trust movement. The first community level land trust was probably the one organized in 1964 in the town of Madison. There are now at least 120 in number in Connecticut and growing. The value of citizens acting collectively and collaboratively is illustrated by Brenneman's account of the struggle in the 1960s to prevent the Rykar Development Corporation from developing 277 acres of the Great Meadows Marsh at the mouth of the Housatonic River in Stratford and Milford. Alarmed by the corporation's proposal to build a port facility on the wetland, a coalition of governmental and nongovernmental organizations came together to protect the marsh. The area is now a unit of the Stewart B. McKinney National Wildlife Refuge.

Ethical and Moral

As we round out this exercise in retrospection, two other aspects must be examined: the extent to which the pragmatic use of natural resources has been a fundamental part of the concept of New England conservation, and the occasions in which a seeming contradiction with preservation has arisen in the course of the region's conservation history. They fall into the areas of ethical and moral concerns.

In the introductory chapter to this book, we remind the reader that New Englanders have long been accustomed to having natural resources put to work on their behalf. Indeed, these personal, economic objectives have been encountered many times both before and after the turn of the twentieth century, and often taken to extremes. This has produced a tug-of-war between organizations seeking economic advancement and environmental preservation — also among those operating at national versus state and local levels. An important part of the story of New England conservation is how it differs from that of conventional environmentalism elsewhere.

New Hampshire makes this point quite explicitly in the mission statement of its much respected Society for the Protection of New Hampshire Forests, describing its organization as an association seeking to perpetuate forests "through their wise use" and their complete reservation only in "places of special scenic beauty." This has enabled the society to work successfully with a spectrum of economic interests that include elements of forest industry and tourism development. Despite the society's long history of pro-environment activism, this almost genetic preference for collaboration over confrontation, described as the result of a Yankee culture literally rooted in the land, is an illustration of what has made New England conservation distinctive from others during the twentieth century.

In Maine, similar differences in values have fueled the present debate over the merits of a new national park suggested for the North Woods. A determined proponent has been the Massachusetts-based Restore: the North Woods. The recent case (November 1997) of the rejected Compact for Maine Forests, described by Thomas Urquhart, an alternative to the effort to ban clearcutting entirely advocated by Green Party spokesperson Jonathan Carter and the Sierra Club, only proved further that Maine voters are still conflicted over matters of preservation and use.

When one looks at New England's extensive heritage of intellectual thought, there seems to be clear evidence of an acceptance of the pragmatic use of natural resources from earliest times.

Even nature icon Henry David Thoreau displayed more than a passing interest in how his neighbors managed their forestland. We made the point before, and make it again, that what Thoreau was advocating was a state of wildness, not wilderness — a condition that permits reasonable accommodation with other uses of the land.

For that reason, we find Massachusett's Quabbin Reservoir and Rhode Island's Scituate Reservoir today serving as quasi-natural areas in addition to their principal function as sources of water supply for their metropolitan areas. New Hampshire's Pittsford Reservoir, built in 1937 for purposes of flood control and power generation when water for log drives was no longer needed, is another such example. New England Power Company's system of headwater reservoirs on the Connecticut and Deerfield Rivers similarly serve multiple purposes, operating under negotiated agreements with fisheries and recreational interests as well as supplying peaking power for the New England electrical grid.

George Perkins Marsh, in *Man and Nature*, was quite explicit about the value of managed, natural resources. Although his view of the future—that responsible stewardship of land could solve most of the environmental problems—has not held up well in practice, it still generates the primary rationale for the current interest in better addressing New England's extensive heritage of land in private hands.

Even our region's most recent intellectual icon, the remarkable Rachel Carson, honed her advocacy for nature in the course of a career as a professional biologist. As recent biographer Mark Hamilton Lytle has written (2007), the lady he characterized as *The Gentle Subversive* agreed with Thoreau's appreciation of the sensibility of science and his concept of humankind "as an inhabitant or part and parcel of Nature." Her view of nature, and the moral and ecological vision she conveyed in her extraordinary writings, came in part from New England where she spent as much time as she could at her cottage on Southport Island on the coast of Maine.

A SAMPLING OF OPINION

On November 19, 2007, sixty recognized New England conservation leaders, representing over a thousand years of experience, were convened at the New England Center in Durham, NH to review the draft regional history of land conservation in New England and contribute suggestions as to its possible findings, recommendations, challenges, and responses. In a remarkable outpouring of ideas, the participants both validated the underlying objective of the conservation history project—to prepare a written record to inform and guide subsequent conservation leaders—and concluded that, regardless of the unremitting press coverage of coming environmental disasters, the story of how the region had recovered from the environmental abuses of the late nineteenth and early twentieth centuries provided a measure of hope that New England can once again meet the challenges of the future. A fuller account of the presentations and colloquy is contained in the summary proceedings distributed to all of the participants. Selected observations and suggestions are provided below.

For example, much criticized was the abject failure of land-use planning over the years and its consequences for the environment—especially the absence of a commonly accepted regional vision that would set basic goals for land use and conservation in New England. Despite this lack, much actual land conservation has occurred, but the twentieth-century practice of drawing lines around an area and simply buying it seems unlikely to be feasible for much longer. Described as currently "besotted with large projects," the region was urged not to overlook the incremental value of small groups of people doing things for themselves where they live and work. The audience was reminded that conservation as a movement was born out of small, special places.

A second failure identified was the current vacuum of political leadership relating to conservation. The potential for improvement is there, the participants asserted, but there is a pressing need to jumpstart a modern equivalent of the interconnected private, state, regional, and national leadership that initiated the conservation movement at the dawn of the twentieth century. Impeding such progress is what one participant described as the "creeping professionalism" occurring within many organizations and agencies and the tensions arising between the traditional volunteer and the professional.

Another common observation was the need to "democratize" conservation in order to make it more relevant and appealing, and to find new ways to engage ordinary people in its effectuation. The conference was reminded that conservation was originally part of the early twentieth century social reform movement. Then and now, conservationists were not faces one would ordinarily see on the street – thus a need to reach very different audiences and even to "translate" the conservation message into forms more relevant to others.

That observation led to a general discussion about the likely near-term drivers of such a transformation and the need for "new frontiers of connectivity" with other important goals of society such as climate change amelioration, energy, transportation, economic development, jobs, education, housing, and social equity. The economic benefits of land conservation, in particular, were singled out as a major potential "hook" to capture support for conservation in the future. A special plea was made not to overlook the role private owners of land could play in this transformation, the assertion being that much as New England has been historically an innovator of private conservation, so too could it emerge as a pioneer for citizen-supported global conservation.

Finally, where will the future stewards of the environment come from? it was asked. The way they arose originally was the answer — by people perceiving a personal — even an economic — interest in the land and engaging with others in remedial and management activities.

In postpresentation and postconference communications, additional thoughtful comments were received.

David K. Leff, for example, former deputy commissioner of the Connecticut Department of Environmental Protection, offered the opinion that we must materially change the focus of conservation if we are to build on the progress of the twentieth century. This means broadening its base of support beyond an audience that is largely affluent and nondiverse with a perceived image of gray-haired folks in hiking boots. More involvement of children in ecology would be a good first step in recognition of the fact that the indifference of a single generation could arrest land conservation and reverse many of today's hard-won gains.

Making better use of the world of new technology, he said, is another challenge for the twenty-first century. That means recognizing the inevitable draw of the Internet and putting it to work for conservation. More and better use of electronic communications could help us convince people that meaningful opportunities for conservation exist nearby, not just in the vast, snow-capped mountain ranges far from sight. We must also persuade others that land conservation incorporates benefits into the daily lives of even ordinary people.

Finally, the concept of conservation as a wise investment, not just an amenity, needs to be hammered home, Leff observed. This means paying more attention to working landscapes and ways in which conservation and development can operate in tandem. More than anything, it will require recognition that the natural environment is a "green" infrastructure meriting care and attention comparable to that of the built infrastructure of roads, utilities, buildings, dams, and bridges.

Echoing and enlarging upon these thoughts were the observations of Richard E. Barringer, Maine's former director of state planning and commissioner of conservation, who noted that the current transportation system throughout New England now provides access to places that could only be imagined a century ago. Travel by passenger vehicle has become the dominant influence on the landscape for at least three-quarters of a century — on the one hand shaping growth and development and, on the other, molding the public's view of conservation as an antigrowth and antidevelopment movement. Coupled with the increasing, competitive pressure on local property taxation, the result has been a twenty-first century heritage of unplanned sprawl. The inadequacy of substate regional planning and governance mechanisms has only made the situation worse.

Another challenge for the future, Barringer observed, is to address the diminished role of the federal government in land conservation since the days of the Land and Water Conservation Fund. More times than not, state, local, and private organizations have had to make up the difference. The problem appears to be one not just of declining federal dollars, but the need for a fundamental redefinition of the federal interest in land conservation. As documented in this book's chapters, the range of potential federal tools, techniques, experience, planning and, if necessary, threats of intervention available to be used is simply enormous. Many have been employed successfully in the past and could be again. On their part, conservationists need to move beyond the single objective of acquiring federal funds for land acquisition. Given the pressure from many quarters on the federal budget, they will need to craft other ways to reengage the federal government meaningfully in New England land conservation.

Native New Englander Robert W. McIntosh, associate regional director of the National Park Service and the former Northeast director of the federal Heritage, Conservation, and Recreation Service, reminded the group that the federal conservation ethic had originated in the West. Working cooperatively with states in the East, and with nongovernmental organizations, has been a learning experience for the federal agencies that is still ongoing. Thus, he said, New Englanders should remain self-reliant, inventive, and resourceful in their approaches to the federal government.

An especially thoughtful, self-styled "meditation" on land conservation came later from Durham conference participant Richard Oliver Brooks, planning and coastal zone specialist for Rhode Island, Connecticut, and Vermont and the founding director of the Vermont Environmental Law Center. In reviewing the history of land conservation nationally, Brooks noted two past strategies: one based on acquisition or regulation designed to carve out and protect small areas from the larger forces affecting the use of land; and a second, national strategy of large-scale set-asides for national forest and parks accompanied by regulatory efforts. Both have helped set the tone for land conservation in New England.

Brooks noted that conservation has many meanings — some still in transition. There is, of course, the classic concept of preservation, but also the recognition in recent years of the need to protect the basic services that nature provides. Thus, modern conservation speaks not only to the preservation of natural places and the assurance of an uninterrupted stream of nature-based resources and values, but also the importance of such places to those who involve themselves on their behalf. Rather than a discreet movement, it represents the beginnings of what many hope will be a major cultural change — some say a "green revolution."

Finally, Brooks observed, the key questions for New England conservation in the future might be summarized as follows. Can it maintain its central identity without fragmenting in different directions? Will the scales of the emerging national and global issues take away New England's relative importance as a regional movement? And, ultimately, might it be time for New England conservation, paraphrasing Robert Frost, to consider taking a road "less-traveled"? With Brooks's perceptive challenges ringing in our ears, let us start with a set of impressions we have gained in the course of our two-year assessment of New England land conservation.

OUR VIEWS

The project has served to remind us of what the New England environment was like in the early days of the twentieth century — in the words of one of our authors "stressed and battered." The turn of the twentieth century may well have been the lowest point for New England's natural resources. At least we hope so.

Viewing today's attractive interplay of woods, fields, and human settlements, interspersed with an abundance of water sources, makes it inconceivable that the face of New England would be anything but natural. But history tells us otherwise. The individual state overviews provide a list of conditions we would not tolerate today. These include accounts of the great North Woods ravaged by the human-induced effects of indiscriminate pulp and timber operations. We encounter streams and other water bodies choked with the effluent from mills and human habitations. We find inland fish and wildlife species, and the once rich resources of the coast, if not extinguished by overharvesting, at least at the very edge of survival.

We have noted the classic, early quilt of individual farm ownerships, surrounded by patchworks of working forests and dependent wildlife, abandoned in favor of better economic prospects for their owners in the new mill villages or the settlements of western America, and simply left to grow fallow until a developer came along. Two of our authors, using the analogy of a lighted map of lands in agricultural use, graphically described the demise of the region's farms as simply "blinking out" as the years advanced.

Another of our cooperators made the interesting point that New England conservation is leading us more and more back to Old England—a rural landscape of permanently secured farms in a setting of publicly accessible footpaths and natural corridors. As Connecticut Greenways Committee Co-chair Susan Merrow observed in 1992, these connections "link the places we live with the places we love." In contrast to the earlier days when landscape change just happened, we are beginning to make conscious choices as to what should go or remain.

But more than anything, we find New Englanders unaware of where we have been and how far we have come. Television journalist Tom Brokaw, once producing a special on the Appalachian Trail in western Connecticut, abruptly redirected the filming to the backdrop of seeming verdant hills and valleys when he learned that those areas, within a few short decades, had once been stripped bare for railroad ties and charcoal. And Rhode Island's Donald Zinn once talked about the need to "unmudify" the public's general thinking about natural resources.

The lessons here are both alarming and comforting. New England has a history of abuse of the land, as well as its conservation. Devastation has occurred, not just from greed and arrogance, but consistent with nature's way of species—in this instance humankind—competing for food, light, space, survival, and prominence among a number of other claimants. The comforting part of the New England story is that despite the abuse, in time the land healed itself.

So those who perceive similar outrages should take heart. Doom is not necessarily around the corner. The energetic optimism of the early part of the century has not yet taken on today's familiar ring of desperation. With improved awareness, mistakes can and have been corrected and avoided. Yet, modern conservationists should pay heed to the warning flags being flown by the region's scientific community that the environment's past ability to recover may have been significantly impaired by the accumulation of insults it has suffered.

The other oddly comforting conclusion to draw from the state and regional histories is how often a disaster has inspired a commitment to conservation. We see this in the fight for a new system of eastern national forests to correct the extensive abuse of overcutting and fire. We see it in the continuing battle to control systemic pollution, reduce emissions, and manage energy production and use. We see it in the modern struggle to prevent development and curb sprawl. One is led to wonder whether near disaster is a necessary prerequisite of conservation action. We hope not.

While we are proud as New Englanders of such a fine record in conservation, we make no pretense of a particular ability or acumen. The record is simply one of our being there first and having to cope with situations at hand as best we could. For the most part, there were no plans or precedents to draw upon. As our authors have pointed out, New England's early leaders just stepped up and took whatever actions they felt were necessary. Thus, for those in other regions, New England should be viewed as the nation's premier experimental laboratory on land conservation whose experience can inform and guide others in the universal search for a secure and productive environment.

In the spirit of Ralph Waldo Emerson's observation that there is properly no history — only biography — our account is composed of a number of stories. From them, we have become convinced that the real message of New England conservation has less to do with ecology and the environment than the ways in which people choose to become involved in its affairs. What this suggests is that despite the rich promise of science and technology, it is the human dimension that is the most important. Deciding what to do should respect the informed judgment of professional managers and scientists, but determining whether or not to do so will always remain the province of society itself.

This brings up the matter of scale. It is clear from our account that the closer the issue is to a group of affected people, the more likely it is to be resolved. From this assertion arises a quandary. Our modern assessment of the environment has brought an awareness of issues at more than local scale — regional, national, and even global in scope. The challenge for the twenty-first century will be to figure out how to adapt the realities of a human-scale system to the needs of the earth as a whole.

Scale questions give rise to another observation. In the early portion of the twentieth century, the people of New England were quite homogenous in terms of background and culture. There was a familiarity that both encouraged and enabled steps to be taken collaboratively. As acculturation occurred, human communities became increasingly complex, often arranged more by educational and economic level than by social circumstance. This has led respected observers of the scene, such as Harvard political scientist Robert Putnam, to caution that the much-sought-after qualities of diversity may be counterproductive in terms of getting something done. The greater the diversity in a community, the less trusting of one another seem to be its inhabitants.

What this means for New England conservation need not be an abandonment of its historic dependency upon community action — only the use of community in other ways. For example, it suggests a need to seek consensus at even smaller scales, such as "neighborwoods," and then find ways to aggregate these components, as necessary, into larger configurations. It also raises the prospect of using environmental determinants, plausible to a range of interests and occupants, to help build the kind of social capital necessary to take collective action. And it also suggests that the use of coalitions and new forms of collaboration, as detailed by our authors, is the course of action likely to work the best in at least the near future.

We have also learned much about ways to resolve problems. Not always are proscriptive actions necessary. In fact, the best solutions are often those inspired by consensus, advanced through coalitions of the convinced, and only backstopped by provisions of program authority or law. New Englanders have learned this truth the hard way, fighting perceived encroachments by others and then replacing them with actions taken on their own. Other regions could well profit from their experience.

We are also much taken by the concept of adaptive management as a way to get things done in a timely fashion. The primary virtue is its avoidance of always having to be right. Adaptive managers choose courses of action based on the most plausible information available at the time. They fully expect the need for fine-tuning later on. To those still exploring the technical and scientific ramifications of an issue, and activists loath to give up their advocacy positions, adaptive management gives them a chance to do something now and reserve the opportunity to do something different later on. It also encourages change at the margins rather than at the core, a stance usually appealing to the generally cautious and conservative American. Happily for us, one of the leading proponents of adaptive management theory, political scientist Kai N. Lee, is a New Englander, now based at Williams College. We would do well to heed his call for the use of both a "compass and a gyroscope" in designing approaches to New England's complex environmental and land-use problems.

Classic in the world of environmentalism is the fight to preserve in the face of ruthless economic aggression. While there is much truth to this assertion, and convincing evidence that such approaches are tactically effective in certain instances, New England has learned that good communication and courteous colloquy can often better carry the day. In the various state accounts are numerous instances where exploiters and environmentalists have reached constructive accord. New England's brand of pragmatism rather than ideology has occasionally caused a schism with conventional environmentalism and some bewilderment among its loyal supporters, but there is encouraging evidence that reaching out to the apparent enemy can often yield substantial benefits. As resource preservation, management, and use become ever more complex, the New England approach seems increasingly the way to go.

Another familiar theme in New England conservation affairs has been the effort to keep things the way they are. As Maine Governor Kenneth Curtis once observed, time has a way of making an area seem even more attractive. The emphasis on the unspoiled ideal, however, does not jibe with the way nature operates. Change is a constant part of the natural process.

The concept of unspoiled nature in New England is really a chimera, because there are few places that have not already been touched by humankind. That is another reason why there is general acceptance of the concept of land being actively managed and little support for the thesis that nature left alone is always the best course of action. Nevertheless, room must be left for efforts to secure special environments, such as remnant old-growth stands, and to ensure that there are natural reserves, lightly touched by human hands, that can serve as research laboratories to determine the best ways to manage land.

The unspoiled is also a slippery slope in another respect. It depends on the mind-set of the observer. For a person conditioned to wilderness, the ideal is quite demanding. Not so the inhabitant of an urban area where undeveloped land is a rare commodity and altogether wild land can be a source of some apprehension. Thus, a challenge for New England and elsewhere is how to build bridges across these differing perceptions of nature and harness the joint help of rural, suburban, and urban constituencies for meaningful conservation action.

What President George H. W. Bush once called "the vision thing" is also worth discussing. To be truthful, there is little evidence of a grand design in the history of conservation in New England during the twentieth century. There has been some evidence of planning, but most of it has occurred after the commitment to do something about the problem has been made. In such circumstances, the plan becomes a pragmatic statement of the rationale for becoming involved and an instrument to obtain the necessary authorizations, personnel, and funds. Simply put, New Englanders ordinarily don't do vision. And although they are usually convincing, they are not always charismatic.

As the years passed, the New England states took action to institutionalize the responsibility for natural resources. As one of our authors observed, institutionalizing the management of resources gave structure to the ongoing dialog of conservation. Connecticut makes claim to having had the first Office of State Forester. Massachusetts, in 2001, took the ultimate step of creating a supersecretary (the Office of Commonwealth Development) to ensure coordination of the state's development, housing, transportation, and environmental programs. No matter what the conservation agency was called, it reflected the ecological reality that if everything is connected to everything else, government agencies dealing with ecological issues should themselves be connected.

Yet a note of caution has crept into the various state accounts. While having public agencies charged with responsibility for public resources has been a generally good thing, nongovernmental entities are increasingly concerned about government's performance. This has led to an increasing burden of legal and legislative precautions and a cumbersome process of citizen appeals. More constructive, it would seem, would be the use of public-private partnerships. One needs not monitor something that contains oneself.

Although gone is the early tradition in Rhode Island and other states of placing conservation responsibilities directly in the hands of standing citizen boards, victim to the quest for more central, political accountability, we believe that government should continue to make use of volunteer citizen advisors. This is what political scientist Alan A. Altshuler, the former director of the Ford Foundation-sponsored Innovation in Government Program, calls "scaffolding," the first step in constructing needed change. Government should be reminded that working collaboratively with nongovernmental interests is not just a matter of history but an opportunity for mutual advantage. The nonprofit sector, on its part, should avoid casting itself as an alternative to government.

In the course of following the trail of twentieth-century conservationists, the matter of the region itself has also arisen. The realities are twofold. First, topographically, there is much to support the concept of New England as a region. The other reality is the extent to which others, for descriptive and even marketing purposes, have used the designation New England in their representations. This has caused us to examine whether there is in fact a viable entity called New England. The answer appears to be no. To be sure, there have been programs and institutions with New England in their title, but few have lasted beyond the immediacy they served and the fund sources that sustained them. All have been objects of suspicion even at the outset. While New Englanders have supported cross-border initiatives on many an occasion, they have tended to do so for purely pragmatic reasons, not out of a sense of loyalty to their region. Home for most still remains localized and intensely personal.

And so the possibility of a new institution whose sole concern would be New England conservation will not appear in our recommendations. Instead, we will dip into the accounts we have collected and use that experience to suggest other ways to improve the conservation and management of natural resources on a regional scale. Our story contains numerous instances where good ideas, under different auspices, have crossed boundaries under the impetus of individual initiative and goodwill.

At the governmental level are the eight historic interstate natural resources compacts that have won formal state and federal approval. The standing New England Governors/Eastern Canadian Premiers Conference operates under a similar but less formal arrangement. The statutory U.S.-Canada International Joint Commission is an example of the principle applied internationally.

As we have seen, there have been at least two examples of state governmental cooperation relating to land conservation at the subregional level: the agreement of the southern New England commissioners of agriculture and natural resources in the late 1950s to meet informally and compare notes on common problems; and a similar move by the northern New England states in the 1980s to engage cooperatively in an assessment of the Northern Forest.

At the nongovernmental level, we have been impressed by Maine's use of Paul Bofinger's experience in setting up its own Forest Society of Maine; Vermont offering Massachusetts and New Hampshire its model of the Vermont Housing and Conservation Fund to help shape joint affordable housing and land conservation initiatives; Connecticut's use of the Massachusetts conservation commission experience as a guide to its own system of municipal conservation and wetlands protection agencies; and Rhode Island's pioneering of coastal zone conservation for New England and the entire nation.

This leads us to suggest some simple and practical ways to advance New England land conservation without the necessity of yet another formal organizational presence.

For example, from the story of the New England River Basins Commission comes the idea of an institutionalized, periodic forum that would enable leaders to share current information, encourage collaborative action, and simply get to know one another better. A good way to test this out would be to celebrate the centennial of the first New England governors-sponsored conference called by Massachusetts Governor Curtis Guild, Jr. in November of 1908 just as national conservation was awakening. Attended by virtually every governor and governor-elect, with an audience of two persons each nominated by individual members of the congressional delegation, the occasion marked the official beginning of New England's own commitment to conservation.

On the nongovernmental side comes the example of the 1937 "Wood Ticks," one of the Benton MacKaye-preferred "notorious aggregations," a voluntary assemblage of New England forest leaders who met annually on Mount Monadnock up until 1950 to take the pulse and brainstorm forest policies for the region. Another example is Rhode Island's Natural Resources Group model of having key individuals meet informally to discuss problems and devise solutions. A regional version of that approach has been the so-called Red Sox, suggested by a conservation organization leader in the 1980s, wherein the heads of state-level nonprofits and the Conservation Law Foundation of New England would meet periodically to plan strategies and reconcile differences in policy positions.

A privately supported initiative to support periodic sabbaticals for present or prospective organizational leaders would also help encourage, generate, and distribute good ideas. As examples, we have been impressed by the story of Kingsbury Browne's law firm sabbatical taken at the Lincoln Institute of land Policy that helped create the concept of the national Land Trust Exchange and sparked a national effort to interject conservation into federal tax policy. Similarly, Charles Foster's Harvard Bullard Fellowship led to the formation of the New England Natural Resources Center. For Richard Barringer, post-doctoral support enabled him to write the *Maine Manifest*, a call to action for effective conservation in Maine. And Russell Brenneman's foundation-supported research and early book on the prospective use of easements for land conservation was a pioneering event for both New England and the nation.

We would also recommend more frequent use by state agency and other organizational heads of the basic administrative authority they all possess to enter into agreements with one another. If used simply for purposes of information exchange, planning, and cooperation, and without any continuing, major commitment of personnel or funds, much can be accomplished without need for formal approval.

At the local level, the Environmental Protection Agency during the 1970s had a popular initiative called the Peer Match Program that paid the travel expenses and encouraged local leaders to consult with their counterparts in other communities on how to expedite the construction of pollution control facilities. For New England land conservation, the need to share views and experience on a timely basis is equally important. We already have many examples of individual leadership causing good ideas to move from one state to another, and we have the national example of the successful "rallies" put on by the Land Trust Alliance that enable a thousand or more local and regional land trust leaders to exchange experiences at these annual sessions. Doing so at local scales would help advance the spirit of cooperation that underlies this book.

From the federal side, we have the case of the Department of the Interior's 1950s special program staff of senior specialists whose only authority was to participate in regional discussions and to report directly to the secretary any actions deemed necessary. Though much disliked by the line units, the regional coordinators' recommendations could only be ignored at the agencies' peril.

The concept of a similar high-level special advisor operating, say, out of the New England Governors Conference, is intriguing. This sort of senior "conservation laureate," carefully chosen to offer a mix of stature and experience, would be given no direct authority, only the standing to attend any meeting—federal, state, or regional—on a conservation matter. From both the access to information, and the connections to the governors, such an individual could become enormously influential and helpful.

The other area that has caught our attention is the matter of communications. The region has come a long way since two men, on horseback, spread the news that the British were coming. Now thanks to computers even school children have instant access to the world's news. Radio and television are in our lives twenty-four hours a day. Regrettably, the sheer volume of information available, the multiplicity of sources, and the absence of a validating mechanism leave users in the dark as to the merits of a representation. New Englanders' response has been the classic one of judging proposals first on the basis of what might be lost rather than gained, and then casting their lot with the individuals or institutions they most know and trust.

As nationally syndicated journalists Neal Peirce and Curtis Johnson have reminded us, we live in a world where the ready availability of accurate and balanced information is crucial. William H. Whyte, in 1962, was among the first to point out that sound resource planning must rest on reliable information. As Peirce and Johnson advised the New England governors in 2006, after a two year study of New England's prospects and impediments, they noted that this "proverbially smart" region of America had better not continue sitting on its hands and letting the quality of its environment decline.

* * *

Although the idea of a special New England conservation information service is beyond the scope of this history, it might be worth a try by others more knowledgeable. For example, we already have an impressive fabric of natural history institutions in our region. We have a commendable system of land-grant and other educational institutions in place. We have Connecticut's government-sponsored Natural Resources Center as a potential state model. And central Natural Heritage and GIS capabilities are available in all of the New England states.

Whether a central information source would be used remains conjectural, but we have been most impressed by story of the role and influence of the conservation journalists in Maine and in Rhode Island, where *The Providence Journal* has long been generous with reporting and editorial time devoted to conservation issues. But as budgets and space tighten up, most of the conventional media sources have been less forthcoming. And when they do carry a conservation story, it tends to reflect a special interest that provides them only one version of the facts. The result is a further contribution to the world of misinformation.

A final area of concern derives from our modest acquaintance with the region's historic leaders. We have dipped into their scholarship, consulted accounts by their contemporaries, and read with interest the few biographies that are available. But, regrettably, we have been unable to travel personally with Thoreau to Maine and Cape Cod; hike with MacKaye the backroads of Shirley, Massachusetts; explore with Carson the tidepools of coastal Maine; or admire with Marsh the reforested slopes of Mount Tom in Vermont. This has made us profoundly concerned about the absence of "witnesses" who can pass on the lessons of our generation of leaders to the next in a humanly knowledgeable way. The solution should be to include in the archival resources emanating from this project a component of audio- and video-supported oral histories taken from the leadership individuals we have identified in our New England conservation history. For many, time is already drawing short.

The final challenge will be to move prudently, patiently, and pragmatically, disregarding the drumbeat from those professing that disaster is just around the corner. As we look back in time, our perspective reveals that even the remarkable has always been possible. Landscape change is now occurring at an ever-increasing pace — much of it unfortunate — but not all is beyond humankind's capacity to design and put in place a set of orderly and constructive remedies. After all, it is not the first time that species have disappeared from the earth, climate has changed, and land and natural resources have been mismanaged and depleted. In the land, we still have a tangible, resilient resource that can respond readily to thoughtful and positive action.

And so our closing assessment is one of hope and high expectations — confidence in our ability to bring about meaningful and responsible change, a sense of excitement about what lies ahead and, above all, an abiding respect for the historic ability New Englanders have had to face the future with equal measures of concern and commitment. Other regions should be so fortunate.

This magnificent etching by New England sportsman-artist A. Lassell Ripley is of the American woodcock, an important indicator species of changes in the environment. Now much reduced in numbers virtually everywhere in New England, as it stirs at dusk to greet the rising moon and perform its remarkable nuptial flight, the bird leaves us with a fitting image of how far we have come in past land conservation, yet how challenging is the future. (Boston Museum of Fine Arts)

POSTLUDE

As this account of land conservation in New England came to an end, the authors remarked on the extraordinary degree of individual leadership their histories had revealed. The editor was encouraged to provide a sample set of mini-biographies of at least some of those responsible for what had transpired.

On the following pages, we offer six such vignettes — one per state. Most of those profiled are no longer with us. Many others would have been selected had there been space available. Those chosen are not ordered in any way, nor is any one deemed to be more important than any other. The intent is to simply showcase the diversity of interests and talents New Englanders have brought to the conservation movement over time and to humanize the written record of their accomplishment.

What the list demonstrates is that land conservation is relevant to virtually everyone. No matter who you are, where you live, or what you become, you can make a genuine difference. We have been heartened by the evidence that the next generation of conservation leaders is already in place and beginning to work effectively. We trust many will be among our readers.

Charles H.W. Foster
Editor

Jean Lande Hennessey
(1927–2004)

Jean Hennessey, the much-beloved doyenne of New Hampshire politics and public affairs, grew up on a farm in the suburbs of Seattle, Washington, the granddaughter of two first-generation Norwegian immigrants and the daughter of an indomitable lady who, in contracting to break wild broncos for the U.S. Cavalry during World War I, was one of the early "horse whisperers," a heritage that Jean would find immensely useful later on as she strove to bring reason and success to the recalcitrant Democratic donkey.

A confidante of philanthropists, party leaders, and presidents, and much sought after to head and advise organizations and agencies, Jean Hennessey was a colleague in the full sense of Felix Frankfurter's "eager collaboration for the larger ends of life." Among her many public service contributions, she was a U.S. Commissioner for the International Joint Commission: U.S. and Canada, a presidentially appointed trustee of the Washington-based Woodrow Wilson International Center for Scholars, and budget director for New Hampshire governor Hugh Gallen.

But Hennessey did more than consort with luminaries. She was one of the early leaders seeking to enhance the status of women, establishing a national program for women in national/corporate philanthropy and practicing what she preached as the first executive director of the New Hampshire Charitable Foundation and Associated Trusts. When she stepped down in 1977, the foundation appropriately established a special *Jean L. Hennessey "Wild Schemes" Fund* in recognition of her spirit of entrepreneurship.

Her efforts on behalf of the environment were unceasing. Among many posts, she served as a founding trustee and chairperson of the New England Natural Resources Center, as a trustee of the Environmental Law Institute and the World Environment Center, and as director of Dartmouth College's Institute on Canada and the United States and senior fellow of its Institute on International Environmental Governance.

But even beyond her extraordinary personal accomplishments, Vassar-smart Jean Hennessey could always be counted upon as a constructive force to bring out the good and the potential in others. Entering a room, she would radiate grace, good humor, and empathy embracing all with her trademark heartfelt, full-bodied laugh. To her immediate family, she was "Maj" (for majesty); to her many associates a warm, generous, persistent, and usually persuasive colleague. Never one for elaborate paeans or plaudits, Jean would have liked the summary tribute offered at the end by one who much admired her: "Gosh! she was fun."

Hennessey, Martha. June 27, 2004. "The Butterfly Effect," eulogy delivered at the Hennessey memorial service, Dartmouth College. Hanover, NH.

Philip Henderson Hoff
(1924–)

Philip Hoff came to Vermont in the early 1950s to practice law. A native of Turners Falls, Massachusetts, he had graduated from Williams College and Cornell University Law School and had served in the Pacific Theater in World War II as a submariner. After having lost his bid for a seat on the Burlington City Council earlier in the year, Hoff was elected to the Vermont House in 1960 as the sole representative of Vermont's largest city. Two years later, he became the first popularly elected Democratic governor in the state's history, serving from 1963–69.

Vermont's archaic administrative structure conferred considerable authority not on the governor but on an intervening layer of boards and commissions, which oversaw many of the most significant governmental functions and were responsible for making the appointments of the relevant department heads. And so Hoff, early in his administration, revived a moribund State Planning Office, won federal entitlements under Section 701 of the Housing Act of 1954, and began to exercise executive leverage over his agencies through the budgeting and appointments approval processes available to his planning and administration departments.

Hoff's administration was further informed by the convergence of two national developments that, for Vermont, proved to be tectonic. One was the U.S. Supreme Court case, Baker v. Carr, which forced the House to reapportion itself, shrinking its size to 150 members and apportioning its representatives by population not simply by municipal origin. The other major event was the advent of President Lyndon

Johnson's Great Society. Hoff's prodding of a newly compliant legislature enabled Vermont to qualify for many federal entitlement programs. "We didn't speak of the 'environment' in that era," Hoff remarked later. "We used the term 'natural resources.'"

Nevertheless, under Hoff's leadership, Vermont led the way in banning billboards, advancing conservation and recreation, and planning growth centers and nucleated settlements, measures that undergirded many of the later concepts embodied in Vermont's pioneering 1970 Land Use and Development Law (Act 250). The Hoff influence was also evident during his term as chairman of the New England Governors Conference, for in 1967 the governors authorized the creation of the New England Regional Commission and the New England River Basins Commission, two federal-state agencies that would later serve the region well.

Charismatic, personable, farsighted, and accomplished, Philip Hoff has left a legacy ranging from significant natural resources conservation to the still-admired Vermont–New York Youth Project, a collaboration between Hoff and New York's Mayor John Lindsay that enabled city children to spend summers in Vermont in programs involving music, dance, sports, and crafts. His empathetic spirit was perhaps best synthesized years later by Vermont's deputy attorney general Gregory McKenzie, who observed: "Phil Hoff cares about people. He even cares about people who don't care about him."

Principal source: Art Ristau, Barre, VT. December 2005.

John Charles Phillips
(1876–1938)

At the turn of the twentieth century, with governmental responsibilities for natural resources but a faint shadow of their prominence later on, much of the nation's conservation leadership was in the hands of what historian John Reiger has termed the "naturalist-sportsman" (*American Sportsmen and the Origins of Conservation*, 1975). These were individuals — primarily men — often from old and wealthy eastern families — whose ready access to money and power made them disproportionately influential.

The naturalist-sportsmen were inspired by the British tradition of pursuing fish and game in the out-of-doors under an accepted code of behavior. They were guided by the interest in science burgeoning throughout the western world. And they were motivated by the urgent need to curb the excesses of plundering and market hunting rampant in the United States in order that fish and wildlife resources could be managed professionally and sustainably.

One such New Englander was Dr. John C. Phillips, who would rise to become an associate curator of birds and mammals at Harvard's Museum of Comparative Zoology, chairman of the Massachusetts Conservation Council, president of the Massachusetts Fish and Game Association, and a founder of the New England Game Conferences, the predecessor of today's popular North American Wildlife and Natural Resources Conferences. Phillips would go on to author more than 100 articles and papers in addition to his respected four volume *Natural History of the Ducks* completed during the period 1922–1926.

Born in 1876 to an old and distinguished Massachusetts family,

John Phillips first acquired credentials in botany and zoology at Harvard's Lawrence Scientific School, then went on to obtain a Harvard medical degree in 1904. Except for his service in World War I and his later command of the 33rd Field Hospital during the Marne and Meuse offensives, he never practiced medicine professionally. He devoted the bulk of his life to the advancement of science and the conservation of natural resources at local, state, regional, national, and international levels. In so doing, he was widely admired as a modest, generous, selfless, talented, and versatile leader whose qualities are now embodied in the prestigious John C. Phillips Medal awarded for meritorious service by the International Union for the Conservation of Nature and Natural Resources (IUCN).

Phillips life was cut short in 1938 when, on a grouse-hunting trip to New Hampshire, he suffered a fatal heart attack. A veritable Who's Who of national and regional conservation leaders turned out to serve as honorary pallbearers. In his honor, Harvard's Museum of Comparative Zoology and Peabody Museum closed for an hour and a half that afternoon. And on a gray day in November, as he was committed to rest in his beloved Wenham woods, a single shot was heard close by, thereby providing a fitting and poignant tribute to the man who had so enjoyed outdoor activities and had contributed so much to their well-being.

Charles H.W. Foster

John Hubbard Chafee
(1922–1999)

John H. Chafee was a big man with a booming voice. When he entered a room, his passion and energy generally made him the center of attention. He was a war hero, a governor, a Navy secretary, and a four-term U.S. Senator. One historian described him as a Yankee with enormous charismatic power, and Chafee used that power to change the way his state, his region, and his nation treated the environment.

In a heavily Democratic state, Chafee, a liberal Republican, battled to get elected governor. He went on to a long career making friends on both sides of the aisle, and among the powerful and the commonplace. One of Chafee's greatest legacies was the collection of beautiful places he left behind. As governor, he created Rhode Island's Green Acres program, institutionalizing the state's acquisition of open space for parks, beaches, and woodlands. As a U.S. Senator, he delighted in obtaining appropriations to expand the federal wildlife refuges along Rhode Island's scenic south coast. He protected thousands of acres. And in the Senate, Chafee played key roles in enacting all of the critical environmental legislation that now protects the country.

"Quite simply, no American has done more for the environment of this country over the past five decades than John Chafee," John DeVillars, regional administrator of the U.S. Environmental Protection Agency, said shortly after Chafee's death in 1999. "He led every significant environmental fight of our generation from drinking water to air quality to endangered species, from rivers to shellfish beds to open space."

Ironically, a few weeks before he died, Chafee told a reporter his biggest accomplishment was sponsoring legislation in 1986 that identified a gritty, forty-six-mile corridor of tired cities and abandoned mills along the Blackstone River as a national heritage corridor. It has since been held up as a model public-private partnership by the National Park Service.

Three years after Chafee's death, a statue was commissioned at Colt State Park, the 460-acre park overlooking Narragansett Bay that Chafee finally persuaded the legislature to acquire after he was elected governor. It depicts Chafee in his forties, still governor, striding off to yet another mission. At the statue's base are words Chafee often used: "What we do today we do for our children and our childrens' children." Chafee's widow, Virginia, said she sometimes visits the park just to look at the statue. "It's so right," she said. "Words can't describe what this means to me."

The statue is seven feet tall. But Chafee was a giant long before.

Lord, Peter B. 2007. *The Providence Journal*. Providence, RI.
McLoughlin, William G. 1978. Rhode Island: a bicentennial history.

Richard Hale Goodwin
(1910–2007)

The career of Richard Hale Goodwin, long-term director of Connecticut College's Connecticut Arboretum and a founder and former president of The Nature Conservancy (TNC), epitomizes the crucial role played by scientists in advancing state, regional, and national land conservation during the twentieth century.

The grandson of John Smith Farlow, an Irish immigrant who made a fortune in railroads, Goodwin grew up in the serene environment of MIT where his father was professor of electrochemistry and dean of graduate students. Despite his subsequent longevity and resolute toughness, he was a sickly child requiring frequent invigoration through numerous camping trips and family excursions in the West.

After enrolling in Harvard and receiving a Ph.D. in botany in 1938, Goodwin accepted a faculty position at the University of Rochester in New York. It was there that his first exposure to natural area preservation occurred—a campaign to secure the 2,000-acre, calcium-rich, Bergen Swamp, which harbored a diverse array of plants and animals.

When he arrived in Connecticut in 1944 to become professor of botany and director of Connecticut College's Connecticut Arboretum, it was only natural that Dick Goodwin would set to work to buffer the Arboretum's ninety-acre holding with adjacent land purchases. What is now called the Bolleswood Natural Area is one of the first such preserves established in Connecticut.

But it was not until he joined The Nature Conservancy, serving two terms as its national president, that his most significant, personal land-conservation effort took place. The Burnham Brook Preserve was one of the TNC's earliest Connecticut preserves, eventually connecting with the Devils Hopyard State Park to form a 2,000-acre reserve protecting an integral part of the Eight Mile River watershed, including Dick and Esther Goodwins' home.

But what about the man himself? An entire generation of colleagues, former students, and friends remains eternally grateful for Dick Goodwin's diverse talents, huge heart, marvelous sense of humor, and healthy understanding of his place on earth. They speak admiringly of his dedication and accomplishment, but also his ready, twinkling smile. Even more applauded is his role as a savvy businessman—the way he employed ingenuity and personal persuasion to construct and finance permanent land-conservation deals in Connecticut and elsewhere.

In one of the many poems in his 1991 autobiography, *A Botanist's Window on the Twentieth Century*, Dick Goodwin speaks of a world in need of those who deeply care. He was in every respect characterizing himself.

Russell L. Brenneman, Westport, CT
David R. Foster, Harvard Forest, Petersham, MA

CONTRIBUTORS AND ADVISORS

RICHARD E. BARRINGER

Following service as commissioner of conservation and state planning director in the administrations of three Maine governors, Dr. Barringer is currently a member of both the public policy and management and community planning and development faculties at the University of Southern Maine's Muskie School of Public Service.

MICHELLE BAUMFLEK

Concerned with the human dimensions of natural resources and currently the non-timber forest products research specialist at the University of Vermont, Baumflek is engaged in identifying culturally and economically important and collaborative planning approaches leading to sustainable use and management of resources throughout the Northern Forest region.

PAUL O. BOFINGER

Retired after nearly thirty-five years as president/forester of the Society for the Protection of New Hampshire Forests, Bofinger now chairs the board of directors of the Northern Forest Center, a nonprofit organization established to assist forest-dependent communities throughout the New England–New York northern forest region.

RUSSELL L. BRENNEMAN

A former corporate lawyer, a longtime officer of the Connecticut Forest & Park Association and the Connecticut League of Conservation

Voters, and an authority on private approaches to the preservation of land, Brenneman has been a conservation advisor to many Connecticut governors and was the first chairman of the Connecticut Resource Recovery Authority.

JAMES C. COLLINS

A lifelong resident of New Hampshire living in the shadow of Mount Cardigan in Orange, NH, and the former acting editor of the alumni magazine at Dartmouth College, Collins is a professional writer and editor now affiliated with *Yankee Magazine* in Dublin, NH.

ROLF DIAMANT

In the course of his thirty year career with the National Park Service, Diamant has worked to develop cooperative conservation strategies for wild and scenic rivers, national heritage areas, partnership parks, and protected areas. He is currently the superintendent of Marsh-Billings-Rockefeller National Historical Park in Woodstock, VT.

CHARLES H.W. FOSTER

A former Massachusetts commissioner of natural resources, secretary of environmental affairs, president of The Nature Conservancy, and dean of the Yale University School of Forestry and Environmental Studies, for the past twenty years Dr. Foster has been an adjunct research fellow at the Belfer Center for Science and International Affairs at the Harvard Kennedy School.

CLARE GINGER

An associate professor in the University of Vermont's Rubenstein School of Environment and Natural Resources, Dr. Ginger's teaching and research interests focus on policy and planning processes where she works with colleagues on projects that integrate interdisciplinary activities in both domestic and international settings.

PERRY R. HAGENSTEIN

Previously a natural resources staff expert and senior policy analyst for the Public Land Law Review Commission in Washington, and later the first executive director and now board chairman of the New England Natural Resources Center, Dr. Hagenstein is a lifetime national associate of the National Research Council/National Academy of Sciences where he has served as a member of and consultant to many of its standing boards and study committees related to environment and natural resources.

Judson D. Hale, Sr.

Born in Boston, raised on a dairy farm in Vanceboro, ME, and now editor-in-chief of the *Old Farmer's Almanac* and *Yankee Magazine*, Hale is in his fiftieth year with the oldest continuously published periodical in North America. The author of *Inside New England* and *The Education of a Yankee*, among many articles and publications, Hale is a recognized and respected observer of the New England scene.

Peter B. Lord

A recent mid-career master's degree graduate in marine affairs from the University of Rhode Island and a recipient in 2005 of the New England Society of Newspapers' master reporter award, Lord has been a staff writer at the *The Providence Journal* for nearly thirty years specializing in environmental issues.

Robert L. McCullough

An associate professor of historic preservation at the University of Vermont with addition credentials in planning and the law, Dr. McCullough is a biographer of Benton MacKaye, the author of the much-acclaimed history of communal forests in New England, and a specialist in community-level transportation-environment relationships.

Robert W. McIntosh

A native New Englander and University of Massachusetts-educated in forestry, McIntosh has been associated with the Department of the Interior and the National Park Service activities in New England for more than forty years. He is currently the NPS's associate regional director with responsibilities for the design, construction, and preservation maintenance in the thirteen-state Northeast Region.

Nora J. Mitchell

Concerned with advancing innovations in collaborative conservation by identifying best practices in areas such as landscape-scale conservation, partnership networks, collaborative leadership, and community engagement, Dr. Mitchell is the founding director of the National Park Service's national Conservation Study Institute and the assistant regional director for conservation studies in its Northeast region.

Richard Ober

Formerly communications director for the Society for the Protection of New Hampshire Forests and for the past seven years the executive director of southwest New Hampshire's Monadnock Conservancy, Ober is currently the vice president for civic leadership at the New Hampshire Charitable Foundation.

Ninian Stein

A graduate of the Yale School of Forestry & Environmental Studies and the recipient of a Ph.D. in anthropology from Brown University, Dr. Stein is now the associate director of the University of Massachusetts (Boston) Environmental Studies Program. She has served as the principal cooperator for Rhode Island, assisting with the taking of six oral histories from prominent RI conservation leaders.

Thomas Urquhart

Educated at Oxford University, associated with local, national, and international conservation organizations for more than thirty years, and a former consultant for the World Wildlife Fund in Gland, Switzerland, and the Organization for Economic Cooperation and Development in Paris, France, Urquhart served for twelve years as the executive director of the Maine Audubon Society. Now a writer specializing in environmental and conservation issues, he is the author of the recent *For the Beauty of the Earth* published by Shoemaker & Hoard.